MW00837471

Lemmas in Olympiad Geometry

Lemmas in Olympiad Geometry

Titu Andreescu

Sam Korsky

Cosmin Pohoata

Library of Congress Control Number: 2016933279

ISBN-10: 0-9885622-3-5
ISBN-13: 978-0-9885622-3-3

9 8 7 6 5 4 3 2 1

www.awesomemath.org

Cover design by Iury Ulzutuev

Preface

This book showcases the synthetic problem-solving methods which frequently appear in modern day Olympiad geometry, in the way we believe they should be taught to someone with little familiarity in the subject. In some sense, the text also represents an unofficial sequel to the recent problem collection published by XYZ Press, *110 Geometry Problems for the International Mathematical Olympiad*, written by the first and third authors; but, the two books can be studied completely independently of each other.

Lemmas in Olympiad Geometry is a project that started in the summer of 2011, when the third author first taught the Geometric Proofs course at the AwesomeMath Summer Camp. Some brief lecture notes were written back then (with the intention of getting expanded), but nothing substantial happened until last summer, when the second author came to the Cornell camp as a teaching assistant for the same course. Ever since, we have all been working together to make the current version of the manuscript possible, and are excited to announce that it is ready.

The work is designed as a medley of the important Lemmas in classical geometry in a relatively linear fashion: gradually starting from Power of a Point and common results to more sophisticated topics, where knowing a lot of techniques can prove to be tremendously useful. We treated each chapter as a short story of its own and included numerous solved exercises with detailed explanations and related insights that will hopefully make your journey very enjoyable. Each chapter is also accompanied by a short list of problems that we have carefully selected. These are problems that we have solved ourselves on our own at some point, and so we are convinced that you are going to appreciate them as well. The last chapter on three dimensional geometry is the only chapter which is not followed by such a list of problems, since we considered it as a bonus section, yet one that has beautiful problems which are also relevant in other subdomains of geometry.

We wish you a pleasant reading and hope that you will enjoy *Lemmas in Olympiad Geometry* as much as we enjoyed writing it.

The authors

Contents

Chapter 1

Power of a Point

One of the most important tools in Olympiad Geometry is the so-called Power of Point Theorem, our first Lemma.

Theorem 1.1. Let Γ be a circle, and P a point. Let a line through P meet Γ at points A and B, and let another line through P meet Γ at points C and D. Then

$$PA \cdot PB = PC \cdot PD.$$

We announce the reader that we will be labeling our Lemmas as Theorems not to follow any convention, but rather to emphasize their importance, since after all they represent the main stars of our show.

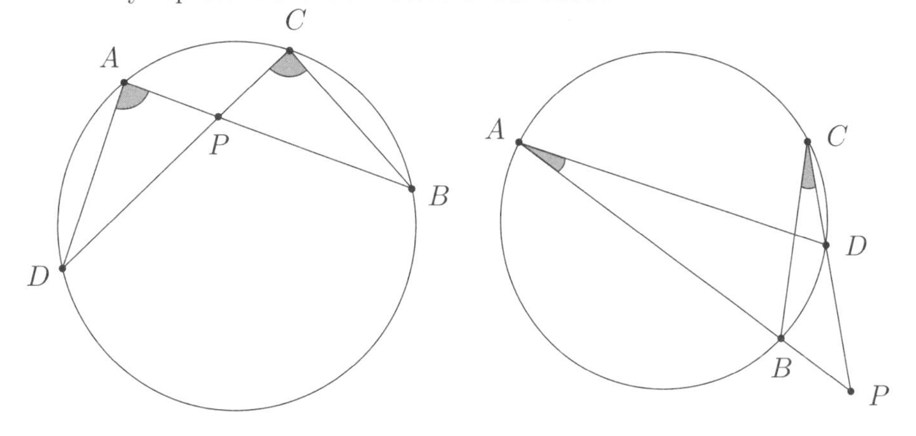

Proof. Of course, there are two configurations to consider here, depending on whether P lies inside the circle or outside the circle. In the case when P lies inside the circle, we have $\angle PAD = \angle PCB$ and $\angle APD = \angle CPB$, so that triangles PAD and PCB are similar; hence

$$\frac{PA}{PD} = \frac{PC}{PB}.$$

Rearranging then yields $PA \cdot PB = PC \cdot PD$.

When P lies outside the circle, we again have $\angle PAD = \angle PCB$ and $\angle APD = \angle CPB$, so again triangles PAD and PCB are similar. We get the same result in this case. □

As a very important special case, when P lies outside the circle and PC is tangent to the circle, we have that

$$PA \cdot PB = PC^2.$$

Conversely, the above represents a very useful criterion for proving concyclities.

Theorem 1.2. Let A, B, C, D be four distinct points. Let the lines AB and CD intersect at P. Assume that either P lies on both line segments AB and CD, or P lies on neither line segment. Then A, B, C, D are concyclic if and only if $PA \cdot PB = PC \cdot PD$.

Proof. Going backwards, the relation $PA \cdot PB = PC \cdot PD$ is equivalent to

$$\frac{PA}{PD} = \frac{PC}{PB},$$

which combined with $\angle APD = \angle CPB$ (which holds in both configurations described above) yields that triangles APD and CPB are similar. Thus, we get that $\angle PAD = \angle PCB$, which in both cases implies that A, B, C, D are concyclic. □

This tells us that no matter what chord XY we take through P (with X, Y on the circle), the value $PX \cdot PY$ is constant. This constant is called the **power of P** with respect to the circle considered. In particular, if $\Gamma(O, R)$ is the circle with center O and radius R, then if we consider the chord XY that passes through the center O (i.e. we choose the diameter of the circle passing through P), we get that

$$PX \cdot PY = \|OP^2 - R^2\|$$

//We say that the points lying on the circle Γ have zero power with respect to Γ!

We emphasize this interplay between products and differences of squares with the following exercise.

Delta 1.1. (IMO 2011 Shortlist) Let $A_1A_2A_3A_4$ be a non-cyclic quadrilateral. Let O_1 and r_1 be the circumcenter and the circumradius of triangle $A_2A_3A_4$. Define O_2, O_3, O_4 and r_2, r_3, r_4 in a similar way. Prove that

$$\frac{1}{O_1A_1^2 - r_1^2} + \frac{1}{O_2A_2^2 - r_2^2} + \frac{1}{O_3A_3^2 - r_3^2} + \frac{1}{O_4A_4^2 - r_4^2} = 0.$$

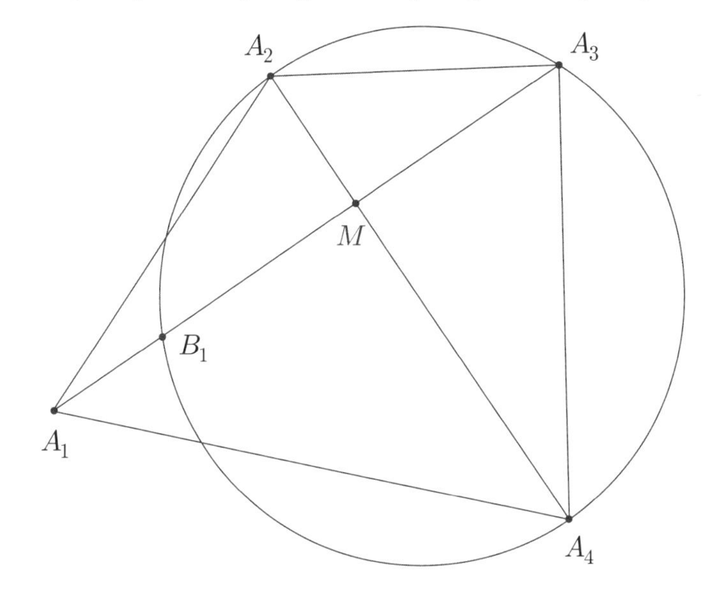

Proof. Let M be the point of intersection of the diagonals A_1A_3 and A_2A_4. On each diagonal choose a direction and let x, y, z, and w be the signed distances from M to the points A_1, A_2, A_3, A_4, respectively. Let ω_1 be the circumcircle of triangle $A_2A_3A_4$ and let B_1 be the second intersection of ω_1 and A_1A_3 (thus, $B_1 = A_3$ if and only if A_1A_3 is tangent to ω_1). Since the expression $O_1A_1^2 - r_1^2$ is the power of the point A_1 with respect to ω_1, we get

$$O_1A_1^2 - r_1^2 = A_1B_1 \cdot A_1A_3.$$

On the other hand, from the equality $MB_1 \cdot MA_3 = MA_2 \cdot MA_4$, we obtain

$$MB_1 = \frac{yw}{z}.$$

Hence, it follows that

$$O_1A_1^2 - r_1^2 = \left(\frac{yw}{z} - x\right)(z - x) = \frac{z - x}{z}(yw - xz).$$

Doing the same thing for the other three expressions, we then get that

$$\sum_{i=1}^{4} \frac{1}{O_iA_i^2 - r_i^2} = \frac{1}{yw - xz}\left(\frac{z}{z - x} - \frac{w}{w - y} + \frac{x}{x - z} - \frac{y}{y - w}\right) = 0,$$

as claimed. This completes the proof. □

By the way, this will not be the only time we will make use of signed distances in this material. Usually, we can assume without loss of generality a certain position of the points in our diagram - however, in problems involving lots of circles, the computations involving the Power of Point Theorem are not the same for all configurations; hence, we often need to take extra care when dealing with signs.

Some warm-up problems now! We begin with another simple interplay between the two formulas for the power of a point.

Delta 1.2. (Euler's Theorem) In a triangle ABC with circumcenter O, incenter I, circumradius R, and inradius r, prove that

$$OI^2 = R(R - 2r).$$

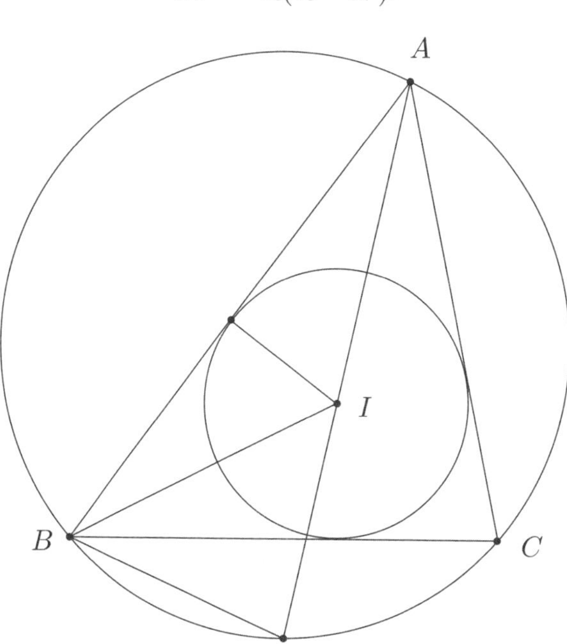

Proof. Let AI meet the circumcircle again at D. In this case, the Power of Point Theorem applied for I yields

$$IA \cdot ID = R^2 - OI^2.$$

Thus, we would like to show that $IA \cdot ID = 2Rr$. First, note that $IA = \frac{r}{\sin \frac{A}{2}}$ (draw the perpendicular from I to AB and apply the Law of Sines in the right triangle that you obtain). Next, note that

$$\angle BID = \angle BAD + \angle ABI = \angle DAC + \angle IBC = \angle DBC + \angle IBC = \angle IBD;$$

hence $ID = BD = 2R\sin\frac{A}{2}$, where the last equality comes from the (extended) Law of Sines in triangle ABD. Hence, we get that

$$IA \cdot ID = \frac{r}{\sin\frac{A}{2}} \cdot 2R\sin\frac{A}{2} = 2Rr,$$

as desired. This completes the proof. □

Note that for any given point P in plane, the above method can be extended to generate an identity for OP^2.

Delta 1.3. Let ABC be an acute-angled triangle and let D be the foot of the A-altitude. Let H be a point on the segment AD. Prove that H is the orthocenter of triangle ABC if and only if $DB \cdot DC = AD \cdot HD$.

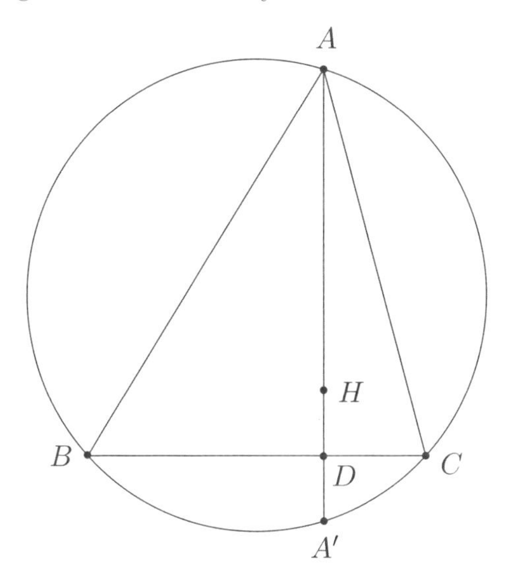

Proof. Let A' be the second intersection of the line AD with the circumcircle of triangle ABC. We know that A' is the reflection of the orthocenter across BC (if not, try angle chasing). Thus, if H is the orthocenter of ABC, then the computing power of D with respect to the circumcircle gives us

$$DB \cdot DC = AD \cdot DA' = AD \cdot HD,$$

as desired. Conversely, we have that $DB \cdot DC = AD \cdot HD$ and also $DB \cdot DC = AD \cdot HA'$ (the power of D with respect to the circumcircle); thus $HD = HA'$, and so H needs to be the orthocenter of ABC, as claimed. □

Although very simple, this proves to be a very useful criterion for showing that a point lying on an altitude of a triangle is the orthocenter. Let's see a couple of problems where this may come in handy.

Delta 1.4. (USA TSTST 2012) In scalene triangle ABC, let the feet of the perpendiculars from A to BC, B to CA, C to AB be A_1, B_1, C_1, respectively. Denote by A_2 the intersection of lines BC and B_1C_1. Define B_2 and C_2 analogously. Let D, E, F be the respective midpoints of sides BC, CA, AB. Show that the perpendiculars from D to AA_2, E to BB_2 and F to CC_2 are concurrent.

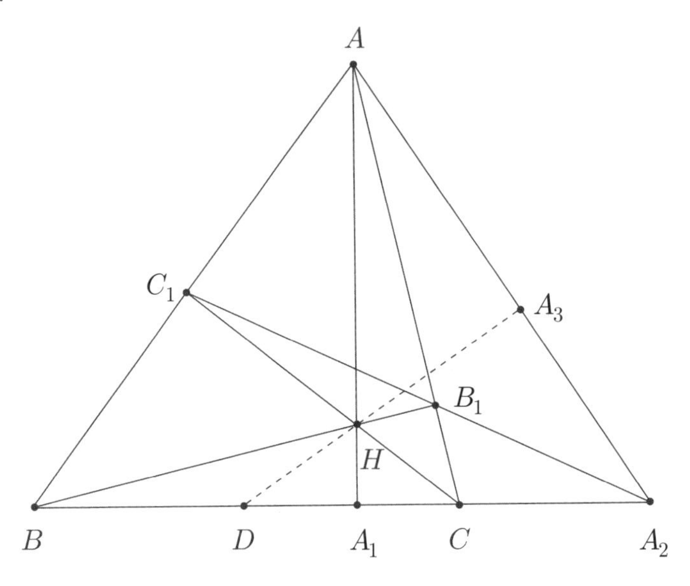

Proof. Let H be the orthocenter of triangle ABC. We claim that H is the desired point of concurrency. Let A_3 be the foot of perpendicular from D to line AA_2. Since $AA_1 \perp BC$ and $DA_3 \perp AA_2$, quadrilateral A_3A_1DA is cyclic. By Power of a Point, we have $A_2C_1 \cdot A_2B_1 = A_2A_3 \cdot A_2A$. Again, by Power of a Point (this time with respect to the nine point circle of triangle ABC) $A_2A_1 \cdot A_2D = A_2C_1 \cdot A_2B_1$, so combining these equations, $A_2C_1 \cdot A_2B_1 = A_2A_3 \cdot A_2A$, implying quadrilateral $A_3C_1B_1A$ is cyclic by **Theorem 1.2**. But H lies on the circumcircle of this quadrilateral, since $HC_1 \perp AB$ and $HB_1 \perp AC$. It follows that $\angle HA_3A = 180° - \angle HB_1A = 90°$, so points D, H, A_3 are collinear. Defining B_3 and C_3 analogously, similar arguments show that points E, H, B_3 and F, H, C_3 are also collinear, so the lines in the problem are concurrent at H as claimed. □

Delta 1.5. (IMO Shortlist 1998) Let I be the incenter of triangle ABC. Let K, L and M be the points of tangency of the incircle of triangle ABC with sides AB, BC, and CA, respectively. The line ℓ passes through B and is parallel to KL. The lines MK and ML intersect ℓ at the points R and S respectively. Prove that $\angle RIS$ is acute.

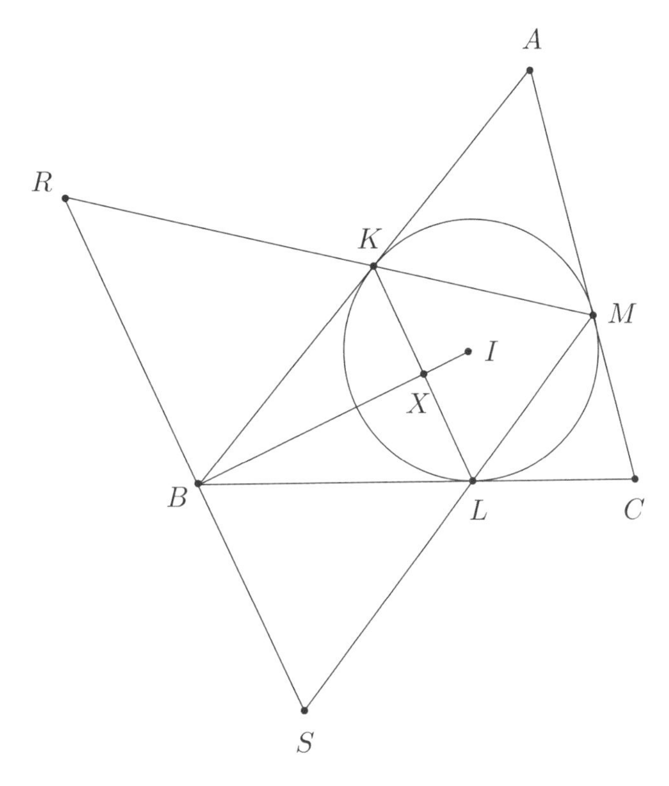

Proof. First note that

$$\angle KRB = \angle MKL = \angle MLC = \angle SLB$$

and

$$\angle RKB = \angle AKM = \angle KLM = \angle LSB$$

Thus, triangle BKS is similar to triangle BRL. This means that $BS \cdot BR = BL^2$. Now let X be the midpoint of segment KL. We have that X lies on the altitude from I to RS and also that $BX = BL \cos \frac{B}{2}$ and $BI = \frac{BL}{\cos \frac{B}{2}}$ which means that $BX \cdot BI = BR \cdot BS$. Hence, by **Delta 1.3**, X is the orthocenter of triangle RIS. But since X is the projection of I onto line KL it's clear that X lies inside of triangle RIS which implies that this triangle is acute as desired. □

//Another way to prove that X is the orthocenter of triangle RIS is to prove that triangle RXS is self-polar with respect to the incircle of triangle ABC.

We continue with a computational problem from the USA Mathematical Olympiad from 1998.

Delta 1.6. (USAMO 1998) Let $\mathcal{C}_1$ and $\mathcal{C}_2$ be concentric circles, with $\mathcal{C}_2$ in the interior of $\mathcal{C}_1$. From a point A on $\mathcal{C}_1$ one draws the tangent AB to $\mathcal{C}_2$ ($B \in \mathcal{C}_2$). Let C be the second point of intersection of AB with $\mathcal{C}_1$, and let D be the midpoint of AB. A line passing through A intersects $\mathcal{C}_2$ at E and F in such a way that the perpendicular bisectors of DE and CF intersect at a point M on AB. Find, with proof, the ratio $\frac{AM}{MC}$.

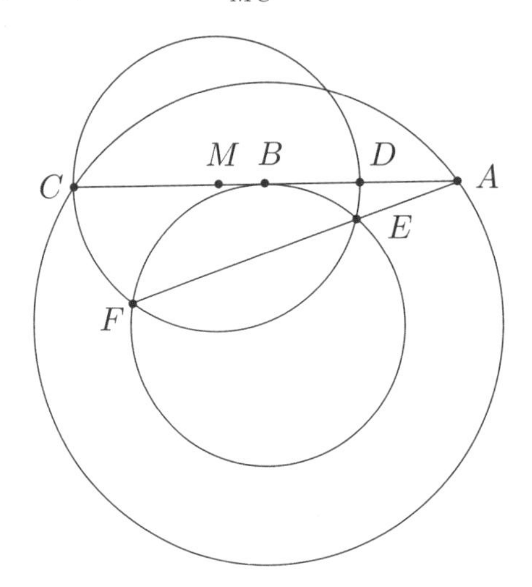

Proof. Let O be the common center of the concentric circles $\mathcal{C}_1, \mathcal{C}_2$. The tangency point B is the midpoint of the chord AC, because AC is perpendicular to the radius OB of the circle $\mathcal{C}_2$, and O is also the center of the circle $\mathcal{C}_1$. The power of the point A with respect to circle $\mathcal{C}_2$ is $AE \cdot AF = AB^2$. But since B is the midpoint of AC and D the midpoint of AB, we have that $AD \cdot AC = \frac{AB}{2} \cdot 2AB = AB^2$ as well. Hence, by **Theorem 1.2**, quadrilateral $CDEF$ is cyclic. The intersection M of the perpendicular bisectors of its diagonals CE, DF is its circumcenter. If this circumcenter is to be on its side CD, it must be the midpoint of this side, hence $DM = MC = \frac{DC}{2}$. Since $DC = \frac{3}{2}AB$, we now have $DM = MC = \frac{3}{4}AB$ and $AM = AD + DM = \frac{AB}{2} + \frac{3}{4}AB = \frac{5}{4}AB$ and so $\frac{AM}{MC} = \frac{5}{3}$. □

We continue with a beautiful IMO problem, where Power of Point can be used in a surprising way.

Delta 1.7. (IMO 2009) Let ABC be a triangle with circumcenter O. The points P and Q are interior points of the sides CA and AB respectively. Let

K, L and M be the midpoints of the segments BP, CQ and PQ. respectively, and let Γ be the circle passing through K, L and M. Suppose that the line PQ is tangent to the circle Γ. Prove that $OP = OQ$.

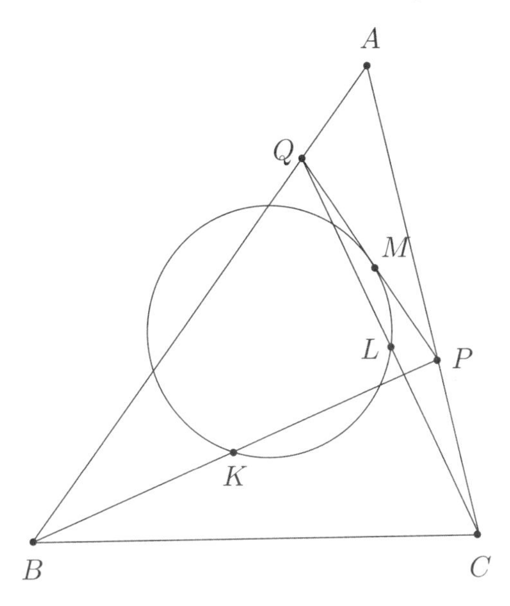

Proof. Since line PQ is tangent to Γ, we have that $\angle QMK = \angle MLK$. Since MK is the P-midline of triangle PQB we have that $MK \parallel AB$ so $\angle QMK = \angle AQM$. Hence, $\angle AQP = \angle MLK$. Similarly we get that $\angle MKL = \angle APQ$, so triangles MKL and APQ are similar. Therefore

$$\frac{AQ}{ML} = \frac{AP}{MK} \Longrightarrow \frac{AP}{BQ} = \frac{AQ}{PC} \Longrightarrow AP \cdot PC = AQ \cdot BQ.$$

Thus, P and Q have the same power with respect to the circumcircle of triangle ABC, so $OP = OQ$ as desired. □

We end this section with a cute result due to Hiroshi Haruki (according to [18]).

Delta 1.8. (Haruki's Lemma) Given two non-intersecting chords AB and CD in a circle and a variable point P on the arc AB remote from points C and D, let E and F be the intersections of chords PC, AB, and of PD, AB, respectively. Prove that the value of

$$\frac{AE \cdot BF}{EF}$$

does not depend on the position of P.

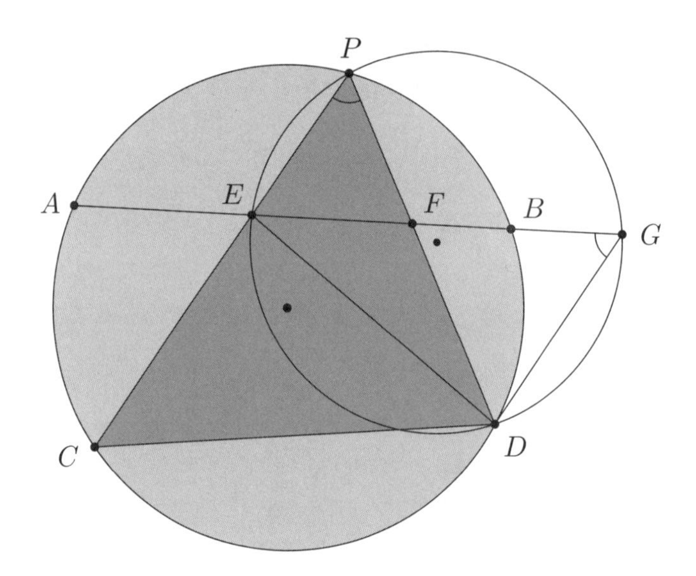

Proof. The proof relies on the fact that the angle $\angle CPD$ is constant. We begin by constructing the circumcircle of triangle PED. Define point G to be the intersection of this circle with the line AB. Note that $\angle EGD = \angle EPD$ as they are subtended by the same chord ED of the circumcircle of triangle PED; these angles remain constant as P varies on the arc AB. Hence, for all positions of P, $\angle EGD$ remains fixed and, therefore, point G remains fixed on the line AB. It follows that BG is constant. On the other hand, by Power of Point, we have that $AF \cdot FB = PF \cdot FD$ and $EF \cdot FG = PF \cdot FD$. Hence,

$$(AE + EF) \cdot FB = EF \cdot (FB + BG),$$

and $AE \cdot FB = EF \cdot BG$. Therefore, we conclude that

$$\frac{AE \cdot BF}{EF} = BG,$$

a constant. □

Haruki's Lemma can be used to give a very short proof of the so-called Butterfly Theorem, a very popular result in projective geometry.

Delta 1.9. (Butterfly Theorem). Let M be the midpoint of chord PQ of a given circle, through which two other chords AB and CD are drawn; AD cuts PQ at X and BC cuts PQ at Y. Then, M is also the midpoint of XY.

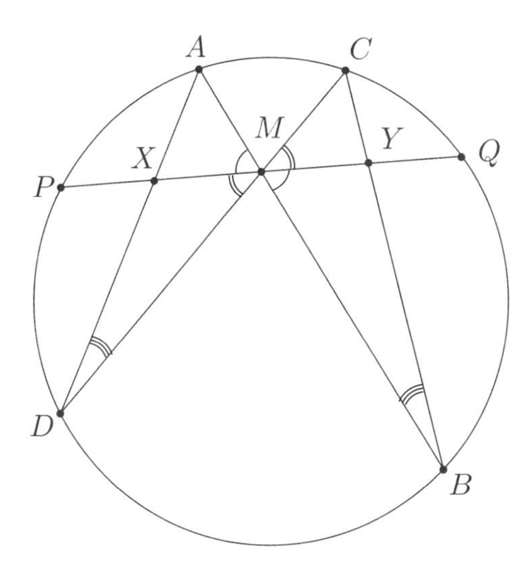

Proof. We think of A and C as being two positions of the variable point traversing the circle. Then, Haruki's lemma tells us that

$$\frac{XP \cdot MQ}{XM} = \frac{MP \cdot YQ}{YM},$$

which, because of $MP = MQ$, is simplified to

$$\frac{XP}{XM} = \frac{YQ}{YM}.$$

Adding 1 to both sides gives

$$\frac{XP + XM}{XM} = \frac{YQ + YM}{YM}.$$

Applying $MP = MQ$ again, we obtain the required $XM = YM$. This completes the proof. □

Assigned Problems

Epsilon 1.1. Let ABC be an acute triangle. Let the line through B perpendicular to AC meet the circle with diameter AC at points P and Q, and let the line through C perpendicular to AB meet the circle with diameter AB at points R and S. Prove that P, Q, R, S are concyclic.

Epsilon 1.2. Let ABC be an acute-angled triangle with circumcenter O and orthocenter H. Prove that

$$OH^2 = R^2(1 - 8\cos A \cos B \cos C).$$

Epsilon 1.3. Let ABC be a triangle and let D, E, F be the feet of the altitudes, with D on BC, E on CA, and F on AB. Let the parallel through D to EF meet AB at X and AC at Y. Let T be the intersection of EF with BC and let M be the midpoint of side BC. Prove that the points T, M, X, Y are concyclic.

Epsilon 1.4. (Kazakhstan MO 2008) Suppose that B_1 is the midpoint of the arc AC, containing B, of the circumcircle of triangle ABC, and let I_b be the B-excircle's center. Assume that the external angle bisector of $\angle ABC$ intersects AC at B_2. Prove that B_2I is perpendicular to B_1I_B, where I is the incenter of ABC.

Epsilon 1.5. (IMO 2000) Two circles Γ_1 and Γ_2 intersect at M and N. Let ℓ be the common tangent to Γ_1 and Γ_2 so that M is closer to ℓ than N is. Let ℓ touch Γ_1 at A and Γ_2 at B. Let the line through M parallel to ℓ meet the circle Γ_1 again at C and the circle Γ_2 again at D. Lines CA and DB meet at E; lines AN and CD meet at P; lines BN and CD meet at Q. Show that $EP = EQ$.

Epsilon 1.6. Let C be a point on a semicircle Γ of diameter AB and let D be the midpoint of the arc AC. Let E be the projection of D onto the line BC and F the intersection of the line AE with the semicircle. Prove that BF bisects the line segment DE.

Epsilon 1.7. Let A, B, C be three points on a circle Γ with $AB = BC$. Let the tangents at A and B meet at D. Let DC meet Γ again at E. Prove that the line AE bisects the segment BD.

Epsilon 1.8. (EGMO 2012) Let ABC be a triangle with circumcenter O. The points D, E, F lie in the interiors of the sides BC, CA, AB respectively, such that DE is perpendicular to CO and DF is perpendicular to BO. (By interior we mean, for example, that the point D lies on the line BC and D is between B and C on that line.) Let K be the circumcenter of triangle AFE. Prove that the lines DK and BC are perpendicular.

Epsilon 1.9. (IMO Shortlist 2013) Let ABC be a triangle with $\angle B > \angle C$. Let P and Q be two different points on line AC such that $\angle PBA = \angle QBA = \angle ACB$ and A is located between P and C. Suppose that there exists an interior point D of segment BQ for which $PD = PB$. Let the ray AD intersect the circumcircle of triangle ABC at $R \neq A$. Prove that $QB = QR$.

Chapter 2

Carnot and Radical Axes

In this section, we will prove a beautiful criterion for perpendicularity. It will also help us justify the existence of the radical axis without using any analytic geometry. The statement goes like this:

Theorem 2.1. Let AB and CD be two segments (not necessarily intersecting). Then, $AB \perp CD$ if and only if

$$AC^2 - AD^2 = BC^2 - BD^2.$$

Obviously, one implication is very easy: the direct implication. We will let you play with the Pythagorean Theorem to settle it. We'll take care of the converse with two proofs. The first one is not that meaningful, as it is rather computational and configuration dependent. Nevertheless, it is pretty straightforward, so we will not omit it.

First Proof. Let us assume that the segments AB and CD intersect. Even though what we are about to do only works for this situation (modulo some signs), the proof for the other case is similar. So, let P be the intersection of the two segments and let $0^\circ \leq \alpha \leq 90^\circ$ be the angle between them. Without loss of generality, $\angle APC = \angle BPC = \alpha$. We know that $AC^2 - AD^2 = BC^2 - BD^2$; however, from the Law of Cosines, we also know that

$$\begin{aligned}
AC^2 &= PA^2 + PC^2 - 2PA \cdot PC \cos\alpha, \\
AD^2 &= PA^2 + PD^2 + 2PA \cdot PD \cos\alpha, \\
BC^2 &= PB^2 + PC^2 + 2PB \cdot PC \cos\alpha, \\
BD^2 &= PB^2 + PD^2 - 2PB \cdot PD \cos\alpha.
\end{aligned}$$

Thus, it follows that

$$-2PA\cos\alpha\cdot(PC+PD)=2PB\cos\alpha\cdot(PC+PD),$$

i.e.

$$2(PA+PB)(PC+PD)\cos\alpha=0,$$

which implies that $\alpha=90^\circ$, as desired. □

Second Proof. This second proof is more insightful, in the sense that it yields the result immediately from an important locus.

Claim. Let CD be a segment in plane. The locus of the points P in plane so that the expression PC^2-PD^2 is constant is a line perpendicular to CD.

Indeed, note that this proves **Theorem 2.1** immediately, since then the condition $AC^2-AD^2=BC^2-BD^2$ means nothing else but the fact that both A and B belong to the locus described above, where the constant is the quantity AC^2-AD^2. Now, let's see how we can prove the claim.

Proof. Let P be a point belonging to the locus and let X be the projection of P on the line CD. Without loss of generality, X lies inside CD. Obviously, if we show that this point X is independent of P, then we are done, since this means that all points P lie on the line perpendicular to CD at X.

Now, to see that X is fixed, we apply the Law of Cosines to get that

$$\begin{aligned}PC^2 &= PD^2+CD^2-2PD\cdot CD\cdot\cos PDC\\ &= PD^2+CD^2-2XD\cdot CD.\end{aligned}$$

Hence, it follows that

$$\text{constant}=PC^2-PD^2=CD^2-2XD\cdot CD.$$

But, clearly, CD and 2 are constants, so the length of the segment XD is also constant; thus, since D is fixed, it follows that X is also fixed (remember, X lies inside CD according to our assumption). This completes the proof. □

This theorem about perpendicularity, as you can imagine, has a lot of interesting implications (and applications) besides helping determine what a radical axis is. We briefly discuss a few first.

Delta 2.1. Prove that medians AA_1 and BB_1 of triangle ABC are perpendicular if and only if $a^2+b^2=5c^2$.

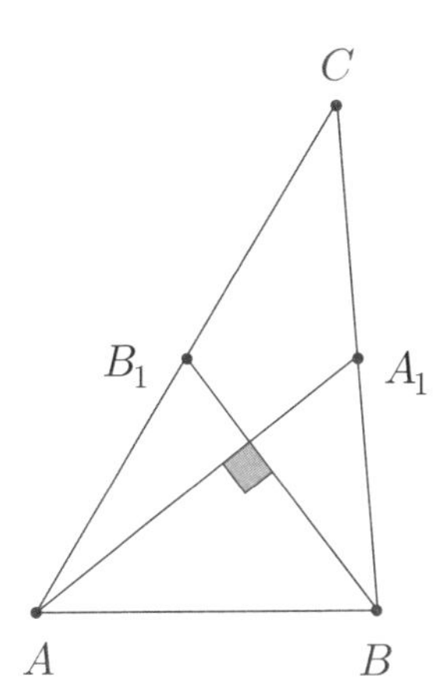

Proof. By **Theorem 2.1**, the medians AA_1 and BB_1 are perpendicular if and only if

$$AB^2 - AB_1^2 = A_1B^2 - A_1B_1^2.$$

This condition rewrites immediately as

$$c^2 - \frac{b^2}{4} = \frac{a^2}{4} - \frac{c^2}{4},$$

which is equivalent to $a^2 + b^2 = 5c^2$ as desired. □

Delta 2.2. Let ABC be a triangle and let X, Y, Z be the midpoints of the arcs BC, CA, AB of the circumcircle of triangle ABC, which do not contain the vertices of the triangle. Prove that the incenter I of triangle ABC is the orthocenter of triangle XYZ.

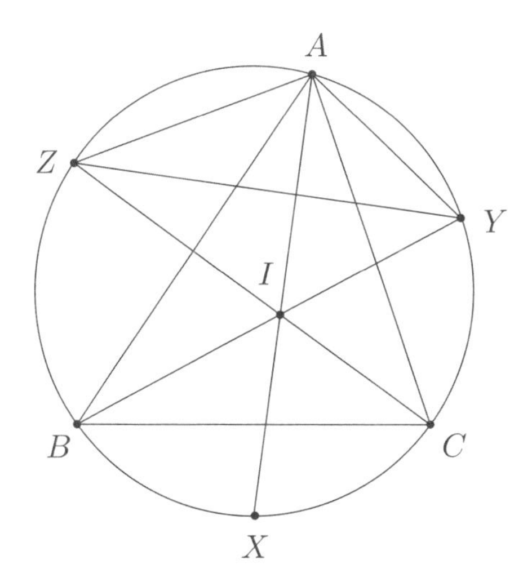

Proof. We clearly just have to prove that $AI \perp YZ$, since afterwards we would just imitate the argument for B and C. But by **Theorem 2.1**, this equivalent to showing that

$$AY^2 - AZ^2 = IY^2 - IZ^2.$$

However, as we showed in **Delta 1.2**, we know that $IY = AY$ and $IZ = AZ$; thus the identity above is apparent. Thus, $AI \perp YZ$ and this completes the proof. □

Now, a more complicated exercise.

Delta 2.3. Let ABC be a triangle and let E and F be the feet of the B and C-internal angle bisectors, respectively. Denote by O the circumcenter of triangle ABC and by I_a the A-excenter of triangle ABC. Prove that $OI_a \perp EF$.

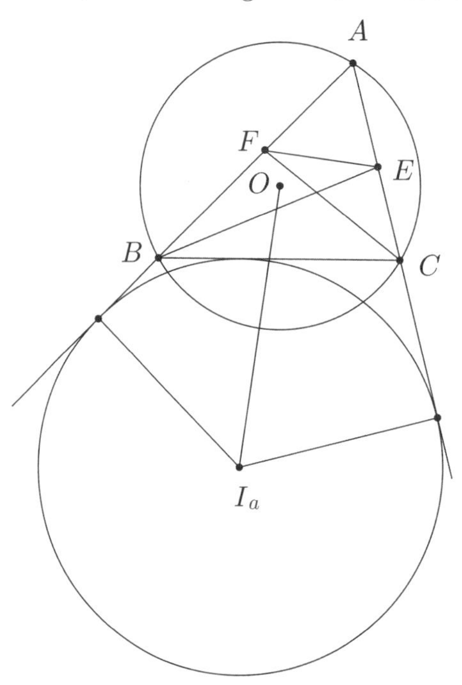

Proof. By **Theorem 2.1**, it is enough to prove that $OF^2 - FI_a^2 = OE^2 - EI_a^2$. So, we'll prove that expression $OF^2 - FI_a^2$ is symmetric with respect to b and c. Let R be the circumradius of triangle ABC and let r_a be the radius of the A-excircle of triangle ABC. From the Law of Cosines in triangle AOF we have

$$\begin{aligned} OF^2 &= AO^2 + AF^2 - 2AO \cdot AF \cos(90^\circ - C) \\ &= R^2 + AF^2 - 2R \cdot AF \sin C \\ &= R^2 + AF^2 - AF \cdot c. \end{aligned}$$

In addition, $FI_a^2 = r_a^2 + (s - AF)^2 = r_a^2 + s^2 - (a+b+c)AF + AF^2$. This implies

$$\begin{aligned} OF^2 - FI_a^2 &= R^2 - AF \cdot c - r_a^2 - s^2 + (a+b+c)AF \\ &= R^2 - r_a^2 - s^2 + AF(a+b) \\ &= R^2 - r_a^2 - s^2 + bc, \end{aligned}$$

and we are done. □

We now move to a very important and beautiful concurrency criterion.

Theorem 2.2. (Carnot's Theorem). Let ABC be a triangle and let M, N, P be points on the sidelines BC, CA, and AB, respectively. Then, the perpendicular lines at M, N, P to BC, CA, and AB, respectively, are concurrent if and only if

$$(MB^2 - MC^2) + (NC^2 - NA^2) + (PA^2 - PB^2) = 0.$$

Proof. Again, we have a very simple implication. Suppose that the perpendicular lines at M, N, P to BC, CA, AB are concurrent at a point X. By **Theorem 2.1**, since $XM \perp BC$, we have that

$$MB^2 - MC^2 = XB^2 - XC^2.$$

And similarly, $NC^2 - NA^2 = XC^2 - XA^2$ and $PA^2 - PB^2 = XA^2 - XB^2$, so, we immediately get that

$$(MB^2 - MC^2) + (NC^2 - NA^2) + (PA^2 - PB^2) = 0.$$

Conversely, assuming that

$$(MB^2 - MC^2) + (NC^2 - NA^2) + (PA^2 - PB^2) = 0,$$

we proceed by contradiction and suppose that the perpendicular lines at M, N, P to BC, CA, AB are not concurrent. In this case, let X be the intersection of the perpendiculars at N and at P to CA and AB respectively and let M' be the projection of X on BC. According to our hopefully false supposition, M' is different from M. The direct implication of Carnot's Theorem that we just proved then yields

$$(M'B^2 - M'C^2) + (NC^2 - NA^2) + (PA^2 - PB^2) = 0,$$

which combined with the above tells us that

$$MB^2 - MC^2 = M'B^2 - M'C^2.$$

By **Theorem 2.1**, this implies that $MM' \perp BC$. And this is clearly a contradiction, since both M and M' lie on BC. Thus, our assumption was false, and the perpendicular lines at M, N, P to BC, CA, and AB, respectively, are concurrent, as claimed. This completes the proof. □

//Do the points M, N, P have to lie on the sidelines of triangle ABC? (Think about **Theorem 2.1**.)

Corollary 2.1. Let A_1, B_1, C_1 be the tangency points of the excircles with centers I_a, I_b, I_c of triangle ABC with the sides BC, CA, and AB, respectively. Then, the lines I_aA_1, I_bB_1, I_cC_1 are concurrent. The concurrency point is usually referred to as the **Bevan point** of triangle ABC.

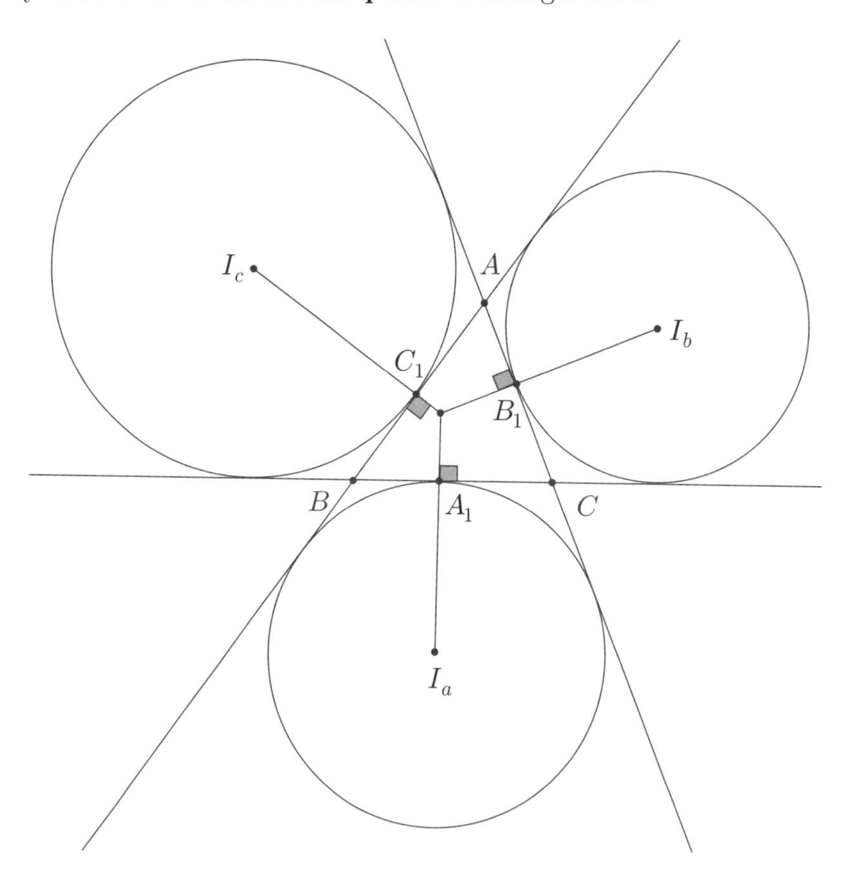

Proof. Obviously, the lines I_aA_1, I_bB_1, I_cC_1 are perpendicular to the sidelines BC, CA, AB of triangle ABC; thus, if we manage to show that

$$(A_1B^2 - A_1C^2) + (B_1C^2 - B_1A^2) + (C_1A^2 - C_1B^2) = 0,$$

then we are done. However, it's easy to verify that $C_1B = B_1C = s - a$, $A_1C = C_1A = s - b$, $A_1B = B_1A = s - c$; thus, Carnot's Theorem implies the desired concurrency. □

Corollary 2.2. (Orthologic Triangles) Let ABC and XYZ be two triangles in the plane. The perpendiculars from X, Y, Z to sides BC, CA, AB respectively concur at a point P. Then, the perpendiculars from A, B, C to sides YZ, ZX, XY respectively concur. This concurrency point and P are referred to as the **orthology centers** of the two triangles, and the triangles are called **orthologic**.

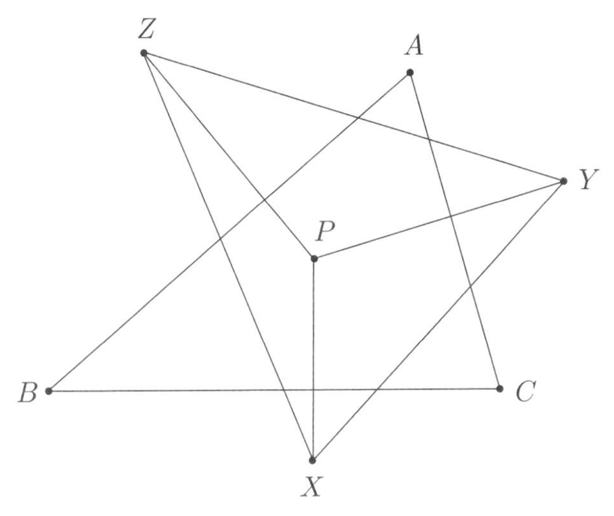

Proof. Using **Theorem 2.1** and then the sidenote after **Theorem 2.2** we have that

$$\begin{aligned}(AY^2 - AZ^2) + (BZ^2 - BX^2) + (CX^2 - CY^2) &= \\ (XC^2 - XB^2) + (YA^2 - YC^2) + (ZB^2 - ZA)^2 &= \\ (PC^2 - PB^2) + (PA^2 - PC^2) + (PB^2 - PA)^2 &= 0\end{aligned}$$

which implies the desired result. □

Delta 2.4. (IMO Shortlist 1987) Find, with proof, the point P in the interior of an acute-angled triangle ABC for which $BL^2 + CM^2 + AN^2$ is a minimum, where L, M, N are the feet of the perpendiculars from P to BC, CA, AB respectively.

Proof. By Carnot's Theorem, we can write

$$\begin{aligned}BL^2 + CM^2 + AN^2 &= BN^2 + CL^2 + AM^2 \\ &= \frac{1}{2}(BL^2 + CL^2 + CM^2 + AM^2 + AN^2 + BN^2).\end{aligned}$$

However, the AM-GM inequality tells us that

$$BL^2 + CL^2 \geq \frac{1}{2}(BL + CL)^2 = \frac{1}{2}BC^2,$$

and similarly,

$$CM^2 + AM^2 \geq \frac{1}{2}CA^2 \text{ and } AN^2 + BN^2 \geq \frac{1}{2}AB^2.$$

It follows that

$$BL^2 + CM^2 + AN^2 \geq \frac{1}{2}(AB^2 + BC^2 + CA^2),$$

and so we have found the minimal value for $BL^2 + CM^2 + AN^2$, as intended. Obviously, the equality holds when P is circumcenter of triangle ABC. This completes the proof. □

Delta 2.5. Let ABC be a triangle and let D, E, F be the feet of the altitudes from A, B, C, respectively. Let X, Y, Z be the midpoints of the segments EF, FD, DE and let x, y, z be the perpendiculars from X, Y, Z to BC, CA, and AB, respectively. Prove that the lines x, y, z are concurrent.

Proof. Let M, N, P be the intersections of the lines x, y, z with the sidelines BC, CA, AB, respectively. In order to effectively use Carnot's Theorem, we need to prove that

$$(MB^2 - MC^2) + (NC^2 - NA^2) + (PA^2 - PB^2) = 0.$$

However, $MX \perp BC$, so by Theorem 2.1, we have that

$$MB^2 - MC^2 = XB^2 - XC^2,$$

and similarly

$$NC^2 - NA^2 = YC^2 - YA^2 \text{ and } PA^2 - PB^2 = ZA^2 - ZB^2.$$

Thus, we want to show that

$$(XB^2 - XC^2) + (YC^2 - YA^2) + (ZA^2 - ZB^2) = 0.$$

But XB and XC are medians in triangles EFB and EFC, so we know that

$$XB^2 = \frac{2h_b^2 + 2FB^2 - EF^2}{4}$$

and

$$XC^2 = \frac{2h_c^2 + 2EC^2 - EF^2}{4}.$$

And so we can write

$$XB^2 - XC^2 = \frac{(h_b^2 - h_c^2) + (FB^2 - EC^2)}{2}.$$

Similarly, we also obtain that

$$YC^2 - YA^2 = \frac{(h_c^2 - h_a^2) + (DC^2 - FA^2)}{2}$$

and

$$ZA^2 - ZB^2 = \frac{(h_a^2 - h_b^2) + (EA^2 - DB^2)}{2}.$$

Thus, we indeed get that

$$\begin{aligned} & (XB^2 - XC^2) + (YC^2 - YA^2) + (ZA^2 - ZB^2) \\ = & -\frac{1}{2}(DB^2 - DC^2) + (EC^2 - EA^2) + (FA^2 - FB^2) \\ = & \ 0 \end{aligned}$$

where the last equality holds because of Carnot's Theorem applied for the concurrent lines AD, BE, CF (which are the altitudes of the triangle ABC). This completes the proof. □

With some more advanced results (namely that the orthocenter of triangle ABC is the incenter of triangle DEF) we can actually prove more - let V be the circumcenter of triangle ABC and I, O, H be the incenter, circumcenter, orthocenter respectively of triangle DEF. Now let P be the midpoint of VH. We will show that these perpendiculars actually concur at point P.

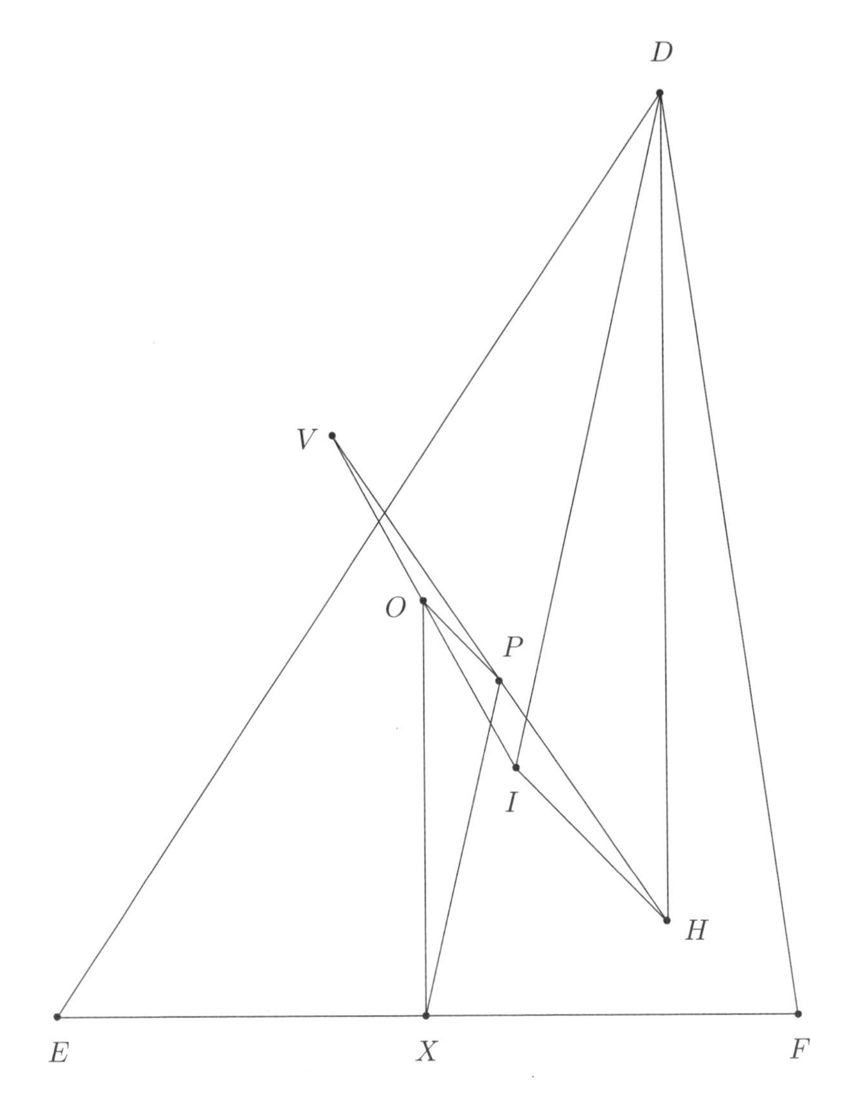

Second Proof. Since $DI \perp BC$ it suffices to show that $XP \parallel DI$. Now since I is the orthocenter of triangle ABC and since O is the nine-point center of triangle ABC we have that V is the reflection of I about O. Therefore segment OP is a midline of triangle VIH, so $OP \parallel IH$ and $IH = 2OP$. Moreover by constructing right triangles and angle chasing it is easy to show that $DH = 2R\cos EDF$ and $OX = R\cos EDF$ where R is the circumradius of triangle DEF. Therefore $DH = 2OX$ and it is clear that $DH \parallel OX$ since both lines are perpendicular to line EF. Now by looking at triangles OPX and HID we find that $XP \parallel DI$ as desired. This completes the proof. □

Delta 2.6. Consider a quadrilateral $ABCD$ inscribed in a circle in which AB is a diameter. Draw the tangents to the circle at A and B. Let E be the midpoint of segment CD. Draw the perpendicular from the midpoint of segment AD to AE and extend this perpendicular to meet the tangent at A at M. Similarly, draw the perpendicular from the midpoint of segment BC to BE and extend this perpendicular to meet the tangent at B at N. Prove that MN is parallel to CD.

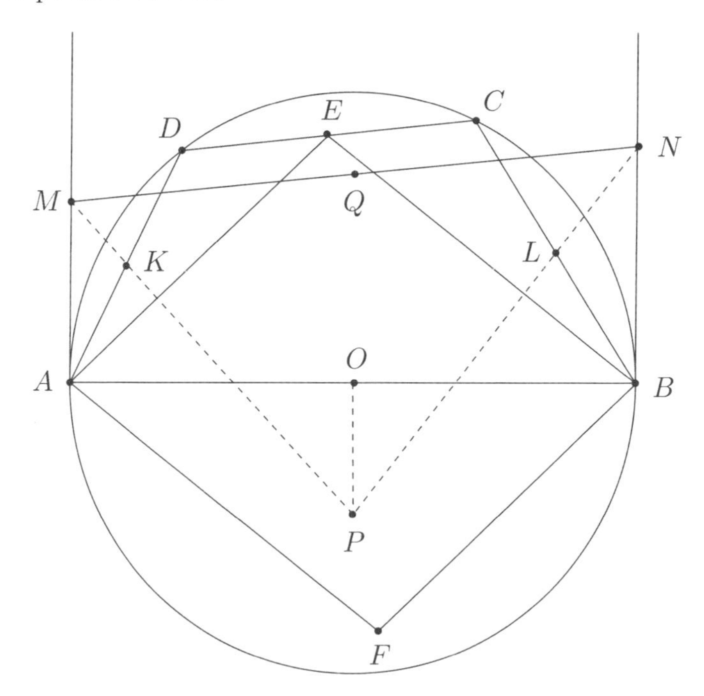

Proof. First, note that the circumcenter O of quadrilateral $ABCD$ is the midpoint of AB. Now, let K, L be the midpoints of segments DA and BC. We have that

$$AO^2 - BO^2 + BL^2 - EL^2 + EK^2 - AK^2 = \frac{1}{4}(BC^2 - BD^2 + CA^2 - DA^2) = 0.$$

Thus, by Carnot's Theorem applied for triangle ABE, the perpendiculars from O, L, K to AB, BE, EA respectively are concurrent at a point, say, P. Let Q be the midpoint of MN and let F be the reflection of E over O. Note that $AEBF$ is parallelogram and triangles PMN and AEF are similar, as their corresponding sides $PM \perp AE, PN \perp AF$ and corresponding medians $PQ \perp AO$ are perpendicular. It follows that $MN \perp EF$. But EF is the perpendicular bisector of segment CD, hence, we conclude that $MN \parallel CD$. This completes the proof. □

Finally, we move to radical axes. We've mentioned them before... but what are they?

Definition. Given two nonconcentric circles $\mathcal{C}_1$ and $\mathcal{C}_2$ with centers O_1 and O_2 respectively and radii r_1 and r_2 respectively, then the **radical axis** of $\mathcal{C}_1$ and $\mathcal{C}_2$ is the locus of points P in the plane that have equal powers with respect to the two circles.

It turns out that this locus is a line perpendicular to O_1O_2. But why? Well, we can justify this easily with the claim from the second proof of **Theorem 2.1**. Indeed, we are asking about the locus of the points P that have the same power with respect to $\mathcal{C}_1$ and $\mathcal{C}_2$, i.e. the locus of points P such that $PO_1^2 - r_1^2 = PO_2^2 - r_2^2$, or equivalently, so that $PO_1^2 - PO_2^2 = r_1^2 - r_2^2$, which is a constant. Thus, P has to lie on a fixed line perpendicular to O_1O_2 as desired.

When two circles intersect at two points, say, X and Y, then obviously their radical axis is the line XY (since both X and Y have power 0 with respect to the two circles). Similarly, when the circles are tangent at some point T, their radical axis is the internal tangent at T with respect to the two circles. However, what happens if the circles are disjoint? How can we draw their radical axis? Well, we need one more concept in order to answer this.

Definition. The **radical center** of three circles γ_1, γ_2, γ_3 is the concurrency point of the radical axes of the three pairs of circles (γ_1, γ_2), (γ_2, γ_3) and (γ_3, γ_1). Why are these lines concurrent? The reason is simple, given the definition of the radical axis. Just take P to be the intersection of the first two radical axes. By definition, P has equal powers with respect to γ_1, γ_2 and γ_3, so P needs to lie on the third radical axis as well - hence the concurrency. This gives the following very easy construction for the radical axis of two non-intersecting circles.

Construction. Let γ_1, γ_2 be two non-intersecting circles. Draw a third circle γ_3 which intersects both γ_1 and γ_2, each at two points, say X, Y on γ_1 and P, Q on γ_2. Then, lines XY and PQ intersect at some point, say R. This point represents the radical center of γ_1, γ_2, γ_3; thus, the radical axis of γ_1 and γ_2 is simply the perpendicular from R to the line joining the centers of the two circles.

Let's see these concepts in use with the following nice result, which can be remembered as some sort of Lemma on its own.

Delta 2.7. Let M and N be points on the lines AB and AC. Prove that the common chord of the circles with diameters CM and BN (the radical axis, that is!) passes through the orthocenter H of triangle ABC.

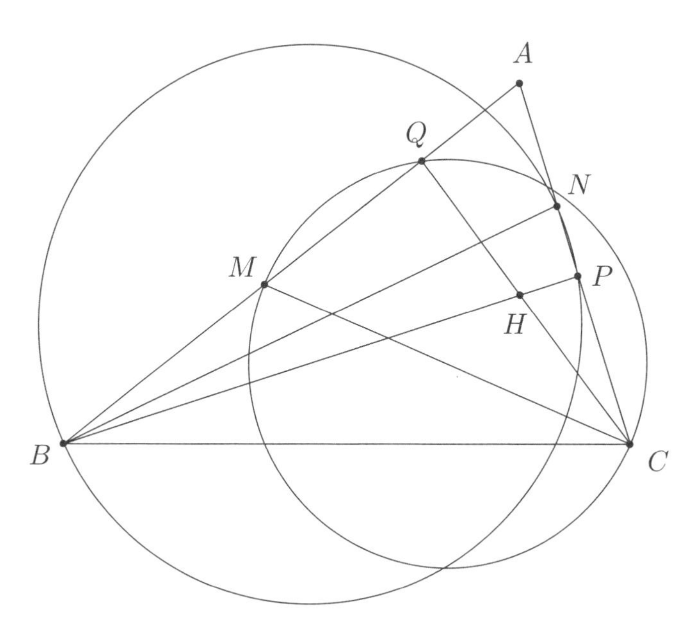

Proof. Let P and Q be the feet of the altitudes from B and C, respectively. Obviously the circle γ_1 with diameter BN passes through P; similarly, the circle γ_2 with diameter CM passes through Q. Thus, the power of the orthocenter H with respect to γ_1 is $HB \cdot HP$, whereas the power of H with respect to γ_2 is $HC \cdot HQ$. Hence, it follows that H lies on the radical axis, since $HB \cdot HP = HC \cdot HQ$ holds because $BCPQ$ is cyclic. This completes the proof. □

Delta 2.8. (IMO 2009) Let H be the orthocenter of an acute-angled triangle ABC. The circle Γ_A centered at the midpoint of BC and passing through H intersects the sideline BC at points A_1 and A_2. Similarly, define the points B_1, B_2, C_1 and C_2. Prove that six points A_1, A_2, B_1, B_2, C_1 and C_2 are concyclic.

Proof. Note that H lies on both Γ_B and Γ_C, so it needs to lie on their radical axis. Moreover, the line passing through the centers of Γ_B and Γ_C is the $A-$midline of triangle ABC, thus it is parallel to BC; hence, it follows that the radical axis of Γ_B and Γ_C is the $A-$altitude of triangle ABC. In particular, A lies on this radical axis, so it has equal powers with respect to Γ_B and Γ_C. This implies that

$$AB_1 \cdot AB_2 = AC_1 \cdot AC_2,$$

so by power of a point the points B_1, B_2, C_1, C_2 lie on a circle Ω_A.

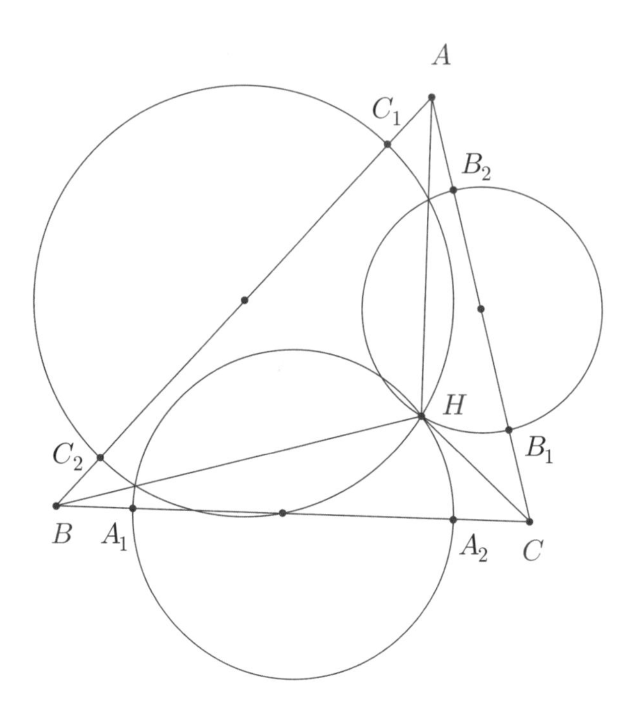

Similarly, we get that the points C_1, C_2, A_1, A_2 lie on a circle Ω_B, and A_1, A_2, B_1, B_2 lie on a circle Ω_C. If at least two of these circle coincide, then we are clearly done, since that would mean all six points A_1, A_2, B_1, B_2, C_1 and C_2 are concyclic. So let's suppose for sake of contradiction that this is not the case, and that they are all different. Then, BC is the radical axis of Ω_B and Ω_C, CA is the radical axis of Ω_C and Ω_A, and AB is the radical axis of Ω_A and Ω_B. This is obviously a contradiction, since the sidelines of triangle ABC are not concurrent! This completes the proof. □

Delta 2.9. Let ABC be a triangle and let D and E be points on sides AB and AC, respectively, such that $DE \parallel BC$. Let P be any point interior to triangle ADE, and let F and G be the intersections of DE with the lines BP and CP, respectively. Let Q be the second intersection point of the circumcircles of triangles PDG and PFE. Prove that the points A, P, and Q are collinear.

Proof. Let the circumcircle of triangle DPG meet line AB again at M, and let the circumcircle of triangle EPF meet line AC again at N. Assume the configuration where M and N lie on sides AB and AC respectively (the arguments for the other cases are similar). We have

$$\angle ABC = \angle ADG = 180^\circ - \angle BDG = 180^\circ - \angle MPC,$$

so $BMPC$ is cyclic.

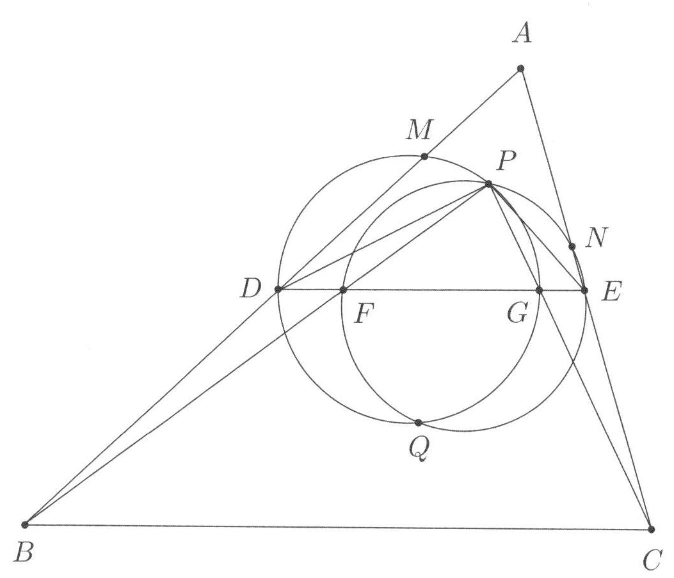

Similarly, $BPNC$ is cyclic as well. So $BCNPM$ is cyclic. Hence, $\angle ANM = \angle ABC = \angle ADE$, so points M, N, D, E are concyclic. By power of a point, this means that $AD \cdot AM = AE \cdot AD$, therefore A has equal power with respect to the circumcircles of triangles DPG and EPF, and thus A lies on the line PQ, their radical axis, as desired. □

The next problem appeared in the 2012 International Mathematical Olympiad as Problem 5 and proved to be quite challenging for numerous strong contestants. We shall give two proofs.

Delta 2.10. (IMO 2012) Let ABC be a triangle with $\angle BCA = 90°$, and let D be the foot of the altitude from C. Let X be a point in the interior of the segment CD. Let K be the point on the segment AX such that $BK = BC$. Similarly, let L be the point on the segment BX such that $AL = AC$. Let M be the point of intersection of AL and BK. Show that $MK = ML$.

We give two proofs! The first one coincides with the very simple official solution, which proved to be pretty hard to figure out.

First Proof. Let k_1 be the circle with center A and radius AC and let k_2 be the circle with center B with radius BC. Obviously $L \in k_1$ and $K \in k_2$. Furthermore, let E and F be the second intersection of the line BX with k_1 and AX with k_2, respectively. Finally, let C' be the second intersection of k_1 and k_2 - it's clear that C' lies on the line CD. From the Power of a Point Theorem in k_1 and k_2, we get

$$EX \cdot XL = CX \cdot XC' = KX \cdot XF,$$

so $EKLF$ is cyclic.

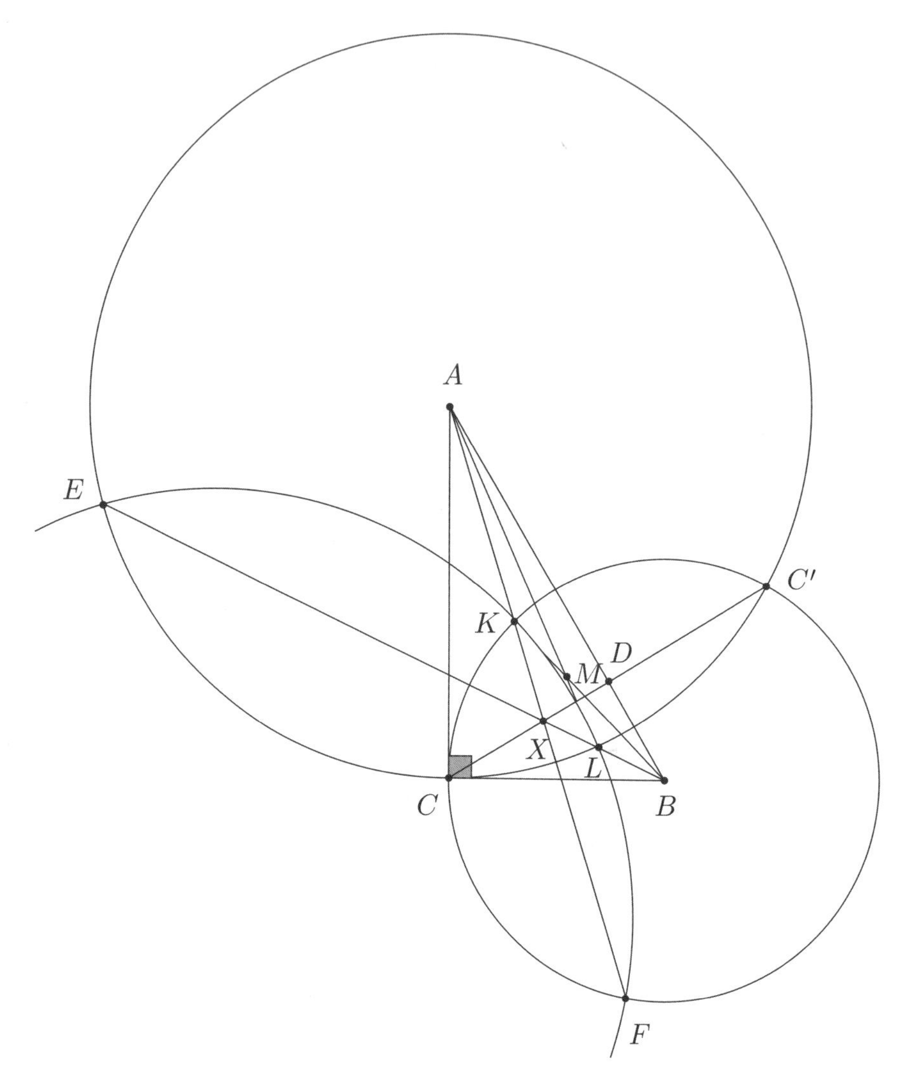

As line BC is tangent to k_1, $EB \cdot LB = CB^2 = KB^2$, so BK is tangent to the circumcircle of KLE. Analogously, AL is tangent to the circumcircle of KLF, and as $EKLF$ is cyclic, the lines AL and BK are then tangent to the same circle; therefore, it follows that $MK = ML$. This completes the first proof. □

The second solution is not natural at all. However, it involves Carnot's Theorem, so that's why we want to include it!

Second Proof. Let circumcircle of triangle ADL intersect the line DC again at U. Then, $\angle AUD = \angle ALD$. Also, $AL^2 = AC^2 = AD \cdot AB$, and so, $\angle AUD = \angle LBD = \angle XBD$. Now, this implies that triangles UAD and BXD are similar, hence

$$\frac{UD}{BD} = \frac{AD}{XD}.$$

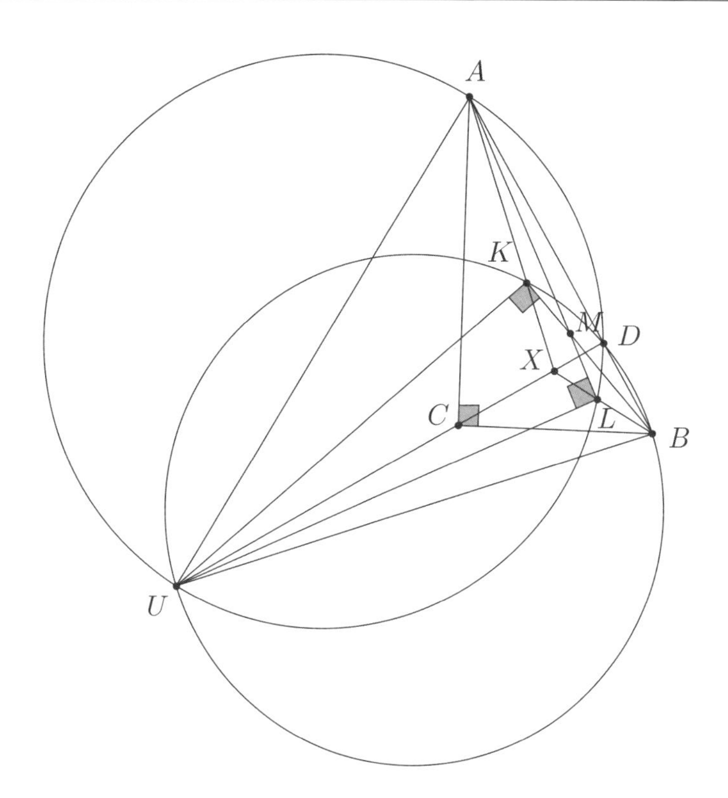

Therefore, triangles UDB and AXD are also similar, and so $\angle BUD = \angle DAX$. However, we can similarly deduce that $\angle DAX = \angle DKB$, and so $BDKU$ is cyclic. But now circles $ADLU$ and $BDKU$ have AU and BU as diameters, thus $\angle ALU = \angle BKU = 90°$. Furthermore, U also lies on CD; therefore, the perpendiculars from K, L, D to BM, AM, AB concur at a point, which is U. So, Carnot's Theorem tells us that

$$BK^2 - KM^2 + ML^2 - LA^2 + AD^2 - DB^2 = 0,$$

which from the equalities $AL = AC$, $BK = BC$, and

$$AD^2 - DB^2 = (AD^2 + DC^2) - (DC^2 + DB^2) = AC^2 - CB^2,$$

immediately yields $MK = ML$, as desired. This completes the second proof and ends the discussion. $\square$

We conclude this section with three more problem involving radical centers. The first one comes from the International Mathematical Olympiad from 1995.

Delta 2.11. (IMO 1995) Let A, B, C, and D be four distinct points on a line, in that order. The circles with diameters AC and BD intersect at X and Y. The line XY meets BC at Z. Let P be a point on the line XY other than Z.

The line CP intersects the circle with diameter AC at C and M, and the line BP intersects the circle with diameter BD at B and N. Prove that the lines AM, DN, and XY are concurrent.

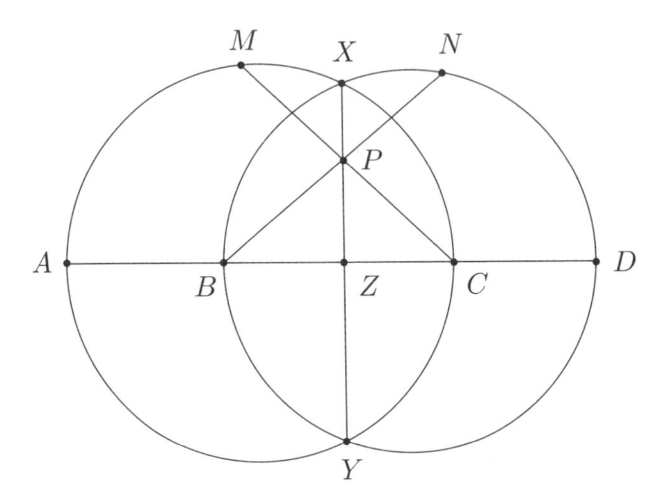

Proof. By power of a point, we have

$$PM \cdot PC = PX \cdot PY = PN \cdot PB,$$

so the points B, C, M, N are concyclic. Note that $\angle AMC = \angle BND = 90^\circ$ hence

$$\angle MND = 90^\circ + \angle MNB = 90^\circ + \angle MCA = 180^\circ - \angle MAD.$$

Therefore, the points A, D, N, M are concyclic. Since AM, DN, XY are the three radical axes of the circumcircles of $AMXC$, $BXND$, and $AMND$, they concur at the radical center of these three circles. This completes the proof. □

Delta 2.12. (Virgil Nicula, Cosmin Pohoata) Let d be an arbitrary line outside a given circle ω with center O. Denote by A the foot of the perpendicular from O to d and consider a mobile point M on ω for which X, Y are the intersections of the circle with diameter AM with ω, d respectively. Prove that the line XY passes through a fixed point.

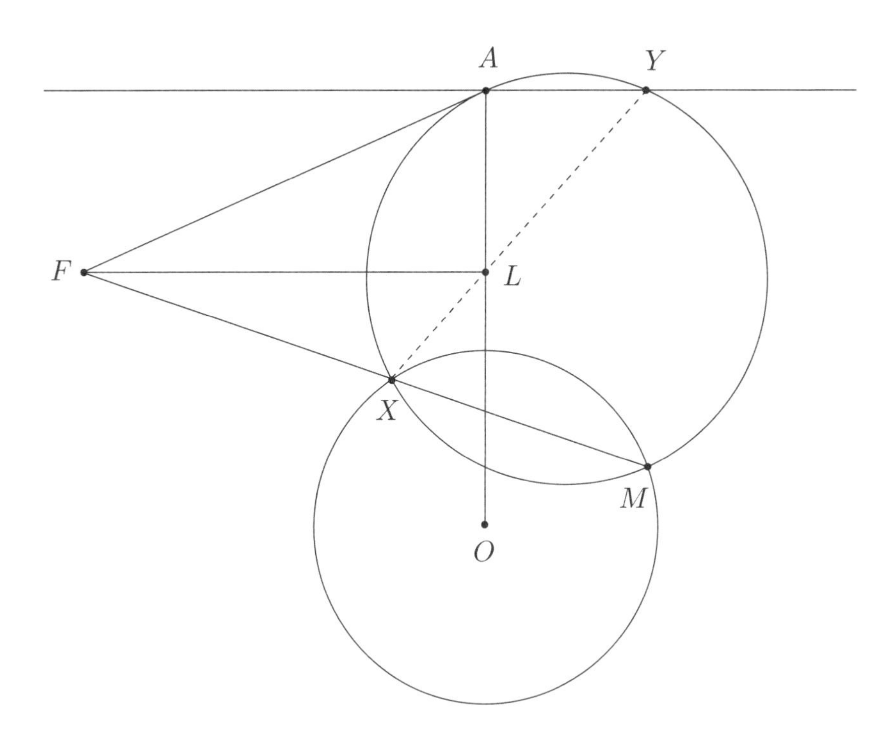

Proof. Consider the line γ the tangent at A to the circle of diameter AM and the intersections $F = \gamma \cap XM$ and $L = OA \cap XY$. Since $\angle FMA = \angle LYA$ and $\angle YAL = \angle FAM = 90°$, we deduce that the triangles $\triangle LAY$ and $\triangle FAM$ are similar. Thus $\angle ALY = \angle AFM$, which implies that the quadrilateral $AFXL$ is cyclic. But since $\angle AXF = 180° - \angle AXM = 90°$, we deduce that $\angle ALF = 90°$ as well which immediately implies that $FL \parallel d$. Now, because γ is the radical axis of the circle of diameter AM and the degenerated circle A, and XM is the radical axis of the circle of diameter AM and ω, it follows that $F = \gamma \cap XM$ is the radical center of the three mentioned circles; so, F is on the radical axis of the degenerated circle A and ω, which is fixed. Thus L is fixed as well. As L lies on XY, L is our desired fixed point. This completes the proof. □

Delta 2.13. (Warut Suksompong and Potcharapol Suteparuk, IMO 2013) Let ABC be an acute-angled triangle with orthocenter H, and let W be a point on the side BC. Denote by M and N the feet of the altitudes from B and C, respectively. Denote by ω_1 the circumcircle of BWN, and let X be the point on ω_1 which is diametrically opposite to W. Analogously, denote by ω_2 the circumcircle of triangle CWM, and let Y be the point on ω_2 which is diametrically opposite to W. Prove that X, Y and H are collinear.

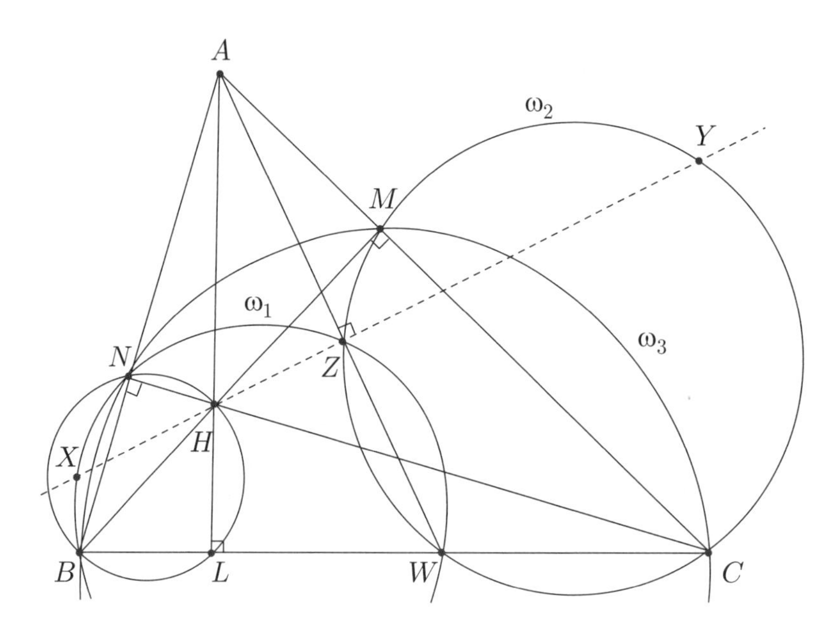

Proof. Let L be the foot of the altitude from A, and let Z be the second intersection of circles ω_1 and ω_2. We show that X, Y, Z and H lie on the same line. It's clear that points B, C, N, M lie on the circle with diameter BC - denote this circle by ω_3. Observe that the line WZ is the radical axis of ω_1 and ω_2; similarly, BN is the radical axis of ω_1 and ω_3, and CM is the radical axis of ω_2 and ω_3. Hence $A = BN \cap CM$ is the radical center of the three circles, and therefore WZ passes through A. Since WZ and WY are diameters in ω_1 and ω_2, respectively, we have $\angle WZX = \angle WZY = 90°$, so the points X and Y lie on the line through Z, perpendicular to AZ. Now, notice that

$$\angle NZM = 360° - \angle NZW - \angle MZW = \angle B + \angle C = 180° - \angle A$$

so Z lies on the circumcircle of triangle AMN. But this is the circle with diameter AH, so we have that $\angle AZH = 90°$. Therefore H also lies on the line through Z perpendicular to AZ, so the proof is complete. □

Assigned Problems

Epsilon 2.1. (Romania District Olympiad 2005) Let I, G be the incenter and the centroid of a non-isosceles triangle ABC. Prove that IG is perpendicular to the sideline BC if and only if $AB + AC = 3BC$.

Epsilon 2.2. Let ABC be a triangle with $AB < AC$. Let X and Y be points on the rays CA and BA so that $CX = AB$ and $BY = AC$. Prove that $OI \perp XY$.

Epsilon 2.3. Let ABC be a triangle and let D, E be points on AB and AC so that $DE \| BC$. Let P be an arbitrary point inside triangle ABC and let PB, PC meet DE at points F and G, respectively. Let O_1, O_2 be the circumcenters of triangles PDG and PEF. Prove that $AP \perp O_1O_2$.

Epsilon 2.4. (Romania JBMO TST 2010) Let I be the incenter of a scalene triangle ABC and denote by γ, δ the circles with diameters IB and IC, respectively. If γ', δ' are the mirror images of γ, δ in IC and IB, prove that the circumcenter O of triangle ABC lies on the radical axis of γ' and δ'.

Epsilon 2.5. In triangle ABC points E and F lie on sides AC and BC such that segments AE and BF have equal length, and circles formed by A, C, F and by B, C, E, respectively, intersect at point C and another point D. Prove that that the line CD bisects $\angle ACB$.

Epsilon 2.6. (ARO 2005) w_B and w_C are excircles of a triangle ABC. The circle w'_B is symmetric to w_B with respect to the midpoint of AC, the circle w'_C is symmetric to w_C with respect to the midpoint of AB. Prove that the radical axis of w'_B and w'_C halves the perimeter of ABC.

Epsilon 2.7. (IMO Shortlist 1998) Let ABC be a triangle with incenter I and circumcircle ω. Let D and E be the second intersection points of ω with AI and BI, respectively. The chord DE meets AC at a point F, and BC at a point G. Let P be the intersection point of the line through F parallel to AD and the line through G parallel to BE. Suppose that the tangents to ω at A and B meet at a point K. Prove that the three lines AE, BD and KP are either parallel or concurrent.

Epsilon 2.8. (USAMO 2009) Given circles ω_1 and ω_2 intersecting at points X and Y, let ℓ_1 be a line through the center of ω_1 intersecting ω_2 at points P and Q and let ℓ_2 be a line through the center of ω_2 intersecting ω_1 at points R and S. Prove that if P, Q, R and S lie on a circle, then the center of this circle lies on line XY (Hint: make sure to deal with edge cases!).

Epsilon 2.9. (ELMO Shortlist 2013) Let ABC be a scalene triangle with circumcircle Γ, and let D,E,F be the points where its incircle meets BC, AC, AB respectively. Let the circumcircles of triangles AEF, BFD, and CDE meet Γ a second time at X, Y, Z respectively. Prove that the perpendiculars from A, B, C to AX, BY, CZ respectively are concurrent.

Epsilon 2.10. (IMO Shortlist 2011) Let $ABCD$ be a convex quadrilateral whose sides AD and BC are not parallel. Suppose that the circles with diameters AB and CD meet at points E and F inside the quadrilateral. Let ω_E be the circle through the feet of the perpendiculars from E to the lines AB, BC and CD. Let ω_F be the circle through the feet of the perpendiculars from F to the lines CD, DA and AB. Prove that the midpoint of the segment EF lies on the line through of ω_E and ω_F.

Epsilon 2.11. Let ABC be a triangle and let D be the foot of the A-internal angle bisector. Let γ_1 and γ_2 be the circumcircles of triangles ABD and ACD and let P, Q be the intersections of AD with the common external tangents of γ_1 and γ_2. Prove that $PQ^2 = AB \cdot AC$. Also, find a converse!

Chapter 3

Ceva, Trig Ceva, Quadrilateral Ceva

Ceva's Theorem is a result which translates concurrencies into (trigono)metric identities. Not only does it give us a criterion for establishing when three lines are concurrent - and we all know that a lot of contest problems ask us precisely this! - but also allows us in certain more complicated settings, where the conclusion is not necessarily the concurrency of three lines, to establish connections between dispersed applications of the Law of Sines around the diagram. We will see this through some examples, of course! But later. First, let us prove Ceva's Theorem.

Theorem 3.1. (Ceva's Theorem) Let ABC be a triangle and let A_1, B_1, C_1 be points on the sides BC, CA, AB of triangle ABC. Then, AA_1, BB_1, CC_1 are concurrent if and only if

$$\frac{A_1B}{A_1C} \cdot \frac{B_1C}{B_1A} \cdot \frac{C_1A}{C_1B} = 1.$$

Let us first assume that we proved the direct implication. That is, let us say we know that if AA_1, BB_1, CC_1 are concurrent, then

$$\frac{A_1B}{A_1C} \cdot \frac{B_1C}{B_1A} \cdot \frac{C_1A}{C_1B} = 1.$$

Let us see how can we prove the converse using this! This is actually very easy. Assume we have three points A_1, B_1, C_1 on the sides BC, CA, AB such that

$$\frac{A_1B}{A_1C} \cdot \frac{B_1C}{B_1A} \cdot \frac{C_1A}{C_1B} = 1.$$

We want to show that the lines AA_1, BB_1, CC_1 are concurrent. Well, we argue by contradiction; suppose that this is not the case, and take the intersection P of the lines BB_1 and CC_1. Let $A_2 = AP \cap BC$. By the direct implication, since the lines AA_2, BB_1, CC_1 are concurrent, we have that

$$\frac{A_2B}{A_2C} \cdot \frac{B_1C}{B_1A} \cdot \frac{C_1A}{C_1B} = 1.$$

Combining this with the previous identity, we get that $\frac{A_1B}{A_1C} = \frac{A_2B}{A_2C}$. Hence, given that points A_1, A_2 both lie in the interior of BC, we conclude that $A_1 = A_2$, contradiction! Therefore, lines AA_1, BB_1, CC_1 are concurrent, proving the converse.

Returning to the direct implication, we now give two proofs. The first one is rather classical and can be seen in most geometry textbooks.

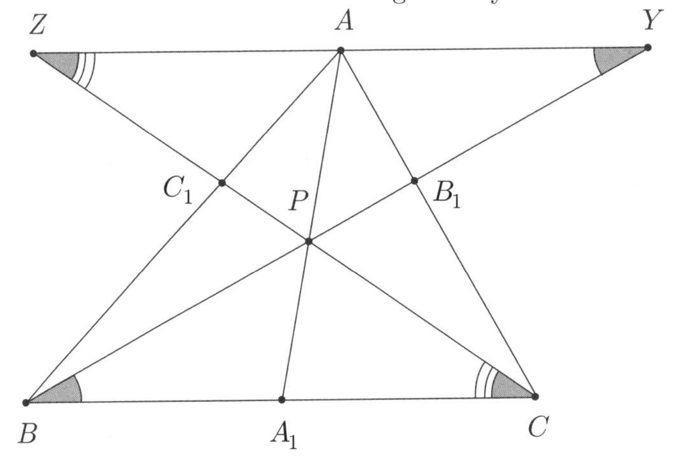

First Proof. Consider the line ℓ parallel to BC which passes through the vertex A and let Y, Z be the intersections of ℓ with BB_1 and CC_1, respectively. Also, let P be the concurrency point of lines AA_1, BB_1, CC_1. We have that $\frac{B_1C}{B_1A} = \frac{BC}{AY}$ (from the similarity of triangles AYB_1 and CBB_1) and $\frac{C_1A}{C_1B} = \frac{AZ}{BC}$ (from the similarity of triangles AZC_1 and BCC_1). Therefore, we get that

$$\begin{aligned} \frac{A_1B}{A_1C} \cdot \frac{B_1C}{B_1A} \cdot \frac{C_1A}{C_1B} &= \frac{A_1B}{A_1C} \cdot \frac{BC}{AY} \cdot \frac{AZ}{BC} \\ &= \frac{A_1B}{A_1C} \cdot \frac{AZ}{AY}. \end{aligned}$$

But $\frac{A_1B}{AY} = \frac{A_1C}{AZ} = \frac{A_1P}{PA}$ (from the similarities of triangles A_1PB and APY

and of triangles A_1PC and APZ). Hence, we conclude that

$$\begin{aligned}\frac{A_1B}{A_1C}\cdot\frac{B_1C}{B_1A}\cdot\frac{C_1A}{C_1B} &= \frac{A_1B}{A_1C}\cdot\frac{AZ}{AY}\\ &= \frac{A_1B}{AY}\cdot\frac{AZ}{A_1C}\\ &= 1.\end{aligned}$$

This settles the first proof.

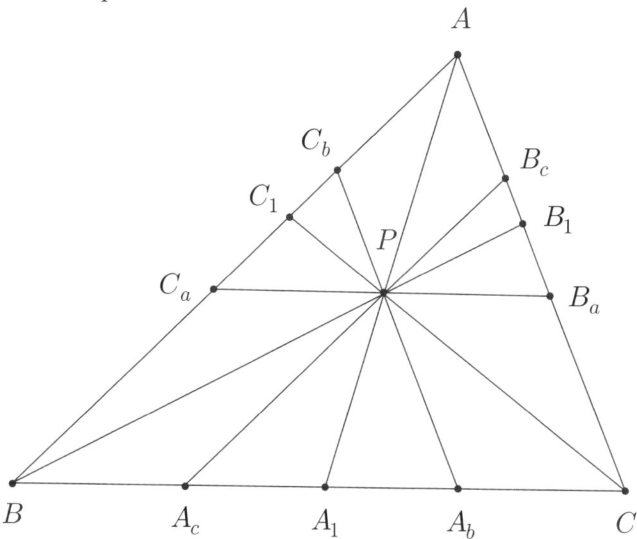

Second Proof. Consider the parallel lines ℓ_a, ℓ_b, ℓ_c to BC, CA, and AB, respectively, which pass through the concurrency point P of the lines AA_1, BB_1, CC_1. Let B_a, C_a be the intersections of ℓ_a with CA, AB, A_b, C_b the intersections of ℓ_b with BC, AB, and last but not least, A_c, C_c the intersections of ℓ_c with BC, CA. First, note that $\frac{A_1B}{A_1C} = \frac{PC_a}{PB_a}$ (explain this to yourselves if not fully clear!). Then, observe that $\frac{PC_a}{BC} = \frac{PC_1}{CC_1}$ (from the similarity of triangles C_1C_aP and C_1BC) and $\frac{PB_a}{BC} = \frac{PB_1}{BB_1}$ (from the similarity of triangles B_1PB_a and B_1BC). Thus, we get that

$$\begin{aligned}\frac{A_1B}{A_1C} &= \frac{PC_a}{PB_a}\\ &= \frac{PC_1}{CC_1} : \frac{PB_1}{BB_1}.\end{aligned}$$

Similarly, we obtain that $\frac{B_1C}{B_1A} = \frac{PA_1}{AA_1} : \frac{PC_1}{CC_1}$ and $\frac{C_1A}{C_1B} = \frac{PB_1}{BB_1} : \frac{PA_1}{AA_1}$. And so we can conclude that

$$\begin{aligned}\frac{A_1B}{A_1C}\cdot\frac{B_1C}{B_1A}\cdot\frac{C_1A}{C_1B} &= \left(\frac{PC_1}{CC_1} : \frac{PB_1}{BB_1}\right)\cdot\left(\frac{PA_1}{AA_1} : \frac{PC_1}{CC_1}\right)\cdot\left(\frac{PB_1}{BB_1} : \frac{PA_1}{AA_1}\right)\\ &= 1.\end{aligned}$$

This wraps up the second proof and thus settles Ceva's Theorem. □

We haven't used trigonometry at all, but as we shall very soon see, this kind of computation will very often lead to identities with sines (when things are not so neat).

Now, Ceva's theorem allows us to establish the existence of many important points in triangle geometry. Consider a triangle ABC...

Corollary 3.1. (The Centroid) Let M, N, P be the midpoints of the sides BC, CA, AB of triangle ABC. Then AM, BN, CP are concurrent at G, the centroid of ABC.

Proof. This is easy!

Corollary 3.2. (The Orthocenter) Let D, E, F be the feet of the altitudes from A, B, C. Then, AD, BE, CF are concurrent at H, the orthocenter of ABC.

Proof. This is slightly trickier, yet... $\frac{DB}{DC} = \frac{c\cos B}{b\cos C}$ etc.

Corollary 3.3. (The Gergonne point) Let A_1, B_1, C_1 be the point of tangency of the incircle with the triangle's sides. Then AA_1, BB_1, CC_1 are concurrent at Γ, the Gergonne point.

Proof. Hint only: Recall that $AB_1 = AC_1 = s - a$, $BC_1 = BA_1 = s - b$, $CA_1 = CB_1 = s - c$, where s is the semiperimeter of triangle ABC.

Corollary 3.4. (The Nagel point) Let A_2, B_2, C_2 be the point of tangency of the excircles with the triangle's sides. Then AA_2, BB_2, CC_2 are concurrent at N_a, the Nagel point.

Proof. Again hint only: Recall that $A_2B = s-c$, $A_2C = s-b$, $B_2C = s-a$, $B_2A = s - c$, $C_2A = s - b$, $C_2B = s - a$.

//With the notations from above, we also have that $MA_1 = MA_2$, $NB_1 = NB_2$, and $PC_1 = PC_2$. Remember this!

As a matter of fact, motivated by the above, let us see a first instance of Ceva's Theorem in a contest-like problem.

Delta 3.1. Let ABC be a triangle and let P be a point in its interior. Let X_1, Y_1, Z_1 be the intersections of AP, BP, CP with BC, CA, and AB, respectively. Furthermore, let X_2, Y_2, Z_2 be the reflections of the points X_1, Y_1, Z_1 over the midpoints of BC, CA, and AB, respectively. Prove that the lines AX_2, BY_2, CZ_2 are concurrent. This concurrency point is called **the isotomic conjugate of P with respect to triangle ABC.**

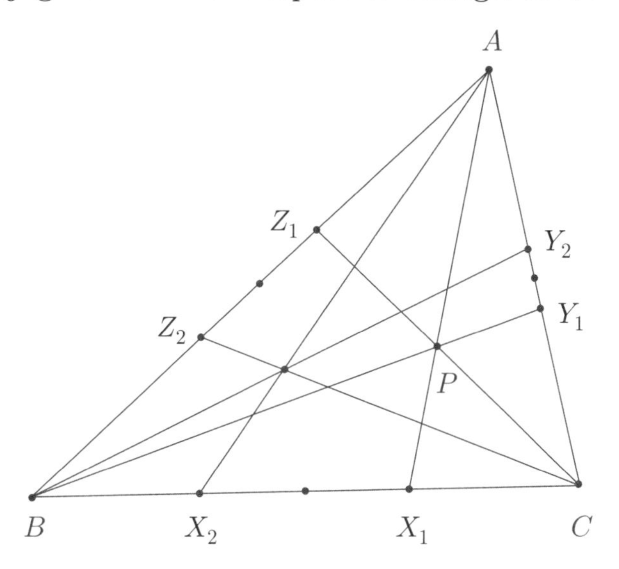

Proof. We have that $X_1B = X_2C$ and $X_1C = X_2B$ (since the segments X_1X_2 and BC share the same midpoint). Hence, $\frac{X_2B}{X_2C} = \frac{X_1C}{X_1B}$, and similarly

$$\frac{Y_2C}{Y_2A} = \frac{Y_1A}{Y_1C} \text{ and } \frac{Z_2A}{Z_2B} = \frac{Z_1B}{Z_1A}.$$

Therefore,

$$\begin{aligned} \frac{X_2B}{X_2C} \cdot \frac{Y_2C}{Y_2A} \cdot \frac{Z_2A}{Z_2B} &= \frac{X_1C}{X_1B} \cdot \frac{Y_1A}{Y_1C} \cdot \frac{Z_1B}{Z_1A} \\ &= \left(\frac{X_1B}{X_1C} \cdot \frac{Y_1C}{Y_1A} \cdot \frac{Z_1A}{Z_1B}\right)^{-1} \\ &= 1, \end{aligned}$$

where the last equality holds because of the direct implication of Ceva's Theorem, since the lines AX_1, BY_1, CZ_1 are concurrent at P. Thus, the **converse** of Ceva's theorem allows us to conclude that the lines AX_2, BY_2, CZ_2 are concurrent! □

This is the first and last time we are going to refer to the components of Ceva's Theorem separately (the direct implication and the converse); we will just say "by Ceva's Theorem", so you always should keep in mind which side of the result we are applying!

//With this new terminology, we can now say that the Gergonne point and the Nagel point of a triangle are isotomic conjugates. There are also some other nice pairs of isotomic conjugates within a triangle, but let's not worry about them at this point.

A very similar result is the following (we will leave it as an exercise): Before attempting it, be sure you know what the power of a point with respect to a circle is.

Delta 3.2. Let ABC be a triangle and let P be a point in its interior. Let X_1, Y_1, Z_1 be the intersections of AP, BP, CP with BC, CA, and AB, respectively. Let the circumcircle of triangle $X_1Y_1Z_1$ intersect the sides of BC, CA, and AB again at X_3, Y_3, and Z_3, respectively. Prove that the lines AX_3, BY_3, CZ_3 are concurrent.

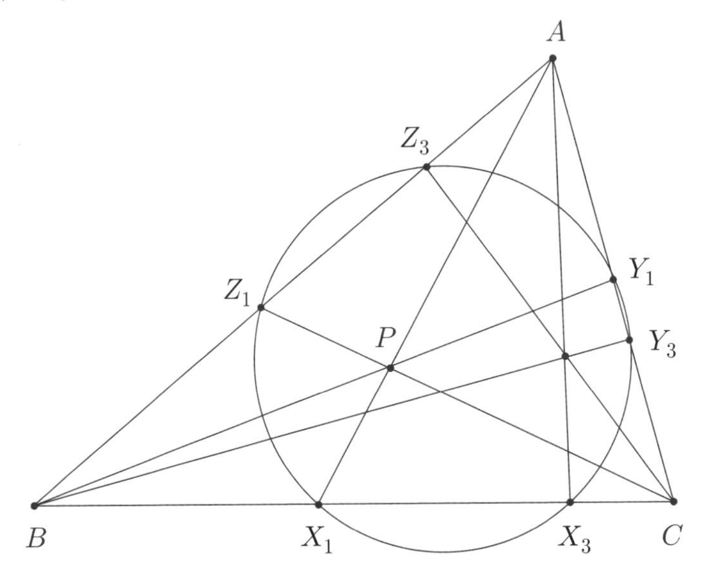

Notice again that this situation generalizes a configuration that we might have seen in our everyday geometry problems: the nine-point/Euler circle. (If you haven't heard about it at this point, do not worry; we will cover it later on in this material when we will prove some theorems about isogonal conjugates.). If P is, say, the orthocenter of triangle ABC, then the circle passing through the feet of the altitudes X_1, Y_1, Z_1 is precisely the nine-point circle of ABC, so it intersects the sides again at the midpoints X_3, Y_3, Z_3 of BC, CA, AB, respectively. Obviously, in this case, the lines AX_3, BY_3, CZ_3 are concurrent (at the centroid of triangle ABC - as we have previously seen).

Delta 3.3. Let $ABCD$ be a trapezoid with $AD \parallel BC$. Lines AB and CD meet at P, segments AC and BD meet at Q. Prove that M, N, P, Q are collinear, where M and N are the midpoints of sides AD and BC. (Hint: Why are M, Q, P collinear? And why are M, Q, N collinear?)

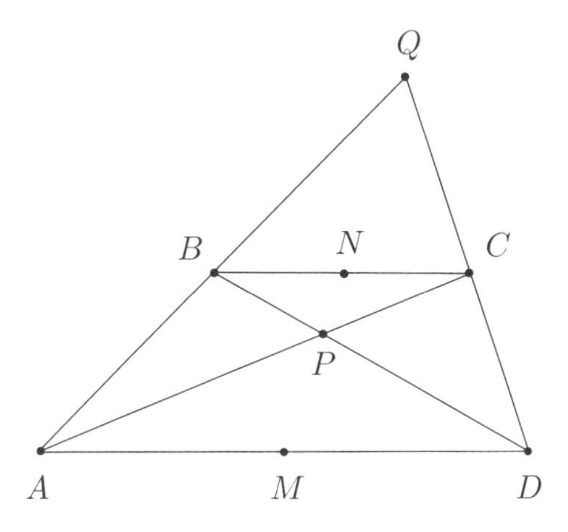

Proof. As hinted, we want to show first that the points M, Q, P are collinear, and then that the points M, Q, N are collinear. Indeed, because $AD \| BC$, we have that $\frac{BP}{BA} = \frac{CP}{CD}$; thus, since

$$1 \cdot \frac{CD}{CP} \cdot \frac{BP}{BA} = \frac{MA}{MB} \cdot \frac{CD}{CP} \cdot \frac{BP}{BA} = 1,$$

Ceva's Theorem yields that the lines QM, AC, BD are concurrent, i.e. the points M, Q, P are collinear. The collinearity of M, Q, N is immediate from similar triangles. Indeed, take line MQ and intersect it with BC, and let N' be the intersection point. Triangles MQD and $N'QB$ are similar, so $\frac{MD}{N'B} = \frac{MQ}{QN'}$; however, triangles MQA and $N'QA$ are also similar, so $\frac{MQ}{QN'} = \frac{MA}{N'C}$; thus, we have that

$$\frac{MD}{N'B} = \frac{MA}{N'C},$$

and since $MA = MD$, it follows that N' is the midpoint of BC, so $N' = N$, which completes the proof. □

Delta 3.4. (USAMO 2003) Let ABC be a triangle. A circle passing through A and B intersects the segments AC and BC at D and E, respectively. Lines AB and DE intersect at F, while lines BD and CF intersect at M. Prove that $MF = MC$ if and only if $MB \cdot MD = MC^2$.

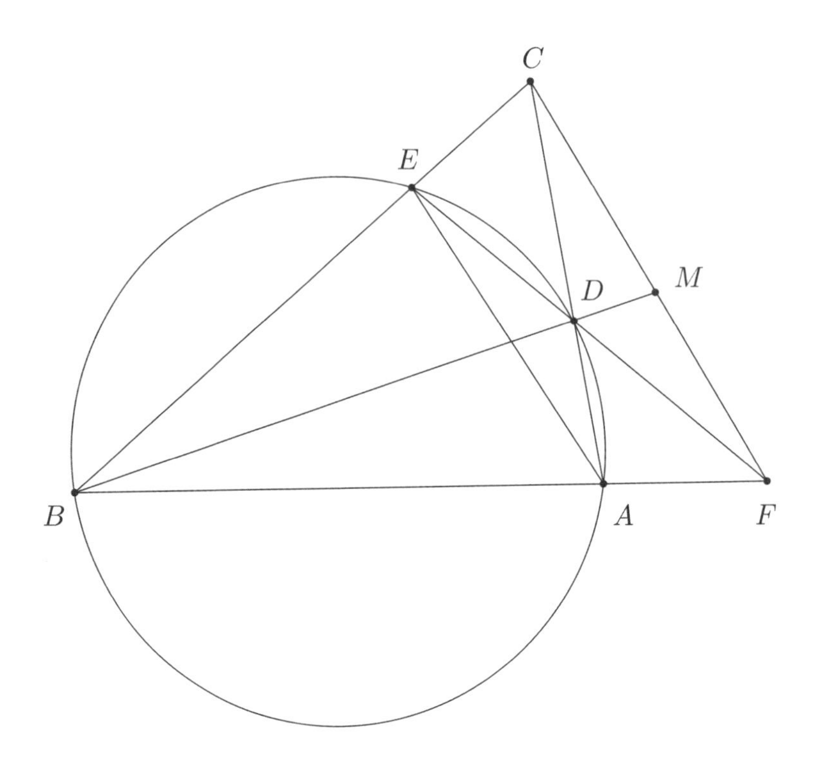

Proof. By Ceva's Theorem on triangle BCF we have that

$$\frac{MF}{MC} \cdot \frac{EC}{EB} \cdot \frac{AB}{AF} = 1$$

so

$$MF = MC \iff \frac{EC}{EB} = \frac{AF}{AB} \iff EA \parallel CF$$

but if $EA \parallel CF$ then $\angle MCD = \angle DAE = \angle DBC$ which would imply that line MC is tangent to the circumcircle of triangle BCD. But by Power of a Point, that tangency is equivalent to $MC^2 = MB \cdot MD$, as desired. □

Now, it's time for Trigonometric Ceva!

Theorem 3.2. (Ceva's Theorem - Dr. Trig form) Let ABC be a triangle and let A_1, B_1, C_1 be points on the sides BC, CA, AB of triangle ABC. Then, AA_1, BB_1, CC_1 are concurrent if and only if

$$\frac{\sin A_1AB}{\sin A_1AC} \cdot \frac{\sin C_1CA}{\sin C_1CB} \cdot \frac{\sin B_1BC}{\sin B_1BA} = 1.$$

Proof. The proof uses the original version of Ceva's Theorem! As a matter of fact, the two are equivalent. But how can we get those sines? Law of Sines, of course! First, note that

$$\frac{A_1B}{AB} = \frac{\sin A_1AB}{\sin AA_1B} \text{ and } \frac{A_1C}{AC} = \frac{\sin A_1AC}{\sin AA_1C}$$

(by the Law of Sines applied twice, in triangles AA_1B and AA_1C, respectively).

Now, observe that $\sin AA_1B = \sin AA_1C$ since one angle is the supplement of the other; hence, by dividing the two relations, we get that

$$\frac{A_1B}{AB} : \frac{A_1C}{AC} = \frac{\sin A_1AB}{\sin A_1AC}.$$

And look - we got in the RHS the ratio we need in our statement! Similarly we get that

$$\frac{\sin C_1CA}{\sin C_1CB} = \frac{C_1A}{CA} : \frac{C_1B}{CB} \quad \text{and} \quad \frac{\sin B_1BC}{\sin B_1BA} = \frac{B_1C}{BC} : \frac{B_1A}{BA},$$

and so we conclude that

$$\begin{aligned} & \frac{\sin A_1AB}{\sin A_1AC} \cdot \frac{\sin C_1CA}{\sin C_1CB} \cdot \frac{\sin B_1BC}{\sin B_1BA} \\ = & \left(\frac{A_1B}{AB} : \frac{A_1C}{AC}\right) \cdot \left(\frac{C_1A}{CA} : \frac{C_1B}{CB}\right) \cdot \left(\frac{B_1C}{BC} : \frac{B_1A}{BA}\right) \\ = & \frac{A_1B}{A_1C} \cdot \frac{B_1C}{B_1A} \cdot \frac{C_1A}{C_1B}. \end{aligned}$$

And this holds for any three points A_1, B_1, C_1 on the sides BC, CA, AB. Thus, via Ceva's Theorem, we can conclude that the lines AA_1, BB_1, CC_1 are concurrent if and only if

$$\frac{\sin A_1AB}{\sin A_1AC} \cdot \frac{\sin C_1CA}{\sin C_1CB} \cdot \frac{\sin B_1BC}{\sin B_1BA} = 1,$$

as claimed. □

//Remember the relation

$$\frac{A_1B}{AB} : \frac{A_1C}{AC} = \frac{\sin A_1AB}{\sin A_1AC},$$

or equivalently

$$\frac{A_1B}{A_1C} = \frac{AB}{AC} \cdot \frac{\sin A_1AB}{\sin A_1AC}.$$

In fact, this holds for any point A_1 on the line BC. It is *extremely* useful and we will use it numerous times in this material. Since it is so important, we isolate it as an exercise.

Delta 3.5. (The Ratio Lemma) Let ABC be a triangle and let A_1 be a point on the line BC. Then, prove that

$$\frac{A_1B}{A_1C} = \frac{AB}{AC} \cdot \frac{\sin A_1AB}{\sin A_1AC}.$$

Corollary 3.5. (The Angle-Bisector Theorem) Let D be a point on the sideline BC of triangle ABC. Then $\frac{DB}{DC} = \frac{AB}{AC}$ if and only if AD is an angle bisector of angle A (either internal or external).

Now, notice that the we can actually generalize the statement of Trig Ceva. First, note that in the statement of **Theorem 3.2** we just have the angles of the type $\angle A_1AB$ and $\angle A_1AC$; thus, we don't really need A_1 to lie on BC. Hence, what can we infer? Well, that A_1, B_1, C_1 can be any points in plane! More precisely, general Trig Ceva states that given any points A_1, B_1, C_1 in plane, then AA_1, BB_1, CC_1 are concurrent if and only if

$$\frac{\sin A_1AB}{\sin A_1AC} \cdot \frac{\sin C_1CA}{\sin C_1CB} \cdot \frac{\sin B_1BC}{\sin B_1BA} = 1.$$

Do remember this!

Now, some simple corollaries justifying the existence of some important triangle centers!

Corollary 3.6. (The Incenter) The internal angle bisectors of the angles of the triangle ABC are concurrent at I, the incenter of ABC.

Proof. What is there to prove? If AA_1, BB_1, CC_1 are the internal angle bisectors of ABC, what's the value of $\frac{\sin A_1AB}{\sin A_1AC}$?

Corollary 3.7. (The Symmedian/Lemoine point) The reflections of the medians in the corresponding internal angle bisectors are concurrent at K, the symmedian point of ABC.

Proof. Let M, N, P be the midpoints of the sides BC, CA, AB, respectively, and let X, Y, Z be the intersections of the sides of ABC with the reflections of lines AM, BN, CP in their corresponding internal angle bisectors - by the way, the lines AX, BY, CZ are called the symmedians of triangle ABC. They have a collection of very interesting properties which we will see very soon. Now, we see that $\frac{\sin XAB}{\sin XAC} = \frac{\sin MAC}{\sin MAB}$ by construction; thus, we get that

$$\begin{aligned}\frac{\sin XAB}{\sin XAC} \cdot \frac{\sin ZCA}{\sin ZCB} \cdot \frac{\sin YBC}{\sin YBA} &= \left(\frac{\sin MAB}{\sin MAC} \cdot \frac{\sin PCA}{\sin PCB} \cdot \frac{\sin NBC}{\sin NBA}\right)^{-1} \\ &= 1,\end{aligned}$$

where the latter holds because the medians AM, BN, CP are concurrent at the centroid G of triangle ABC!

And even more generally:

Delta 3.6. Let ABC be a triangle and let P be a point in its interior. Let X_1, Y_1, Z_1 be the intersections of AP, BP, CP with BC, CA, and AB, respectively. Furthermore, let AX_2, BY_2, CZ_2 be the reflections of lines AX_1, BY_1, CZ_1 in the corresponding internal angle bisectors of ABC. Then, these reflections are concurrent and their concurrency point is called the **isogonal conjugate of P with respect to triangle ABC.**

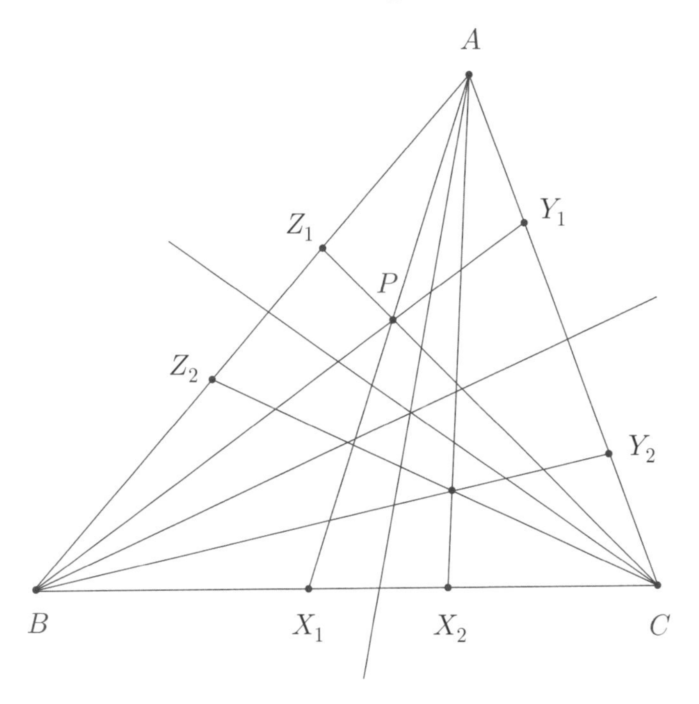

Proof. We have that $\angle X_1AB = \angle X_2AC$ and $\angle X_2AB = \angle X_1AC$ so $\frac{\sin X_1AB}{\sin X_1AC} = \frac{\sin X_2AC}{\sin X_2AB}$. Multiplying this equation with similar results yields

$$\begin{aligned}\frac{\sin X_2AC}{\sin X_2AB} \cdot \frac{\sin Y_2BA}{\sin Y_2BC} \cdot \frac{\sin Z_2CB}{\sin Z_2CA} &= \left(\frac{\sin X_1AC}{\sin X_1AB} \cdot \frac{\sin Y_1BA}{\sin Y_1BC} \cdot \frac{\sin Z_1CB}{\sin Z_1CA}\right)^{-1} \\ &= 1,\end{aligned}$$

where the latter holds because the cevians AX_1, BY_1, CZ_1 are concurrent at P. Hence, by Trig Ceva, the proof is complete. □

We will return to isogonal conjugates very soon, as they have numerous interesting properties that are exploited in various contests around the world. For now, let's just see a few more applications of Ceva.

Delta 3.7. Points A_1, B_1, C_1 are chosen on the sides BC, CA, AB, respectively of a triangle ABC. Denote by G_a, G_b, G_c are the centroids of triangles AB_1C_1, BC_1C_1, CA_1B_1, respectively. Prove that the lines AG_a, BG_b, CG_c are concurrent if and only if lines AA_1, BB_1, CC_1 are concurrent.

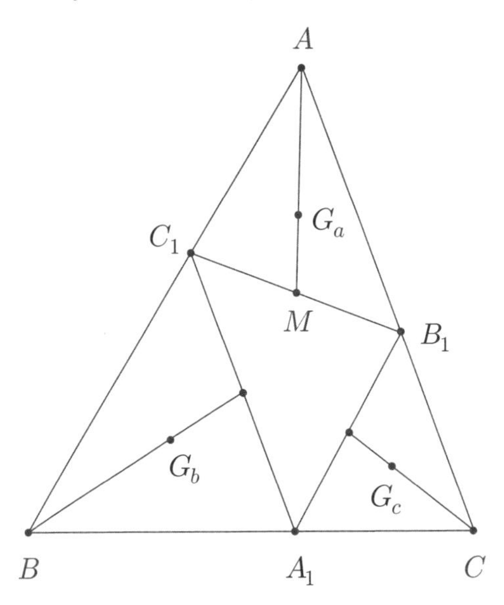

Proof. Let M be the midpoint of B_1C_1. Then, the Ratio Lemma yields that

$$\frac{AB_1}{AC_1} = \frac{\sin C_1AM}{\sin B_1AM} = \frac{\sin C_1AG_a}{\sin B_1AG_a}$$

Doing the same for the rest and multiplying yields that

$$\frac{\sin C_1AG_a}{\sin B_1AG_a} \cdot \frac{\sin B_1CG_c}{\sin A_1CG_c} \cdot \frac{\sin A_1BG_b}{\sin C_1BG_b} = \frac{AB_1}{AC_1} \cdot \frac{A_1C}{BC_1} \cdot \frac{BC_1}{A_1B}.$$

Thus, one of these is equal to one if and only if the other one is equal to one as well. By the converses of Trig Ceva and regular Ceva, we see that AA_1, BB_1, and CC_1 concur if and only if AG_a, BG_b, and CG_c concur as desired. $\square$

Delta 3.8. Let D be the foot of the altitude from A in triangle ABC and let M, N be points on the sides CA, AB such that the lines BM and CN intersect on AD. Prove that AD is the bisector of angle $\angle MDN$.

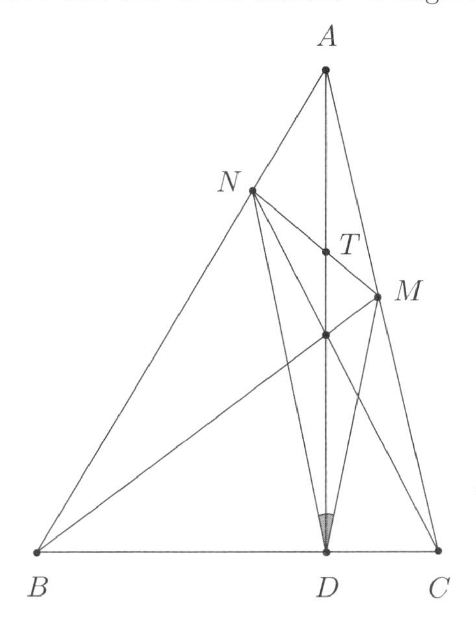

Proof. Let T be the intersection of AD with MN. By the Ratio Lemma, we have that

$$\frac{TM}{TN} = \frac{AM}{AN} \cdot \frac{\sin DAC}{\sin DAB}.$$

The same Ratio Lemma, however, tells us that

$$\frac{DC}{DB} = \frac{AC}{AB} \cdot \frac{\sin DAC}{\sin DAB}.$$

Hence, we get that

$$\frac{TM}{TN} = \frac{AM}{AN} \cdot \frac{DC}{DB} \cdot \frac{AB}{AC} = \frac{AM}{AN} \cdot \frac{DC}{DB} \cdot \frac{\sin C}{\sin B},$$

where the last equality holds because of the Law of Sines applied in triangle ABC.

On the other hand from the Law of Sines, applied twice in triangles BDN and CDM, we get that

$$DM = CM \cdot \frac{\sin C}{\sin CDM} = CM \cdot \frac{\sin C}{\sin (90^\circ - MDA)} = CM \cdot \frac{\sin C}{\cos MDA},$$

and similarly

$$DN = BN \cdot \frac{\sin B}{\cos NDA};$$

thus,

$$\frac{DM}{DN} = \frac{CM}{BN} \cdot \frac{\sin C}{\sin B} \cdot \frac{\cos NDA}{\cos MDA}.$$

But the lines AD, BM, CN are concurrent, so from Ceva's Theorem we know that

$$\frac{DB}{DC} \cdot \frac{MC}{MA} \cdot \frac{NA}{NB} = 1, \text{ i.e. } \frac{DC}{DB} \cdot \frac{AM}{AN} = \frac{CM}{BN}.$$

Hence, we get that

$$\frac{TM}{TN} = \frac{CM}{BN} \cdot \frac{\sin C}{\sin B} = \frac{DM}{DN} \cdot \frac{\cos MDA}{\cos NDA}.$$

But the Ratio Lemma gave us that

$$\frac{TM}{TN} = \frac{DM}{DN} \cdot \frac{\sin MDA}{\sin NDA};$$

thus $\tan MDA = \tan NDA$, and so the angles $\angle MDA$ and $\angle NDA$ are equal, as claimed. □

Delta 3.9. Conversely, in the same configuration as above, but this time considering the points M, N to be defined on CA, AB such that $\angle MDA = \angle NDA$, prove that the lines AD, BM, CN are concurrent.

The "if and only if" statement coming from these two implications is known in literature as Blanchet's Theorem. See, for example, [19], [37]. We will come back to this in a separate Chapter/Appendix as well and see some very beautiful applications. For the moment, we continue with a few more applications of Ceva.

Delta 3.10. Let ABC be an arbitrary triangle, and let D, E, F be any three points on the lines BC, CA, AB such that the lines AD, BE, CF concur. Let the parallel to the line AB through the point E meet the line DF at a point Q, and let the parallel to the line AB through the point D meet the line EF at a point T. Then, the lines CF, DE and QT are concurrent.

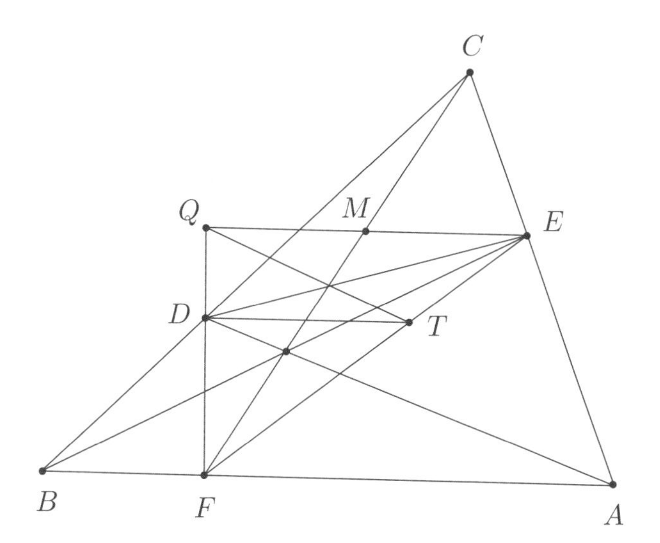

Proof. We draw D, E, F in the interiors of segments BC, CA, AB for convenience. The proof can be adapted for the other cases very simply. We proceed by hoping that Ceva's Theorem will help us. More precisely, if we let M to be the intersection of CF and EQ, note that the concurrency of CF, DE, QT is equivalent to proving that

$$\frac{MQ}{ME} \cdot \frac{TE}{TF} \cdot \frac{DF}{DQ} = 1$$

(by Ceva in triangle FQE). However, $TD \| EQ$, so it suffices to show that $MQ = ME$, as triangles FDT and FQE are similar. Now, in order to evaluate $\frac{MQ}{ME}$ we use the Ratio Lemma, of course! We have that

$$\frac{MQ}{ME} = \frac{FQ}{FE} \cdot \frac{\sin CFQ}{\sin CFE} = \frac{\sin FEQ}{\sin FQE} \cdot \frac{\sin CFQ}{\sin CFE},$$

where the last equality holds because of the Law of Sines in triangle FQE. However, $\angle FQE = \angle BFD$ and $\angle FEQ = \angle AFE$ because $EQ \| AB$; thus

$$\frac{MQ}{ME} = \frac{\sin BFD}{\sin AFE} \cdot \frac{\sin CFQ}{\sin CFE}.$$

However, the Ratio Lemma again tells us that

$$\frac{DB}{DC} = \frac{FB}{FC} \cdot \frac{\sin BFD}{\sin CFQ} \text{ and } \frac{EC}{EA} = \frac{FC}{FA} \cdot \frac{\sin CFE}{\sin AFE}.$$

Thus, it follows that

$$\begin{aligned} \frac{MQ}{ME} &= \frac{\sin BFD}{\sin AFE} \cdot \frac{\sin CFQ}{\sin CFE} \\ &= \frac{DB}{DC} \cdot \frac{EC}{EA} \cdot \frac{FA}{FB} \\ &= 1, \end{aligned}$$

where the last equality holds because of Ceva's Theorem, as the lines AD, BE, CF are concurrent! This completes the proof. □

Delta 3.11. (Cevian Nest Theorem) Let D, E, F be points on sides BC, CA, AB respectively of a triangle ABC. Also let X, Y, Z be points on sides EF, FD, DE respectively of triangle DEF. Consider the three triples of lines (AX, BY, CZ), (AD, BE, CF), and (DX, EY, FZ). If any two of these triples are concurrent, the third one is as well.

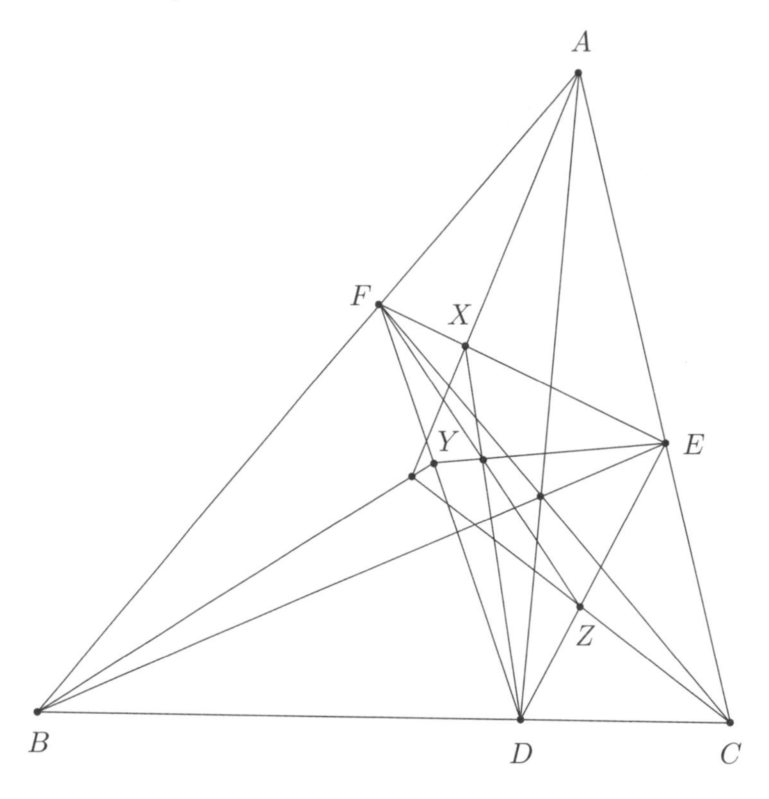

Proof. We will prove that if the triples (AD, BE, CF) and (DX, EY, FZ) are concurrent, so is the triple (AX, BY, CZ) (the other directions follow similarly). By the Ratio Lemma we obtain $\frac{FX}{EX} = \frac{AF}{AE} \cdot \frac{\sin XAF}{\sin XAE}$. Multiplying this and similar results we obtain

$$\frac{\sin XAF}{\sin XAE} \cdot \frac{\sin YBD}{\sin YBF} \cdot \frac{\sin ZCE}{\sin ZCD} = \left(\frac{FX}{EX} \cdot \frac{DY}{FY} \cdot \frac{EZ}{DZ}\right) \cdot \left(\frac{AE}{AF} \cdot \frac{BF}{BD} \cdot \frac{CD}{CE}\right) = 1$$

after two applications of Ceva's Theorem. Then by Trig Ceva, we are done. □

Finally, let's see something for quadrilaterals that resembles the Trig Ceva we proved for triangles!

Theorem 3.3. (Quadrilateral Ceva) If $ABCD$ is a convex quadrilateral which has diagonals that intersect at P, then

$$\frac{\sin PAD}{\sin PAB} \cdot \frac{\sin PBA}{\sin PBC} \cdot \frac{\sin PCB}{\sin PCD} \cdot \frac{\sin PDC}{\sin PDA} = 1.$$

Surprisingly, this is very easy to prove.

Proof. By the Law of Sines applied in triangles PAB, PBC, PCD, PDA, we get that

$$\frac{PA}{PB} = \frac{\sin PBA}{\sin PAB} \text{ and } \frac{PB}{PC} = \frac{\sin PCB}{\sin PBC} \text{ and } \frac{PC}{PD} = \frac{\sin PDC}{\sin PCD} \text{ and } \frac{PD}{PA} = \frac{\sin PAD}{\sin PDA},$$

respectively. Hence, by multiplying these relations, we obtain the desired identity. □

This result is also easy to obtain using the Ratio Lemma in the obvious fashion. Now, Quadrilateral Ceva is very useful when dealing with (hidden) quadrilaterals and the angles determined by their sides with their diagonals. Let's see an example.

Delta 3.12. (IMO 2009) Let ABC be a triangle with $AB = AC$. The angle bisectors of $\angle CAB$ and $\angle ABC$ meet the sides BC and CA at D and E, respectively. Let K be the incenter of triangle ADC. Suppose that $\angle BEK = 45°$. Find all possible values of $\angle CAB$.

Proof. Let $I = BE \cap AD$. Note that I is the incenter of triangle ABC. Letting $x = \angle EBC$ and angle chasing we have that $\angle EIC = 2x$ and $\angle CID = 90° - x$ and $\angle CEK = 135° - 3x$. By Quadrilateral Ceva on quadrilateral $IECD$ with interior point K we obtain

$$\begin{aligned} & \sin 45° \sin(90° - x) = \sin 2x \sin(135° - 3x) \\ \Longrightarrow \quad & \sin 45° = 2 \sin x \sin(135° - 3x) = \cos(135° - 4x) - \cos(135° - 2x), \end{aligned}$$

which rearranges as

$$\begin{aligned} & \cos 135° + \cos(135° - 4x) = \cos(135° - 2x) \\ \Longrightarrow \quad & 2\cos(135 - 2x)\cos 2x = \cos(135 - 2x). \end{aligned}$$

Consequently, we either have $\cos(135° - 2x) = 0$ or $\cos 2x = \frac{1}{2}$. This means that $2x$ is equal to either $45°$ or $60°$ and so $\angle CAB = 180° - 4x$ is either $60°$ or $90°$. □

Assigned Problems

Epsilon 3.1. (Korea 1997) In an acute triangle ABC with $AB \neq AC$, let V be the intersection of the angle bisector of A with BC, and let D be the foot of the perpendicular from A to BC. If E and F are the intersections of the circumcircle of AVD with CA and AB, respectively, show that the lines AD, BE, CF concur.

Epsilon 3.2. Let ABC be a triangle such that $\angle ABC = 15°$ and $\angle ACB = 30°$. Let D be a point on the line through A perpendicular to AC, such that $AC = AD$, with the condition that BC separates the points A and D. Find the magnitude of $\angle ADB$.

Epsilon 3.3. (Brazilian NMO 1993) Let $ABCD$ a convex quadrilateral with $\angle BAC = 30°$, $\angle CAD = 20°$, $\angle ABD = 50°$ and $\angle DBC = 30°$. If its diagonals intersect at P, prove that $PC = PD$.

Epsilon 3.4. Let $ABCD$ be a convex quadrilateral with $\angle DAC = \angle BDC = 36°$, $\angle CBD = 18°$ and $\angle BAC = 72°$. The diagonals intersect at point P. Determine $\angle APD$.

Epsilon 3.5. (Mathematical Gazette) Let $\triangle ABC$ be an isosceles triangle ($AB = AC$) with $\angle BAC = 20°$. Point D is on side AC such that $\angle DBC = 60°$. Point E is on side AB such that $\angle ECB = 50°$. Find with proof the magnitude of $\angle EDB$.

Epsilon 3.6. (China TST 2014) Let the circumcenter of triangle ABC be O. H_A is the projection of A onto BC. The extension of AO intersects the circumcircle of BOC at A'. The projections of A' onto AB, AC are D, E, and O_A is the circumcenter of triangle DH_AE. Define H_B, O_B, H_C, O_C similarly. Prove that H_AO_A, H_BO_B, H_CO_C are concurrent

Epsilon 3.7. (Romania JBMO TST 2007) Let ABC be a right triangle with $\angle A = 90°$, and let D be a point lying on the side AC. Denote by E reflection of A into the line BD, and by F the intersection point of CE with the perpendicular in D to the line BC. Prove that AF, DE and BC are concurrent.

Epsilon 3.8. (Romania TST 2002) Let ABC be an acute triangle. The segment MN is the midline of the triangle that is parallel to side BC and P is the projection of the point N on the side BC. Let A_1 be the midpoint of the segment MP. Points B_1 and C_1 are constructed in a similar way. Show that if AA_1, BB_1, and CC_1 are concurrent lines, then the triangle ABC is isosceles.

Epsilon 3.9. Denote by AA_1, BB_1, CC_1 the altitudes of an acute triangle ABC, where A_1, B_1, C_1 lie on the sides BC, CA, and AB, respectively. A circle passing through A_1 and B_1 touches the arc AB of its circumcircle at C_2. The points A_2, B_2 are defined similarly.

1. (Tuymaada Olympiad 2007) Prove that the lines AA_2, BB_2, CC_2 are concurrent.
2. (MathLinks Contest 2008) Prove that the lines A_1A_2, B_1B_2, C_1C_2 are concurrent on the Euler line of triangle ABC.

Chapter 4

Menelaus' Theorem

We will now talk about a theorem very similar to Ceva's Theorem - another tool in our "turning Olympiad problems into (trigono)metric identities" toolbox!

Theorem 4.1. (Menelaus' Theorem) Let ABC be a triangle and let $A_1 \in BC$, $B_1 \in AC$, $C_1 \in AB$ so that either exactly one or all three of these points lie outside the segments BC, CA, AB. Then, A_1, B_1, C_1 are collinear if and only if

$$\frac{A_1B}{A_1C} \cdot \frac{B_1C}{B_1A} \cdot \frac{C_1A}{C_1B} = 1.$$

So the same condition as Ceva's Theorem! But be careful about the additional hypothesis about the position of the points A_1, B_1, C_1 with respect to the sides of ABC. As with Ceva, the direct implication is the "hard" part to prove, as the converse can be dealt with by using the same trick we used to prove the converse of Ceva's Theorem. So let's take care of that first.

Assuming the identity (and assuming that we have proved the direct implication), we want the collinearity. Assume furthermore without loss of generality that A_1 lies on the extension of BC, whereas the other two points lie on the corresponding sides. Now, we argue by contradiction, i.e. let's say the points A_1, B_1, C_1 are not collinear. Then, take the intersection of B_1C_1 with BC and call this point A_2. This point needs to lie on the extension of the side BC since both B_1 and C_1 are inside CA and AB, respectively. In this case, the direct implication tells us that

$$\frac{A_2B}{A_2C} \cdot \frac{B_1C}{B_1A} \cdot \frac{C_1A}{C_1B} = 1.$$

But we also know that

$$\frac{A_1B}{A_1C} \cdot \frac{B_1C}{B_1A} \cdot \frac{C_1A}{C_1B} = 1,$$

hence $\frac{A_1B}{A_1C} = \frac{A_2B}{A_2C}$, and considering what we know about their positions, we get that $A_1 = A_2$, which proves the converse.

As for the direct implication, we give again two separate proofs!

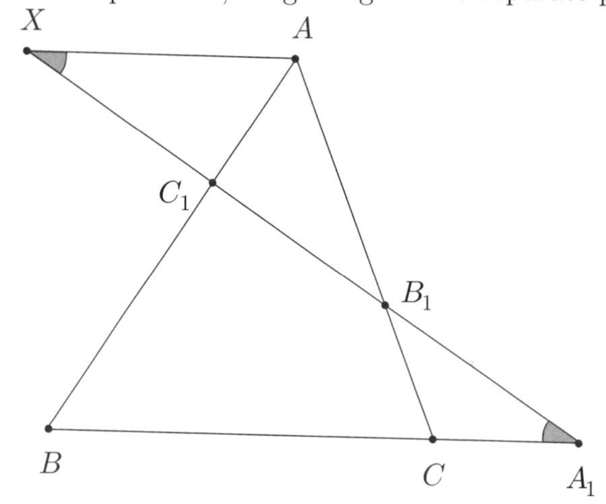

First Proof. Draw the parallel through A to the line BC and let X be the intersection of the line determined by A_1, B_1, C_1 with this parallel. We have that

$$\frac{B_1C}{B_1A} = \frac{A_1C}{AX} \text{ and } \frac{C_1A}{C_1B} = \frac{AX}{A_1B}$$

(from similar triangles). Therefore, we get that

$$\begin{aligned} \frac{A_1B}{A_1C} \cdot \frac{B_1C}{B_1A} \cdot \frac{C_1A}{C_1B} &= \frac{A_1B}{A_1C} \cdot \frac{A_1C}{AX} \cdot \frac{AX}{A_1B} \\ &= 1, \end{aligned}$$

which is precisely what we wanted.

Second proof. Even simpler. Just draw the projections of the vertices A, B, C on the line determined by A_1, B_1, C_1 and denote by h_1, h_2, h_3 the distances from A, B, C to this line. We have that

$$\frac{A_1B}{A_1C} = \frac{h_2}{h_3}, \frac{B_1C}{B_1A} = \frac{h_3}{h_1}, \frac{C_1A}{C_1B} = \frac{h_1}{h_2}$$

(from similar triangles). It follows immediately that their product is 1. □

Now, using this simple result about collinearities, combined with Ceva's Theorem and the other affiliated results from the first section, we can prove very nice and very difficult problems, as we will see! But let's start with some rather direct applications.

Corollary 4.1. The centroid G of triangle ABC divides the medians into nice ratios: if M, N, P are the midpoints of the sides BC, CA, AB, then

$$\frac{AG}{GM} = \frac{BG}{GN} = \frac{CG}{GP} = 2.$$

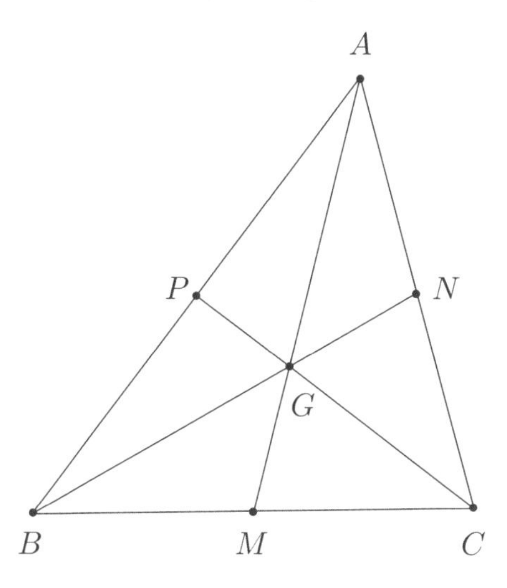

Proof. Let us look at triangle ABM and the collinear points C, G, P lying on the lines BM, MA, and AB, respectively. By Menelaus' Theorem, we have that

$$\frac{CB}{CM} \cdot \frac{GM}{GA} \cdot \frac{PA}{PB} = 1.$$

But M is the midpoint of BC and P is the midpoint of AB, so $\frac{CB}{CM} = 2$ and $\frac{PA}{PB} = 1$. Hence, it follows that $\frac{GM}{GA} = \frac{1}{2}$, i.e. $\frac{AG}{GM} = 2$, as desired. We can do the same thing for the other two ratios. □

Corollary 4.2. (The Euler Line) The centroid G lies on the line OH, where O and H are the circumcenter and orthocenter of triangle ABC respectively. Furthermore, we have that $HG = 2GO$.

Proof. Let M be the midpoint of BC, D the foot of the A-altitude (on BC) and let G' be the intersection of AM with the line OH. We would like to show that G' is precisely G and that it splits HO in the nice ratio from the statement.

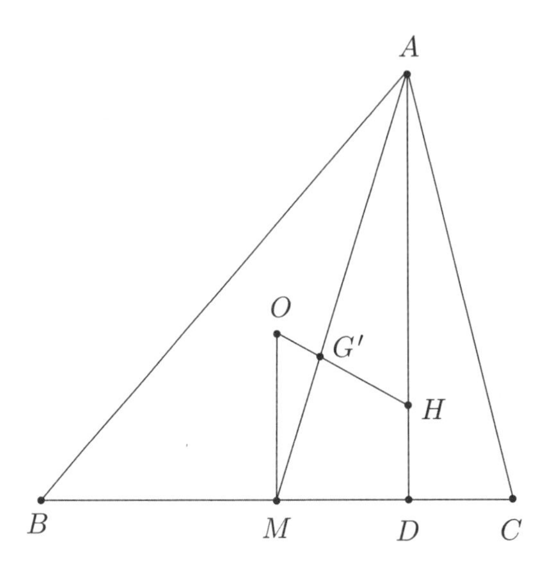

We do this by using what we learned in **Corollary 4.1**; we show that $\frac{AG'}{G'M} = 2$. If we manage to do this, then by the above the coincidence of G and G' follows immediately. But recall that $AH = 2R|\cos A|$ and $OM = R|\cos A|$. Hence, $AH = 2OM$, and since triangles AHG' and MOG' are similar, we get that

$$\frac{AH}{MO} = \frac{AG'}{MG'} = 2,$$

which is precisely what we were looking for to get that $G = G'$. This completes the proof. $\square$

Some applications now!

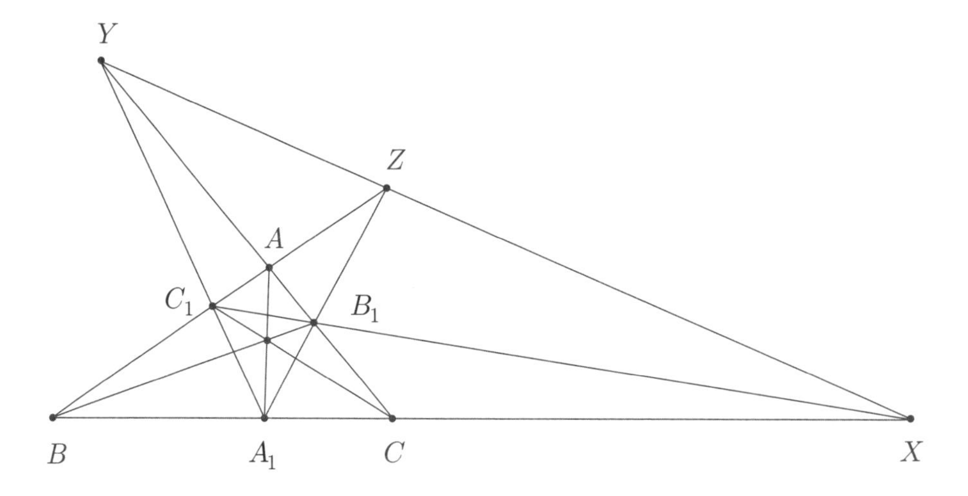

Delta 4.1. Let ABC be a triangle and P be a point in its interior. Let A_1, B_1, C_1 be the intersections of AP, BP, CP with the sides BC, CA, and AB, respectively. Consider the intersections X, Y, Z of BC with B_1C_1, of CA with C_1A_1, and of AB with A_1B_1, respectively. Prove that X, Y, Z are collinear.

Proof. By Menelaus' Theorem applied for triangle ABC and the collinear points B_1, C_1, X, we have that

$$\frac{XB}{XC} \cdot \frac{B_1C}{B_1A} \cdot \frac{C_1A}{C_1B} = 1.$$

Similarly, by applying Menelaus again two more times, we get that

$$\frac{YC}{YA} \cdot \frac{C_1A}{C_1B} \cdot \frac{A_1B}{A_1C} = 1 \text{ and } \frac{ZA}{ZB} \cdot \frac{A_1B}{A_1C} \cdot \frac{B_1C}{B_1A} = 1.$$

Thus, we get that

$$\begin{aligned} \frac{XB}{XC} \cdot \frac{YC}{YA} \cdot \frac{ZA}{ZB} &= \left(\frac{A_1B}{A_1C} \cdot \frac{B_1C}{B_1A} \cdot \frac{C_1A}{C_1B}\right)^{-2} \\ &= 1, \end{aligned}$$

where the last equality holds by Ceva's Theorem, since the lines AA_1, BB_1, CC_1 are concurrent at P. Hence, by Menelaus' Theorem - applied this time for the case when all the three points we want to prove collinear lie on the extensions of our reference triangle's sides, we get that X, Y, Z lie on the same line. □

As we will see in later sections, this problem could also be solved by a single application of Desargues' Theorem!

Delta 4.2. (The Lemoine Line) Let ABC be a triangle and let A_1 be the intersection point of the tangent at A to the circumcircle of ABC with the sideline BC. Similarly, define B_1 and C_1. Prove that A_1, B_1, C_1 are collinear.

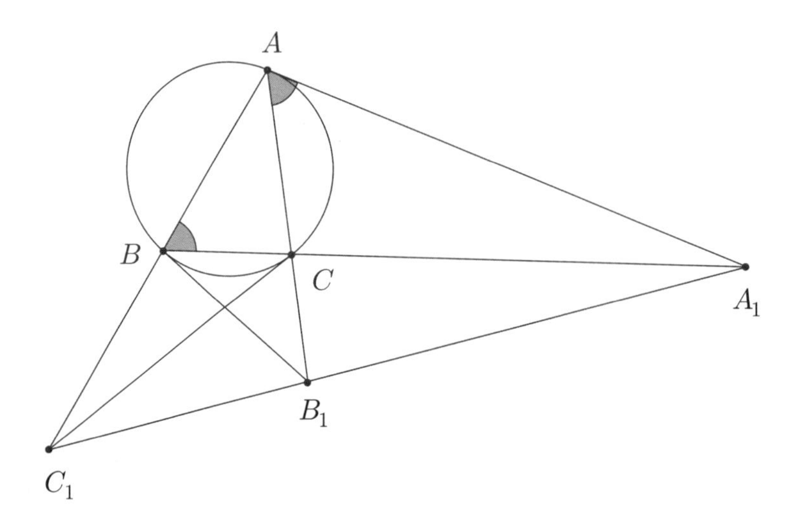

Proof. We would like to use Menelaus' Theorem, so we want to find the ratios $\frac{A_1B}{A_1C}$ etc. So, we use the Ratio Lemma! More precisely, we have that

$$\frac{A_1B}{A_1C} = \frac{AB}{AC} \cdot \frac{\sin A_1AB}{\sin A_1AC}.$$

But the line AA_1 is tangent at A to the circumcircle of ABC, hence $\angle A_1AB = \angle C$ and $\angle A_1AC = \angle C + \angle A = 180° - \angle B$. Thus, we get that

$$\frac{A_1B}{A_1C} = \frac{AB}{AC} \cdot \frac{\sin C}{\sin B} = \frac{AB^2}{AC^2},$$

where the last equality holds because of the Law of Sines applied in triangle ABC.

Similarly, we get that $\frac{B_1C}{B_1A} = \frac{BC^2}{BA^2}$ and $\frac{C_1A}{C_1B} = \frac{CA^2}{CB^2}$; thus

$$\frac{A_1B}{A_1C} \cdot \frac{B_1C}{B_1A} \cdot \frac{C_1A}{C_1B} = 1,$$

and so Menelaus assures the collinearity. □

Delta 4.3. (Quadrilateral Menelaus) Let $A_1A_2A_3A_4$ be a quadrilateral and let d be a line which intersects the sides A_1A_2, A_2A_3, A_3A_4 and A_4A_1 in the points M_1, M_2, M_3, and M_4, respectively. Then,

$$\frac{M_1A_1}{M_1A_2} \cdot \frac{M_2A_2}{M_2A_3} \cdot \frac{M_3A_3}{M_3A_4} \cdot \frac{M_4A_4}{M_4A_1} = 1.$$

Proof. Let the line d intersect A_1A_3 at X. By Menelaus' theorem applied in triangles $A_1A_2A_3$ and $A_1A_3A_4$, we get that

$$\frac{M_1A_1}{M_1A_2} \cdot \frac{M_2A_2}{M_2A_3} \cdot \frac{XA_3}{XA_1} = 1$$

and

$$\frac{M_3A_3}{M_3A_4} \cdot \frac{M_4A_4}{M_4A_1} \cdot \frac{XA_1}{XA_3} = 1.$$

Thus, by multiplying the above two identities, we conclude that

$$\frac{M_1A_1}{M_1A_2} \cdot \frac{M_2A_2}{M_2A_3} \cdot \frac{M_3A_3}{M_3A_4} \cdot \frac{M_4A_4}{M_4A_1} = 1.$$

This completes the proof. □

Delta 4.4. (Polygonal Menelaus) Let d be a line that intersects the sides A_iA_{i+1} of the n-gon $A_1A_2 \ldots A_{n-1}A_n$ in the points M_i, for all $1 \leq i \leq n$ (where $A_{n+1} = A_1$). Prove that

$$\prod_{i=1}^{n} \frac{M_iA_i}{M_iA_{i+1}} = 1.$$

(Hint: Use induction!)

Even though we included it above, Quadrilateral (or Polygonal) Menelaus is not as useful as Quadrilateral Ceva, so we won't dwell on it much. Instead, we turn again to "triangle" Menelaus and look at some more applications.

Delta 4.5. (Van Aubel's Theorem) Let ABC be a triangle and let P be an interior point of this triangle. Let the lines AP, BP, CP meet the sides BC, CA, AB at A', B', C', respectively. Prove that $\frac{PA}{PA'} = \frac{C'A}{C'B} + \frac{B'A}{B'C}$.

Proof. By Menelaus' Theorem applied in triangle ABA' for the collinear points C, P, C', we get that

$$\frac{CB}{CA'} \cdot \frac{PA'}{PA} \cdot \frac{C'A}{C'B} = 1, \text{ i.e. } \frac{PA}{PA'} = \frac{CB}{CA'} \cdot \frac{C'A}{C'B}.$$

This means that

$$\begin{aligned} \frac{PA}{PA'} &= \frac{A'B + CA'}{CA'} \cdot \frac{C'A}{C'B} \\ &= \frac{C'A}{C'B} + \frac{A'B}{CA'} \cdot \frac{C'A}{C'B} \\ &= \frac{C'A}{C'B} + \frac{B'A}{B'C}, \end{aligned}$$

where the last equality holds from Ceva's Theorem, since the lines AA', BB', CC' are concurrent at P. This completes the proof. □

Delta 4.6. (China TST 2012) Let the incircle of triangle ABC touch the sidelines BC, CA, AB at A_1, B_1, C_1, respectively. Let A_2 be the reflection of A_1 over the line B_1C_1 and similarly define B_2 and C_2 (as the reflections of B_1 and C_1 over C_1A_1 and A_1B_1, respectively). Let $A_3 = AA_2 \cap BC$, $B_3 = BB_2 \cap CA$, and $C_3 = CC_2 \cap AB$. Prove that A_3, B_3, C_3 are collinear.

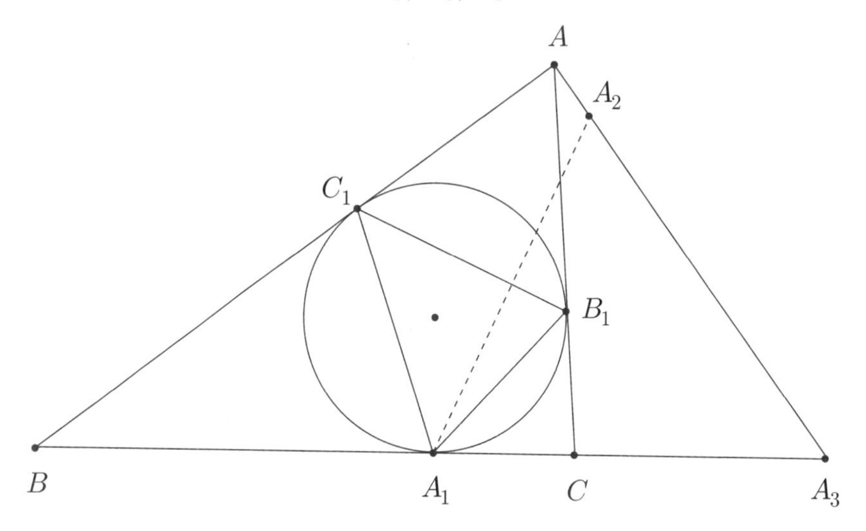

First Proof. We would like to use Menelaus in triangle ABC, so we need to compute the ratios $\frac{A_3B}{A_3C}$ etc... The best way to do it is, as in the proof of the existence of the Lemoine line, by recalling the Ratio Lemma. First, let $\angle BAC = a$, $\angle ABC = b$ and $\angle ACB = c$ and note that

$$\begin{aligned} \angle A_2C_1A &= |\angle A_2C_1B_1 - \angle AC_1B_1| \\ &= |\angle A_1C_1B_1 - \angle AC_1B_1| \\ &= |\,(90^\circ - c/2) - (90^\circ - a/2)\,| \\ &= |a/2 - c/2|. \end{aligned}$$

And similarly, $\angle A_2B_1A = |a/2 - b/2|$. Now let $\alpha = |a/2 - b/2|$, $\beta = |a/2 - c/2|$ and $\gamma = |b/2 - c/2|$. Since $A_2C_1 = A_1C_1$ and $A_2B_1 = A_1B_1$, the Law of Sines tells us that

$$\frac{\sin \angle C_1AA_2}{\sin \angle B_1AA_2} = \frac{\frac{A_2C_1 \cdot \sin\beta}{AA_2}}{\frac{A_2B_1 \cdot \sin\alpha}{AA_2}} = \frac{A_1C_1 \cdot \sin\beta}{A_1B_1 \cdot \sin\alpha}.$$

But again, by the Ratio Lemma,

$$\frac{A_3B}{A_3C} = \frac{AB \cdot \sin \angle C_1AA_2}{AC \cdot \sin \angle B_1AA_2} = \frac{AB \cdot A_1C_1 \cdot \sin\beta}{AC \cdot A_1B_1 \cdot \sin\alpha}.$$

And similarly, using the same argument, we can find that

$$\frac{B_3C}{B_3A} = \frac{BC \cdot A_1B_1 \cdot \sin\alpha}{AB \cdot B_1C_1 \cdot \sin\gamma}$$

and

$$\frac{C_3A}{C_3B} = \frac{AC \cdot B_1C_1 \cdot \sin\gamma}{BC \cdot A_1C_1 \cdot \sin\beta}.$$

Multiplying yields that

$$\frac{A_3B}{A_3C} \cdot \frac{B_3C}{B_3A} \cdot \frac{C_3A}{C_3B} = 1,$$

and so by Menelaus' Theorem, we conclude that A_3, B_3 and C_3 are indeed collinear. □

In fact, we can actually prove more in this configuration. One can show that the line $A_3B_3C_3$ is actually the Euler line of triangle $A_1B_1C_1$. This will be the method of our second proof:

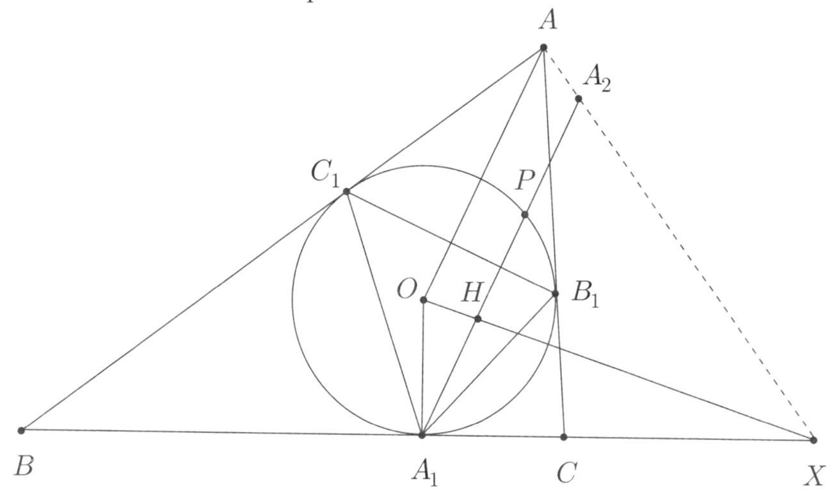

Second Proof. Let O, H be the circumcenter and orthocenter respectively of triangle $A_1B_1C_1$. Also let line A_1H intersect the circumcircle of triangle $A_1B_1C_1$ again at point P. Let line OH intersect line BC at X, and assume without loss of generality that A_1 lies between B and X. And finally, let R be the circumradius of triangle $A_1B_1C_1$. Now, note that by the Ratio Lemma on triangle OA_1H with cevian A_1X we find that

$$\frac{XH}{XO} = \frac{A_1H}{A_1O} \cdot \frac{\sin XA_1P}{\sin XA_1O} = \frac{A_1H \cdot \sin A_1C_1P}{R} = \frac{A_1H \cdot A_1P}{2R^2}$$

where we used the Extended Law of Sines for the last reduction. But we know that $A_1P = A_2H$ and that $A_1H = 2R\cos B_1A_1C_1$ so we have that

$$\frac{XH}{XO} = \frac{A_2H \cdot \cos B_1A_1C_1}{R} = \frac{A_2H}{AO}$$

which since $A_2H \parallel AO$ (since both lines are perpendicular to B_1C_1) implies that X lies on line AA_2. Hence, $X = A_3$. This completes the proof as we can do the same for B_3 and C_3. □

Delta 4.7. Let ABC be a triangle with incenter I. Let D, E, F be the tangency points of the its incircle with BC, CA, AB, respectively. Prove that the circumcircles of triangles AID, BIE, CIF have two common points (in other words, they have one point in common that is different from I).

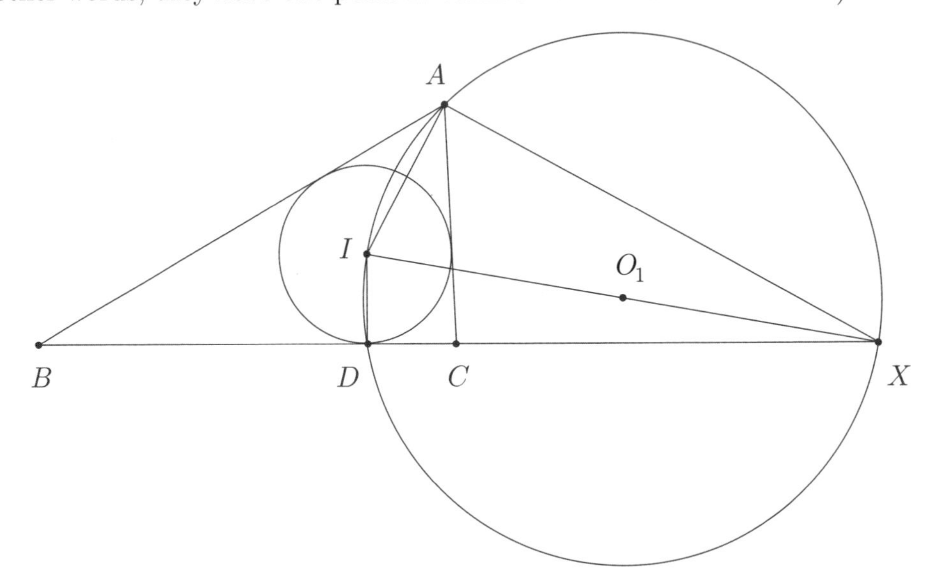

Proof. We want to show that three circles that share a point have one other point in common. So, it suffices to show that the centers of the circles are collinear! (convince yourself why - think about radical axis). Let O_1, O_2, O_3 be the circumcenters of triangles AID, BIE, CIF respectively, and let X, Y, Z be the reflections of I over O_1, O_2, O_3, respectively. Since $\angle XDI = \angle XAI = 90°$ it's clear that X is the foot of the A-external angle bisector of triangle ABC, so by the external angle bisector theorem we have that

$$\frac{XB}{XC} = \frac{AB}{AC}.$$

Similarly, we get that Y and Z lie on CA and AB and furthermore $\frac{YC}{YA} = \frac{BC}{BA}$ and $\frac{ZA}{ZB} = \frac{CA}{CB}$, so Menelaus' Theorem yields that points X, Y, Z are collinear.

But this immediately implies that points O_1, O_2, O_3 all lie on the I-midline of triangle IXY, so we are done. $\square$

Delta 4.8. (USAMO 2012) Let P be a point in the plane of triangle ABC, and ℓ a line passing through P. Let A', B', C' be the points where the reflections of lines PA, PB, PC with respect to ℓ intersect lines BC, AC, AB respectively. Prove that A', B', C' are collinear.

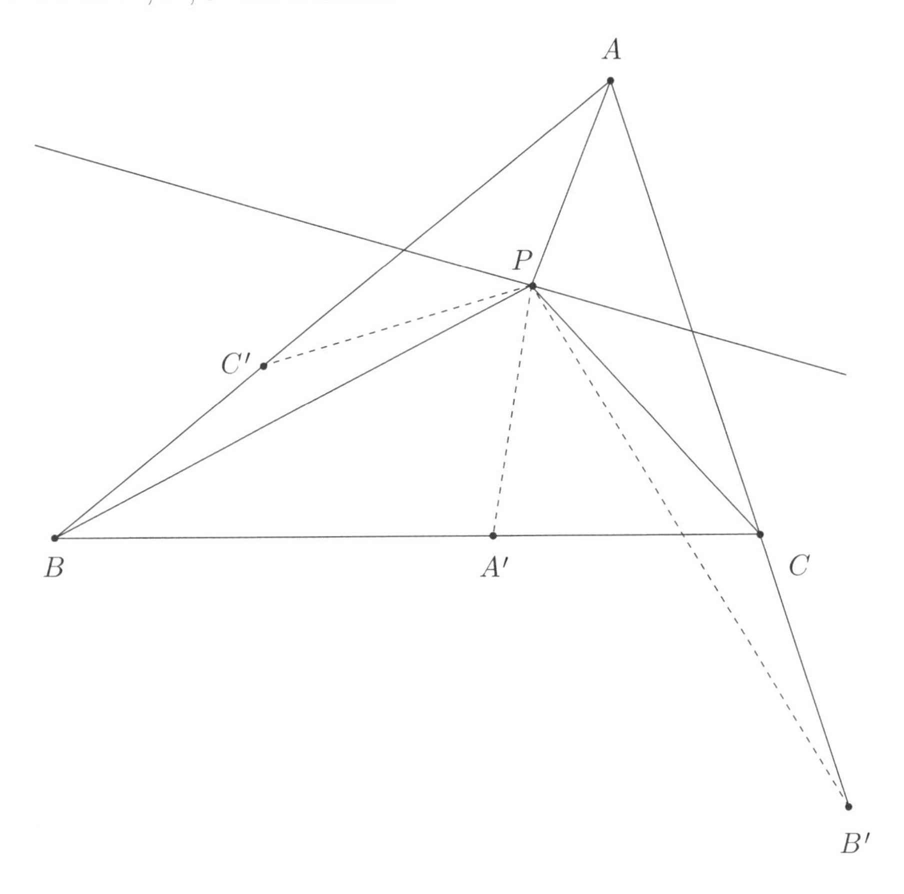

Proof. Assume without loss of generality that the configuration is as shown. Let $\alpha = \angle(\ell, PA) = \angle(\ell, PA')$ and $\beta = \angle(\ell, PB) = \angle(\ell, PB')$ and $\gamma = \angle(\ell, PC) = \angle(\ell, PC')$. Then by the Law of Sines in triangles $A'CP$ and $A'BP$ and $B'AP$ and $B'CP$ and $C'BP$ and $C'AP$ we obtain

$$\frac{A'C}{\sin|\gamma - \alpha|} = \frac{A'P}{\sin BCP}$$

$$\frac{A'B}{\sin|\alpha + \beta|} = \frac{A'P}{\sin CBP}$$

$$\frac{B'A}{\sin|\alpha+\beta|} = \frac{B'P}{\sin CAP}$$

$$\frac{B'C}{\sin|\beta-\gamma|} = \frac{B'P}{\sin ACP}$$

$$\frac{C'B}{\sin|\beta-\gamma|} = \frac{C'P}{\sin ABP}$$

$$\frac{C'A}{\sin|\gamma-\alpha|} = \frac{C'P}{\sin BAP}$$

Multiplying and dividing these equations in the appropriate manner yields

$$\frac{A'B}{A'C} \cdot \frac{B'C}{B'A} \cdot \frac{C'A}{C'B} = \frac{\sin CAP}{\sin BAP} \cdot \frac{\sin ABP}{\sin CBP} \cdot \frac{\sin BCP}{\sin ACP} = 1$$

where the second equality follows from Trig Ceva. Hence, by Menelaus' Theorem on triangle ABC, points A', B', C' are collinear as desired. □

Assigned Problems

Epsilon 4.1. (IMO Shortlist 1995) The incircle of triangle $\triangle ABC$ touches the sides BC, CA, AB at D, E, F respectively. X is a point inside triangle ABC such that the incircle of triangle XBC touches BC at D, and touches CX and XB at Y and Z respectively. Show that E, F, Z, Y are concyclic. (Hint: Prove that EF, YZ and BC are concurrent! To do this, let $T_1 = BC \cap EF$, $T_2 = BC \cap YZ$ and show that $\frac{T_1B}{T_1C} = \frac{T_2B}{T_2C}$.)

Epsilon 4.2. Let Γ be a circle and let B be a point on a line that is tangent to Γ at the point A. The line segment AB is rotated about the center of the circle through some angle to the line segment $A'B'$. Prove that AA' passes through the midpoint of BB'.

Epsilon 4.3. Let P be a point in the interior of a triangle ABC and let D, E, F be the projections of P onto BC, CA, AB respectively. Let X be the point on EF such that $PX \perp PA$ and define Y and Z similarly. Prove that points X, Y, Z are collinear.

Epsilon 4.4. Let ABC be an isosceles triangle with $AC = BC$. Its incircle touches AB in D and BC in E. A line distinct of AE goes through A and intersects the incircle in F and G. Line AB intersects line EF and EG in K and L, respectively. Prove that $DK = DL$.

Epsilon 4.5. (Serbia 2014) Let ABC be a triangle and consider points D, E on sides BC, CA respectively. Let F be the intersection of the circumcircle of triangle CDE and the line parallel from to AB passing through C. Now let G be the intersection of lines AB and FD. Point H is selected on line AB such that $\angle BEG = \angle HDA$. Given that $HE = DG$, prove that Q is on the angle bisector of $\angle BCA$, where Q is the intersection of lines BE and AD.

Epsilon 4.6. (IMO Shortlist 2006) Let ABC be a triangle such that $\angle ACB < \angle BAC < 90^\circ$. Let D be a point on AC such that $BD = BA$. The incircle of ABC touches AB at K and AC at L. Let J be incenter of triangle BCD. Prove that line KL bisects segment AJ.

Epsilon 4.7. Let ABC be a triangle and let O be its circumcenter. Let A_1 be the other end of the diameter of (O) passing through A (in other words, A_1 is the antipode of A with respect to the circumcircle of ABC) and denote by A_2 the reflection of O across BC. Similarly define B_1, B_2, C_1, C_2. Prove that the circumcircles of triangles OA_1A_2, OB_1B_2, OC_1C_2 share two common points. (Hint: Show that the centers are collinear; reread **Delta 4.7.**)

Chapter 5

Desargues and Pascal

This section comes as an annex to the previous one about Menelaus, for the two results presented here are just consequences of Menelaus' Theorem. However, they deserve to be treated separately, as there are many contest problems out there that can be beautifully solved using these two.

First, we deal with Desargues' Theorem, which perfectly merges concurrencies and collinearities.

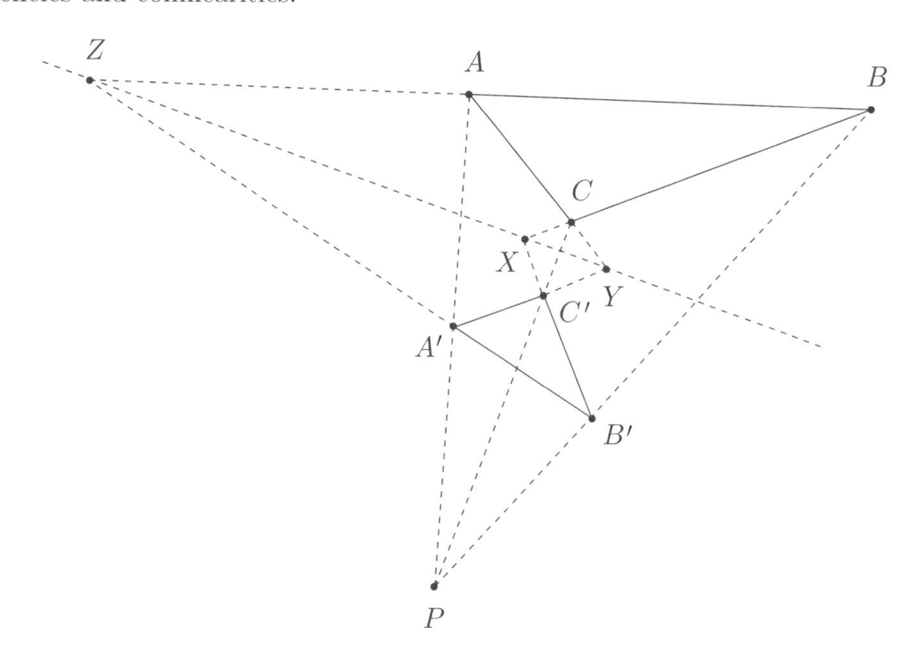

Theorem 5.1. (Desargues' Theorem) Let ABC and $A'B'C'$ be two triangles. Then, the lines AA', BB', CC' are concurrent if and only if the intersections of BC and $B'C'$, of CA and $C'A'$, and of AB and $A'B'$ are collinear.

Note that Desargues translates a question that asks you about a concurrency into a question about collinearity and vice-versa. The whole idea is that maybe the other thing is easier to prove in the configuration you are working with; you will be surprised to see that this actually happens a lot!

Definition. Triangles for which the situation from the statement of Desargues' Theorem holds are called **perspective**. The concurrency point of AA', BB', CC' is then called the **perspector** of the two triangles, whereas the line determined by the three intersections is called the **perspectrix** of triangles ABC and $A'B'C'$.

Proof. Let us first prove the direct implication, i.e. assume that the lines AA', BB', CC' are concurrent at a point, say P. Let X be the intersection of BC and $B'C'$, Y the intersection of CA and $C'A'$, and Z the intersection of AB and $A'B'$. To show that X, Y, Z are collinear, we will use Menelaus in triangle ABC. So, in other words, we want to show that

$$\frac{XB}{XC} \cdot \frac{YC}{YA} \cdot \frac{ZA}{ZB} = 1.$$

Hence, we need to find the ratios $\frac{XB}{XC}$ etc. Now the points B', C', X are collinear, so Menelaus in triangle PBC gives us that

$$\frac{XB}{XC} \cdot \frac{C'C}{C'P} \cdot \frac{B'P}{B'B} = 1.$$

Similarly, Menelaus for C', A', Y and A', B', Z in triangles PCA and PAB, respectively, tells us that

$$\frac{YC}{YA} \cdot \frac{A'A}{A'P} \cdot \frac{C'P}{C'C} = 1 \text{ and } \frac{ZA}{ZB} \cdot \frac{B'B}{B'P} \cdot \frac{A'P}{A'A} = 1.$$

Therefore, multiplying these last three equations, we get that

$$\frac{XB}{XC} \cdot \frac{C'C}{C'P} \cdot \frac{B'P}{B'B} = 1,$$

as desired, so X, Y, Z are indeed collinear.

For the converse, we proceed as follows. Assuming that X, Y, Z are collinear, we actually know that the lines BC, ZY and $B'C'$ are concurrent, i.e. the triangles BZB' and CYC' are perspective; hence, by the direct implication, we get that the intersections $BZ \cap CY = A$, $ZB' \cap YC' = A'$, $BB' \cap CC'$ are collinear, which is equivalent to saying that the lines AA', BB', CC' are concurrent, which proves the converse! This settles the proof of Desargues' Theorem. $\square$

Notice again how we used the direct implication to prove the converse! This is quite a common trick when dealing with concurrencies and collinearities - keep it in mind.

Let's see some examples where we could use this.

Delta 5.1. Take another look at **Delta 4.1**, and laugh.

Delta 5.2. (Moldavian TST 2011) In triangle ABC with $AB < AC$, the point H denotes the orthocenter. The points A_1 and B_1 are the feet of perpendiculars from A and B respectively. The point D is the reflection of C over point A_1. If $E = AC \cap DH$, $F = DH \cap A_1B_1$, and $G = AF \cap BH$, prove that the lines CH, EG and AD are concurrent.

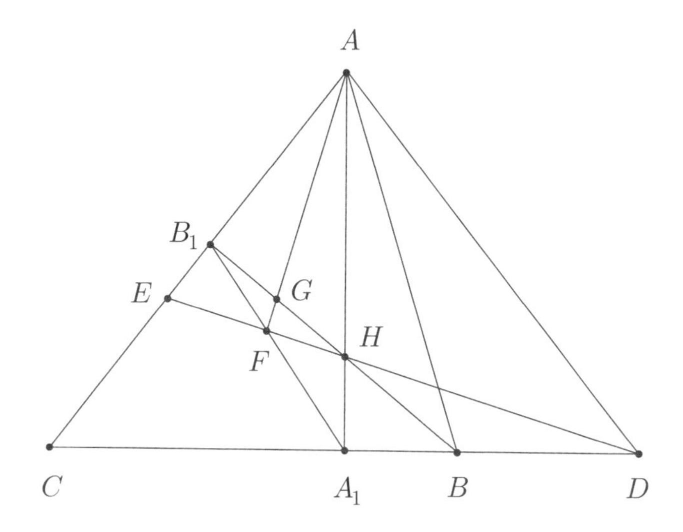

Proof. We see that the points $CE \cap HG = B_1$, $ED \cap GA = F$, $CD \cap HA = A_1$ are collinear on A_1B_1, i.e. the triangles CED and HGA are perspective. Hence, by Desargues' Theorem, it follows that the lines CH, EG, AD are concurrent, as claimed. □

Delta 5.3. (Sharygin 2012) Point D lies on side AB of triangle ABC. Let ω_1 and Ω_1, ω_2 and Ω_2 be the incircles and the excircles (touching segment AB) of triangles ACD and BCD respectively. Prove that the common external tangents to ω_1 and ω_2, Ω_1 and Ω_2 meet on AB.

Proof. Denote the centers of $\omega_1, \omega_2, \Omega_1, \Omega_2$ by I_1, I_2, I_3, I_4, respectively. It's clear that the original problem can be reduced to showing that lines

AB, I_1I_2, I_3I_4 concur. But note that $AI_1 \cap BI_2$ is the incenter of triangle ABC, $AI_3 \cap BI_4$ the C-excenter of triangle ABC, and $I_3I_1 \cap I_2I_4 = C$.

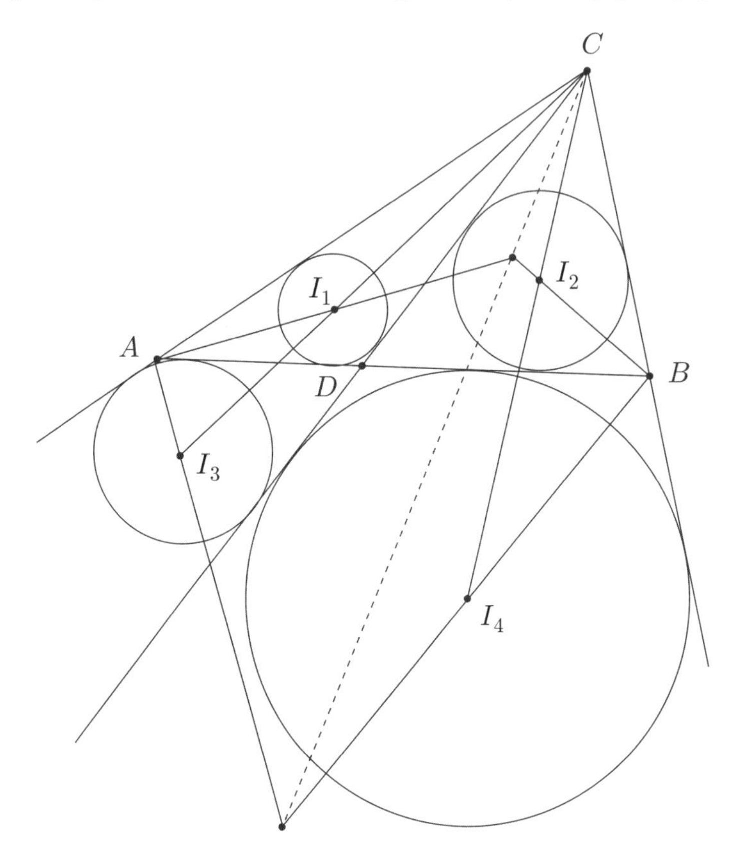

Thus, the three points $AI_1 \cap BI_2, AI_3 \cap BI_4, I_3I_1 \cap I_2I_4$ all lie on the C-internal angle bisector of triangle ABC. Thus, the triangles AI_1I_3 and BI_2I_4 are perspective by Desargues' Theorem. Therefore, the lines AB, I_1I_2 and I_3I_4 are concurrent as desired. □

There are also some trickier Desargues' that one can apply, when, for example, sidelines of the two triangles involved are parallel. This is the case in the following problem.

Delta 5.4. Let $BCXY$ be a rectangle constructed outside triangle ABC. Let D be the foot of the altitude from A lying on BC and let U, V be the intersection points of DY with AB and of DX with AC, respectively. Prove that $UV \| BC$.

Proof. The two triangles ABC and DYX are perspective (as the lines AD, BY, CX are "concurrent" at the point at infinity associated with the direction of the A-altitude).

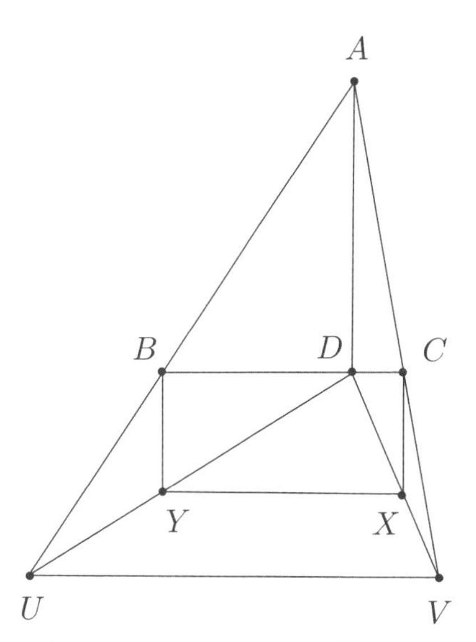

Thus, by Desargues' Theorem, the points $U = DY \cap AB$, $V = DX \cap AC$ and $BC \cap YX$ (the point at infinity on line BC) are collinear, i.e. $UV \parallel BC$. This completes the proof. $\square$

Now, we introduce one of the most useful theorems in Olympiad geometry. Before we give the theorem, it is extremely important to keep in mind that **the converse does not hold!** However, replacing the word "cyclic" with "lying on a conic" makes the following theorem an if and only if statement.

Theorem 5.2. (Pascal's Theorem) Let $ABCDEF$ be a cyclic hexagon (with vertices not necessarily in this order on the circle). Then, the intersections $AB \cap DE$, $BC \cap EF$, $CD \cap FA$ are collinear.

Proof. Let $J = AB \cap DE, L = BC \cap EF, K = CD \cap FA, G = BC \cap FA, H = DE \cap FA$, and $I = BC \cap DE$. By Menelaus's Theorem on triangle GHI with points D, K, C we find that

$$\frac{DI}{DH} \cdot \frac{CG}{CI} \cdot \frac{KH}{KG} = 1$$

By Menelaus' Theorem on the same triangle with points A, J, B and then with points E, L, F we obtain two similar equations and multiplying them together yields

$$\frac{KH}{KG} \cdot \frac{LG}{LI} \cdot \frac{JI}{JH} \cdot \left(\frac{ID \cdot IE}{IB \cdot IC}\right) \cdot \left(\frac{HF \cdot HA}{HD \cdot HE}\right) \cdot \left(\frac{GC \cdot GB}{GF \cdot GA}\right) = 1.$$

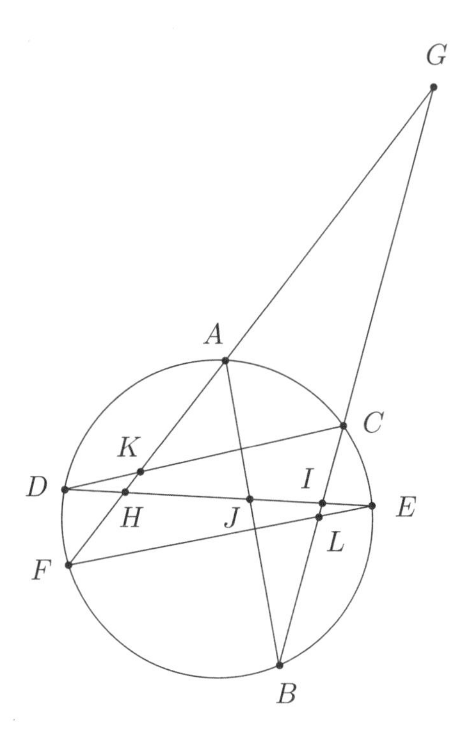

By Power of Point, the expressions in parentheses are each equal to 1 and now Menelaus on triangle GHI with points J, L, K yields the desired collinearity. □

Corollary 5.1. (Pappus's Theorem) Given one set of collinear points A, B, C and another set of collinear points A', B', C', let $X = BC' \cap B'C$ and $Y = CA' \cap C'A$ and $Z = AB' \cap A'B$. Then points X, Y, Z are collinear.

Proof. Note that two intersecting lines are a degenerate conic so applying Pascal's Theorem to degenerate hexagon $AB'CA'BC'$ we immediately obtain the desired result. □

Now, let's see some applications of Pascal; keep in mind that degenerate cases of Pascal are often tremendously useful. For example, we can apply Pascal's theorem for the degenerated hexagon $ABCDEA$, and in this case, we have that the intersections $AB \cap DE$, $BC \cap EA$, and $CD \cap AA$ (which is nothing but the intersection of CD with the tangent at A to the circumcircle of the (degenerate) hexagon $ABCDEA$) are collinear. So, keep your eyes open!

First, a particular case of something we already did with Menelaus (see **Delta 4.2**).

Delta 5.5. Let ABC be a triangle with circumcircle ω and let A_1 be the intersection of the tangent to ω at A and line BC. Define B_1, C_1 similarly. Prove that points A_1, B_1, C_1 are collinear.

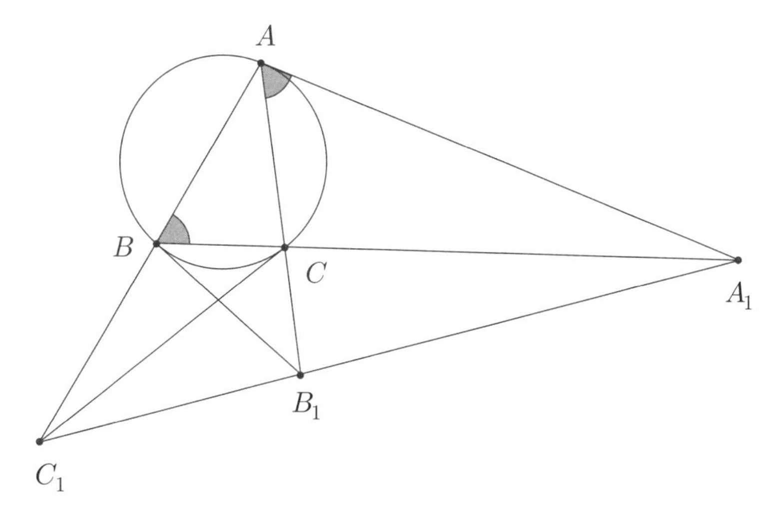

Proof. Indeed, we can apply Pascal's theorem for the cyclic (and degenerate) hexagon $AABBCC$. This gives us that $AA \cap BC$, $BB \cap CA$, $CC \cap AB$ are collinear which is precisely what we need, hence this completes the proof. □

Delta 5.6. (Forum Geometricorum) Let ABC be a triangle and let B_1, C_1 be points on the sides CA, AB respectively. Let Γ be the incircle of triangle ABC and let E, F be the tangency points of Γ with the same sides CA and AB, respectively. Furthermore, draw the tangents from B_1 and C_1 to Γ which are different from the sidelines of triangle ABC and take the tangency points with Γ to be Z and Y, respectively, like in the picture below.

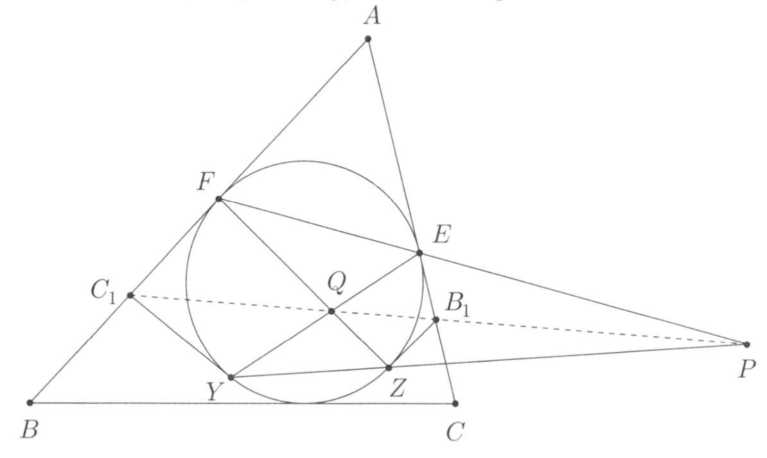

Prove that the lines B_1C_1, EF and YZ are concurrent.

Proof. Let $P = EF \cap YZ$ and let $Q = EY \cap FZ$. Then by Pascal's Theorem on degenerate hexagon $EFFZYY$ we have that points P, Q, C_1 are collinear. Also, by Pascal's Theorem on degenerate hexagon $FEEYZZ$ we have that points P, Q, B_1 are collinear. This implies that points P, B_1, C_1 are collinear which proves the desired result. □

Delta 5.7. (Romania TST 2012) Let γ be a circle and l a line in its plane. Let K be a point on l, located outside of γ. Let KA and KB be the tangents from K to γ, where A and B are distinct points on γ. Let P and Q be two points on γ. Lines PA and PB intersect line l in points R and S respectively. Lines QR and QS intersect γ again at points C and D respectively. Prove that the tangents from C and D to γ are concurrent on line l.

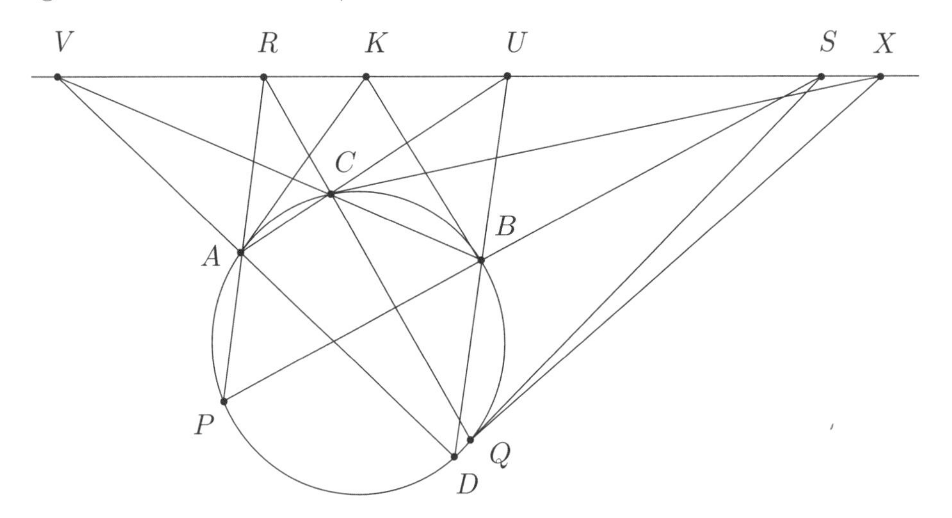

Proof. Let $X = CC \cap DD$, so that we wish to show that X lies on l. Also let $U = AC \cap BD$ and $V = AD \cap BC$. By Pascal's Theorem on $BPADQC$ we have that points R, S, V are collinear which means that V lies on l. By Pascal's Theorem on $AACBBD$ we have that points K, U, V are collinear so U lies on l as well. Now by Pascal's Theorem on $ACCBDD$ we have that points U, V, X are collinear so X lies on l as desired. □

//If one is familiar with poles and polars (**Section 12** of this book), there is a very short solution to this problem, the idea of which is to notice that l is the polar of $AB \cap CD$ with respect to γ.

Delta 5.8. (IMO Shortlist 2007) Let ABC be a fixed triangle, and let X, Y, Z be the midpoints of sides BC, CA, AB, respectively. Let P be a variable point on the circumcircle of triangle ABC. Let lines PX, PY, PZ meet the circumcircle of triangle ABC again at A', B', C', respectively. Assume that the points A, B, C, A', B', C' are distinct, and lines AA', BB', CC' form a triangle. Prove that the area of this triangle does not depend on P.

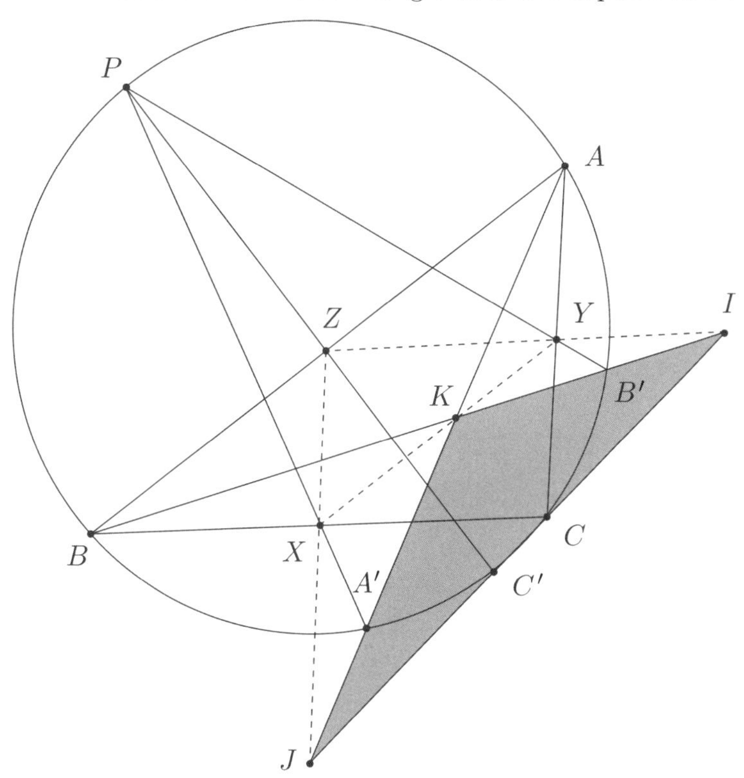

Proof. Without loss of generality, let P be on the arc AB that does not contain C of the circumcircle of triangle ABC. Now let $I = BB' \cap CC'$ and $J = CC' \cap AA'$ and $K = AA' \cap BB'$. By Pascal's Theorem on $ABB'PC'C$ we have that points I, Y, Z are collinear. Similarly we find that points J, Z, X and K, X, Y are collinear. Now since $XJ \parallel CA$ it's easy to see that triangles KJX and KAY are similar so we have that

$$\frac{KX}{KY} = \frac{KJ}{KA}$$

and similarly, since triangle KIY is similar to triangle KBX, we have

$$\frac{KX}{KY} = \frac{KB}{KI}$$

so we have that

$$[IJK] = \frac{KI \cdot KJ \cdot \sin IKJ}{2} = \frac{KA \cdot KB \cdot \sin AKB}{2} = [ABK] = \frac{[ABC]}{2}$$

which obviously implies the desired result. □

Delta 5.9. (IMO 2011 Shortlist) Let ABC be an acute triangle with circumcircle Ω. Let B_0 be the midpoint of AC and let C_0 be the midpoint of AB. Let D be the foot of the altitude from A and let G be the centroid of the triangle ABC. Let ω be a circle through B_0 and C_0 that is tangent to the circle Ω at a point $X \neq A$. Prove that the points D, G and X are collinear.

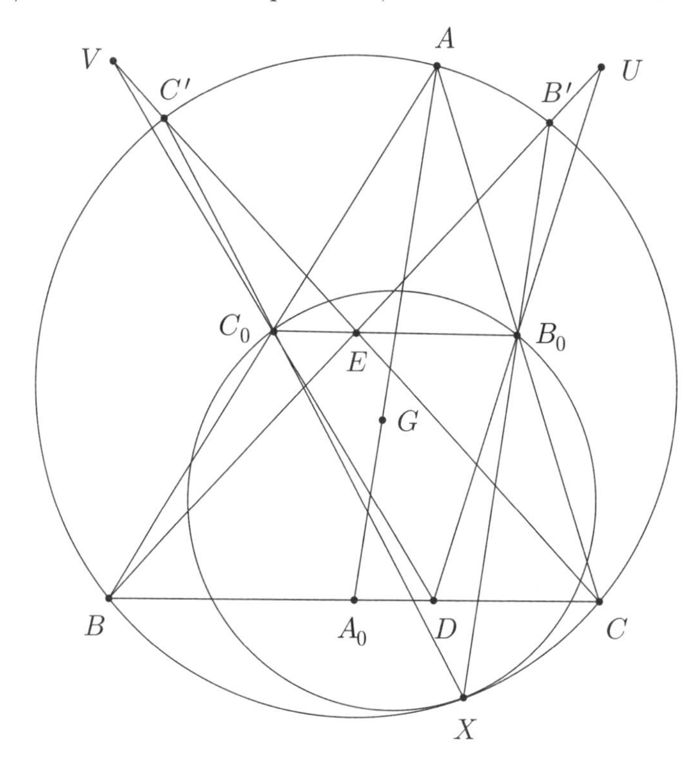

Proof. Let A_0 be the midpoint of BC. Let lines XB_0 and XC_0 meet Ω again at B' and C', and let $E = BB' \cap CC'$. Pascal's Theorem on $ABB'XC'C$ gives that E lies on B_0C_0. Since the homothety carrying ω to Ω carries B_0C_0 to $B'C'$, we have that $B'C' \parallel BC$. Then quadrilateral $B'C'BC$ is an isosceles trapezoid, so E is the foot of the perpendicular from A_0 to B_0C_0. Now since $A_0E \parallel AD$ and since $AD = 2A_0E$ we know that points E, G, D are collinear (this is similar to the proof of the existence of the Euler line). Therefore it suffices to show that lines $B'B_0, C'C_0, ED$ are concurrent.

Since lines BB_0, CC_0, ED concur at G, triangles BCE and B_0C_0D are perspective. Let $U = BE \cap B_0D$ and $V = CE \cap C_0D$. Desargues' Theorem on triangles BCE and B_0C_0D implies $UV \parallel B_0C_0$. By Desargues' Theorem again, this implies that triangle $B'C'E$ and triangle B_0C_0D are perspective, so $B'B_0, C'C_0, ED$ are concurrent as desired. □

We finish with a cute application of the converse of Pascal's Theorem (which in this case turns out to be the converse of Pappus's Theorem). We'll actually provide two proofs to the following famous result, since even though the second proof doesn't utilize Desargue or Pascal, it uses an insanely beautiful claim.

Delta 5.10. (The Newton-Gauss Line) Let ABC be a triangle and let D and E be points on segments BC and CA respectively. Let $F = AB \cap DE$. Show that the midpoints of segments AD, BE, and CF are collinear.

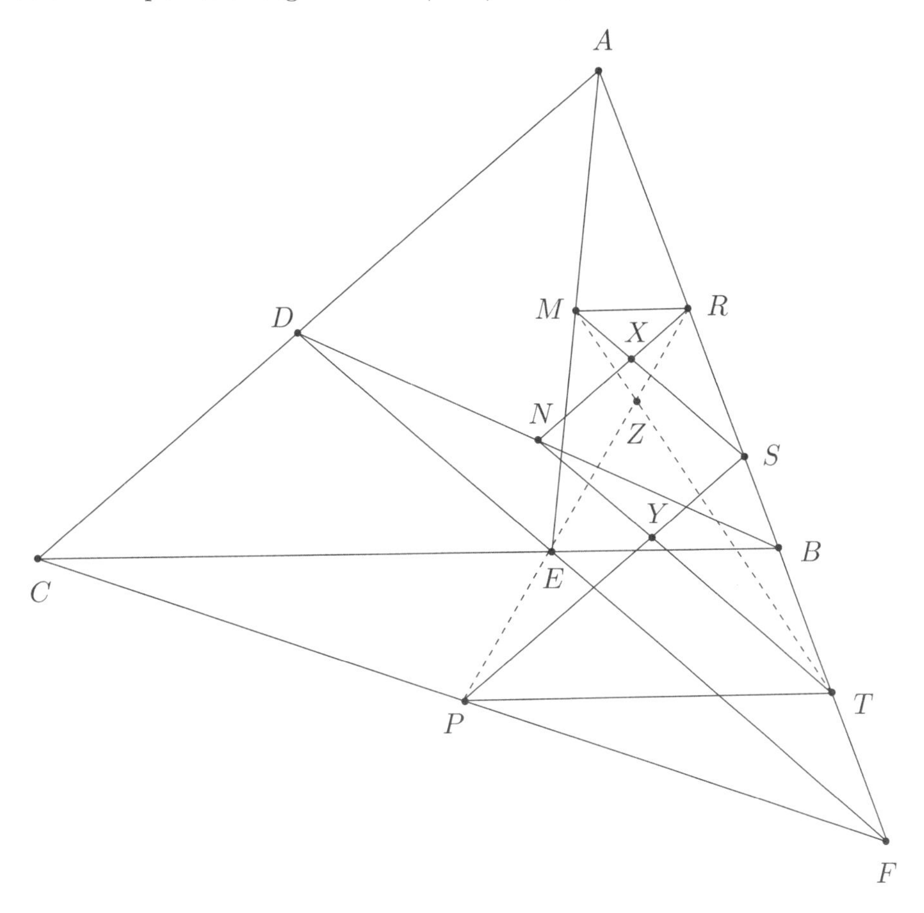

First Proof. Let M, N, P be the midpoints of segments AD, BE, CF respectively and let R, S, T be the midpoints of segments AC, AE, CE respectively. It's easy to see that $MR \parallel PT \parallel BC$ and $NR \parallel PS \parallel AB$ and $NT \parallel MS \parallel DE$ so if we let $X = MS \cap NR$ and $Y = NT \cap PS$ then triangles MXR and TYP are perspective (their perspectrix is the line at infinity). Hence by Desargues' Theorem, if we let $Z = MT \cap PR$, then points X, Y, Z are collinear. Now by the converse Pascal's Theorem on degenerate hexagon $MTNRPS$ we have that these 6 points lie on a conic. But since three of these points are collinear it's clear that the only possible conic is the degenerate conic of two lines, hence points M, N, P are collinear as desired. □

Second Proof. We begin with a claim about a useful locus:

Claim. (Leon Anne's Theorem) Let $ABCD$ be a quadrilateral. Then the locus of points P such that $[ABP]+[CDP]=[BCP]+[DAP]$, where $[XYZ]$ denotes the signed area of triangle XYZ (positive if the triangle intersects the interior of quadrilateral $ABCD$ and negative otherwise), is a line.

Proof. Interpret the problem on the Cartesian plane. The area of a triangle with a fixed base and a moving apex is clearly a linear function of the Cartesian coordinates of the apex, hence the desired result.

Returning to the problem, since M is the midpoint of AD we have that $[ACM]=[DCM]$ and $[AFM]=[DFM]$ so M lies on the Leon-Anne line of quadrilateral $ACDF$. Similarly we find that N and P lie on this line, which completes the second proof. □

//With Leon Anne's Theorem in mind, try showing that if a quadrilateral $ABCD$ has an inscribed circle with center I then points M, I, N are collinear, where M and N are the midpoints of segments AC and BD respectively.

Assigned Problems

Epsilon 5.1. Let $ABCD$ be a convex quadrilateral. Let the parallel line through A to BD meet CD at F and let the parallel through D to AC meet AB at E. If M, N, P, Q denote the midpoints of the segments BD, AC, DE, AF, prove that the lines MN, PQ, AD are concurrent.

Epsilon 5.2. On the circumcircle ω of triangle ABC, two points D, E are situated. AD and AE intersect BC at X and Y, respectively. Let D', E' be the reflections of D, E across the perpendicular bisector of BC. Prove that $D'Y, E'X$ intersect on ω.

Epsilon 5.3. (APMO 2008) Let Γ be the circumcircle of a triangle ABC. A circle passing through points A and C meets the sides BC and BA at D and E, respectively. The lines AD and CE meet Γ again at G and H, respectively. The tangent lines of Γ at A and C meet the line DE at L and M, respectively. Prove that the lines LH and MG meet at Γ.

Epsilon 5.4. (IMO 2010) Given a triangle ABC, with I as its incenter and Γ as its circumcircle, AI intersects Γ again at D. Let E be a point on the arc BDC, and F a point on the segment BC, such that $\angle BAF = \angle CAE < \frac{1}{2}\angle BAC$. If G is the midpoint of IF, prove that lines EI and DG concur on Γ.

Epsilon 5.5. (IMO Shortlist 2004) Let Γ be a circle and let d be a line such that Γ and d have no common points. Further, let AB be a diameter of the circle Γ; assume that this diameter AB is perpendicular to the line d, and the point B is nearer to the line d than the point A. Let C be an arbitrary point on the circle Γ, different from the points A and B. Let D be the point of intersection of the lines AC and d. One of the two tangents from the point D to the circle Γ touches this circle Γ at a point E; hereby, we assume that the points B and E lie in the same half-plane with respect to the line AC. Denote by F the point of intersection of the lines BE and d. Let the line AF intersect the circle Γ at a point G, different from A. Prove that the reflection of the point G in the line AB lies on the line CF.

Epsilon 5.6. Let Γ be a circle and ℓ be a line lying outside Γ. Let $K \in \ell$ and let AB and CD be two chords of Γ passing through K. Take P, Q two points on Γ. Let PA, PB, PC, PD meet ℓ at X, Y, Z, T, respectively, and then let QX, QY, QZ, QT meet again Γ at R, S, U, V. Show that RS and UV meet on ℓ.

Epsilon 5.7. Let ABC be a triangle and let O be its circumcenter. An arbitrary line through O intersects sides AB and AC at points K and L respectively. Let M, N be the midpoints of segments KC, LB respectively. Show that $\angle MON = \angle BAC$.

Epsilon 5.8. (USA TST 2015) Let ABC be a non-equilateral triangle and let M_a, M_b, M_c be the midpoints of the sides BC, CA, AB, respectively. Let S be a point lying on the Euler line. Denote by X, Y, Z the second intersections of M_aS, M_bS, M_cS with the nine-point circle. Prove that AX, BY, CZ are concurrent.

Chapter 6

Jacobi's Theorem

We continue with a useful application of Trig Ceva. The following result is a lemma which is an excellent tool for proving concurrencies. Its name comes from its discoverer, C.F.A. Jacobi. This will be a short section, but we felt that it was important to isolate the following result:

Theorem 6.1. (Jacobi's Theorem) Let ABC be a triangle, and let X, Y, Z be three points in its plane such that $\angle YAC = \angle BAZ$, $\angle ZBA = \angle CBX$ and $\angle XCB = \angle ACY$. Then, the lines AX, BY, CZ are concurrent.

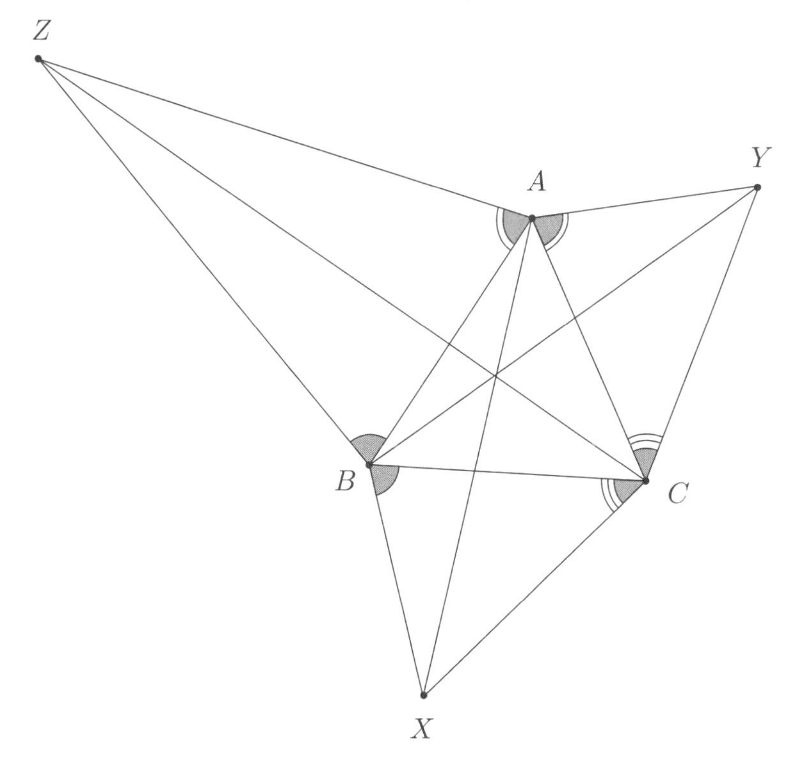

First Proof. The proof is very simple. To avoid complications, we use directed angles taken modulo 180°. Denote by A, B, C, x, y, z the magnitudes of the angles $\angle CAB$, $\angle ABC$, $\angle BCA$, $\angle YAC$, $\angle ZBA$, and $\angle XCB$, respectively. Since the lines AX, BX, CX are (obviously) concurrent (at X), Trig Ceva yields

$$\frac{\sin CAX}{\sin XAB} \cdot \frac{\sin ABX}{\sin XBC} \cdot \frac{\sin BCX}{\sin XCA} = 1.$$

We now notice that

$$\angle ABX = \angle ABC + \angle CBX = B + y, \quad \angle XBC = -\angle CBX = -y,$$

$$\angle BCX = -\angle XCB = -z, \quad \angle XCA = \angle XCB + \angle BCA = z + C.$$

Hence, we get

$$\frac{\sin CAX}{\sin XAB} \cdot \frac{\sin (B+y)}{\sin (-y)} \cdot \frac{\sin (-z)}{\sin (C+z)} = 1.$$

Similarly, we can find

$$\frac{\sin ABY}{\sin YBC} \cdot \frac{\sin (C+z)}{\sin (-z)} \cdot \frac{\sin (-x)}{\sin (A+x)} = 1,$$

$$\frac{\sin BCZ}{\sin ZCA} \cdot \frac{\sin (A+x)}{\sin (-x)} \cdot \frac{\sin (-y)}{\sin (B+y)} = 1.$$

Multiplying all these three equations and canceling similar terms, we get

$$\frac{\sin CAX}{\sin XAB} \cdot \frac{\sin ABY}{\sin YBC} \cdot \frac{\sin BCZ}{\sin ZCA} = 1.$$

And using Trig Ceva once more, we find that the lines AX, BY, CZ are concurrent, which completes the proof. □

//Notice how we began by writing a triviality (the fact that AX, BX, CX were concurrent at X). Good ideas usually come from things like this!

The following proof was given by the user vittasko on the Art of Problem Solving online forum, and we include it because of the surprising way in which radical axes are utilized.

Second Proof. Define three points D, E, F such that $\angle BDC = \angle YAC$, $\angle AEC = \angle ZBA$ and $\angle AFB = \angle XCB$ and denote by ω_1, ω_2, ω_3 the circumcircles of triangles DBC, ECA, FAB respectively. An easy angle chase shows that line AX is the radical axis of circles ω_2 and ω_3 and similarly line BY is the radical axis of circles ω_3 and ω_1 and line CZ is the radical axis of circles

ω_1 and ω_2. Therefore lines AX, BY, CZ concur at the radical center of circles $\omega_1, \omega_2, \omega_3$. This completes the proof. □

We proceed with some corollaries about important triangle centers!

Corollary 6.1. (The Torricelli or First Fermat point) Let XBC, YCA, ZAB be the equilateral triangles erected on the sides of ABC towards the exterior of the triangle. Then, the lines AX, BY, CZ are concurrent at a point that is usually denoted by T (the Torricelli point) or F_+ (the first Fermat point).

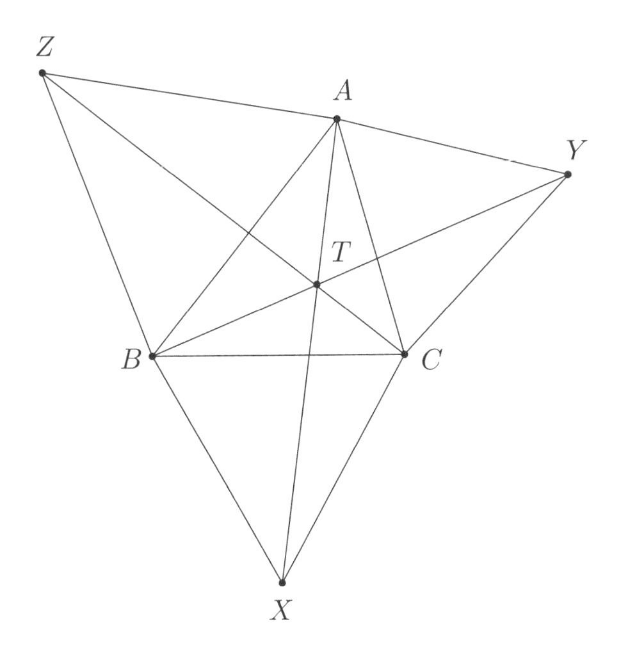

Proof. We have that $\angle YAC = \angle BAZ = 60°$, $\angle ZBA = \angle CBX = 60°$ and $\angle XCB = \angle ACY = 60°$; thus by Jacobi's Theorem the lines AX, BY, CZ are indeed concurrent. □

Corollary 6.2. (The Napoleon point(s)) Let XBC, YCA, ZAB be the equilateral triangles erected on the sides of ABC towards the exterior of the triangle. Furthermore, let O_a, O_b, O_c be the circumcenters of triangles XBC, YCA, ZAB. Then, the lines AO_a, BO_b, CO_c are concurrent at N_+, the first Napoleon point.

Proof. We have that $\angle O_bAC = \angle BAO_c = 30°$, $\angle O_cBA = \angle CBO_a = 30°$ and $\angle O_aCB = \angle ACO_b = 30°$. Jacobi's Theorem does the rest. □

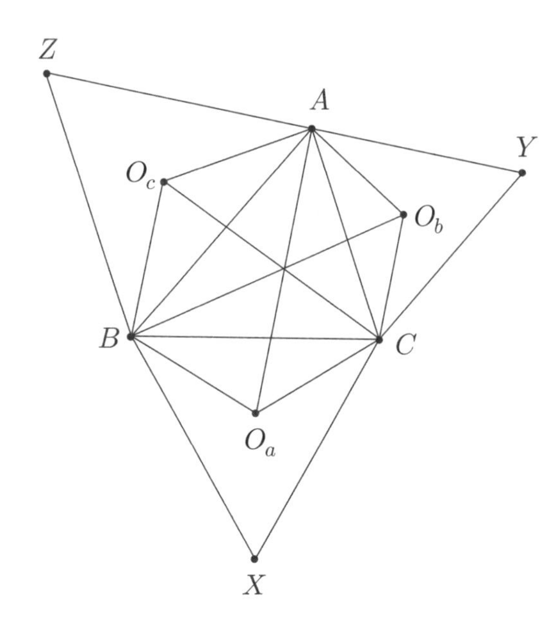

Corollary 6.3. (The Kiepert point(s)) Let XBC, YCA, ZAB be the similar isosceles triangles erected on the sides of ABC towards the exterior of the triangle, having the base angles of magnitude say α. Then, in general, the lines AX, BY, CZ are concurrent at K_α, the Kiepert point of $\alpha \in (0, 90^\circ)$.

As you can see, this generalization is very easy knowing Jacobi's Theorem. Kiepert knew that; yet, what he also knew and proved is that the locus of these concurrency points is very special as the angle α varies in $(0, 90^\circ)$ - it's a hyperbola! The proof however is slightly more complicated; we leave that as **Epsilon 51**.

Now, time for some more relevant applications.

Delta 6.1. (Kariya's Theorem) Let I be the incenter of a given triangle ABC, and let D, E, F be the points where the incircle of ABC touches the sides BC, CA, AB respectively. Now, let X, Y, Z be three points on the lines ID, IE, IF such that the directed segments IX, IY, IZ are equal. Then, the lines AX, BY, CZ are concurrent.

Proof. Being the points of tangency of the incircle of triangle ABC with the sides AB and BC, the points F and D are symmetric to each other with respect to the angle bisector of the angle $\angle ABC$, i.e. with respect to the line BI. Thus, the triangles BFI and BDI are congruent.

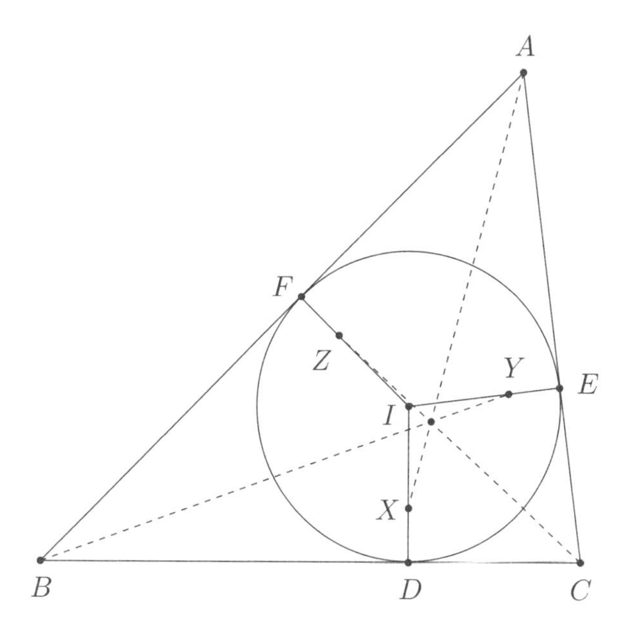

Now, the points Z and X are corresponding points in these two triangles, since they lie on the (corresponding) sides IF and ID of these two triangles and satisfy $IZ = IX$. Corresponding points in inversely congruent triangles form oppositely equal angles, i.e. $\angle ZBF = -\angle XBD$. In other words, $\angle ZBA = \angle CBX$. Similarly, we have that $\angle XCB = \angle ACY$ and $\angle YAC = \angle BAZ$. Note that the points X, Y, Z satisfy the condition for Jacobi's Theorem, and therefore, we conclude that the lines AX, BY, CZ are concurrent. □

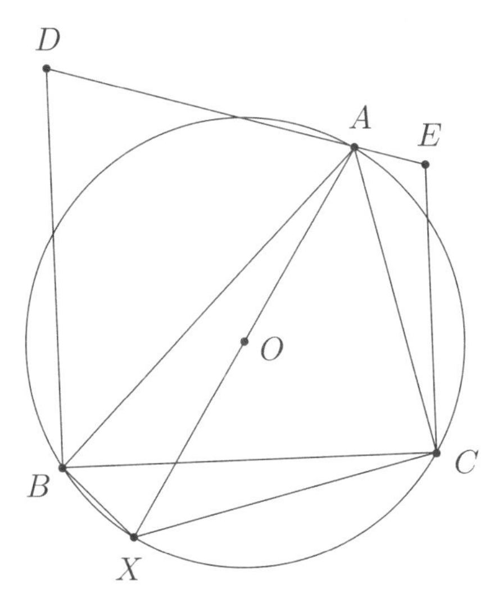

Delta 6.2. Let ABC be a triangle and let the external angle bisector of the angle $\angle BAC$ intersect the lines perpendicular to BC and passing through B

and C at the points D and E, respectively. Prove that the lines BE, CD, AO are concurrent, where O is the circumcenter of ABC.

Proof. Let X be the point diametrically opposite to A on the circumcircle of triangle ABC. Then, AX is a diameter of this circumcircle, so $\angle ABX = 90^\circ$. Thus, using $\angle DBC = 90^\circ$, we get

$$\angle CBX = \angle ABX - \angle ABC = 90^\circ - \angle ABC = \angle DBC - \angle ABC = \angle ABD.$$

Similarly, $\angle BCX = \angle ACE$.

Now, we also have $\angle EAC = \angle BAD$ (since line DE is the A-exterior angle bisector of triangle ABC), hence, Jacobi's Theorem yields that the lines AX, BE and CD concur. Now, the line AX coincides with the line AO (since the segment AX is a diameter of the circumcircle of triangle ABC, and thus it passes through the center O of this circumcircle). Hence, the lines AO, BE and CD concur as desired. □

Delta 6.3. (IMO Shortlist 2001) Let A_1 be the center of the square inscribed in acute triangle ABC with two vertices of the square on side BC. Thus one of the two remaining vertices of the square is on side AB and the other is on AC. Points B_1, C_1 are defined in a similar way for inscribed squares with two vertices on sides AC and AB, respectively. Prove that lines AA_1, BB_1, CC_1 are concurrent.

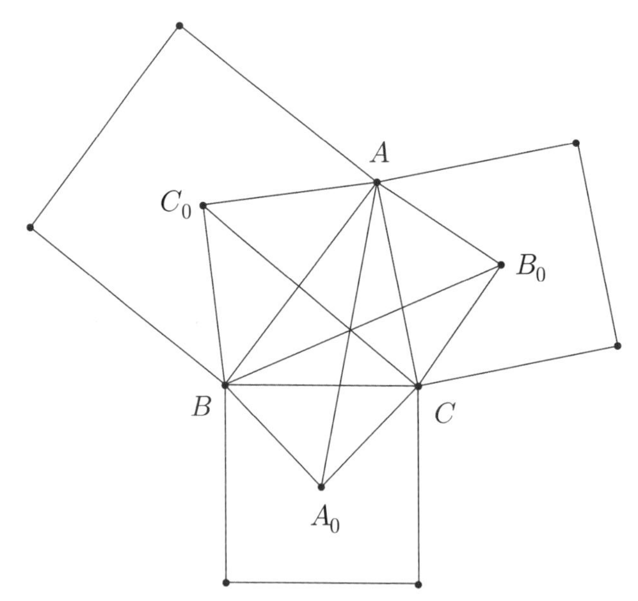

Proof. Let BCX_1X_2 be the square constructed on the side BC in the exterior of the triangle and let A_0 be the center of this square. The points A,

A_1, A_0 are collinear, by homothety. Similarly, if we define B_0, C_0 to be the centers of the squares erected on the sides CA, AB which are in the exterior of ABC, we get that B, B_0, B_1 and C, C_0, C_1 are collinear. But $\angle C_0AB = \angle B_0AC = 45°$, $\angle C_0BA = \angle A_0BC = 45°$, $\angle B_0CA = \angle A_0CB = 45°$, so by Jacobi's Theorem, the lines AA_0, BB_0, CC_0 are collinear, which settles the proof. □

Delta 6.4. (Baltic Way 2009) In a triangle ABC, draw the altitudes AD, BE, CF and let H be its orthocenter. Let O_1, O_2, O_3 be the incenters of triangles EHF, FHD, DHE, respectively. Prove that the lines AO_1, BO_2, CO_3 are concurrent.

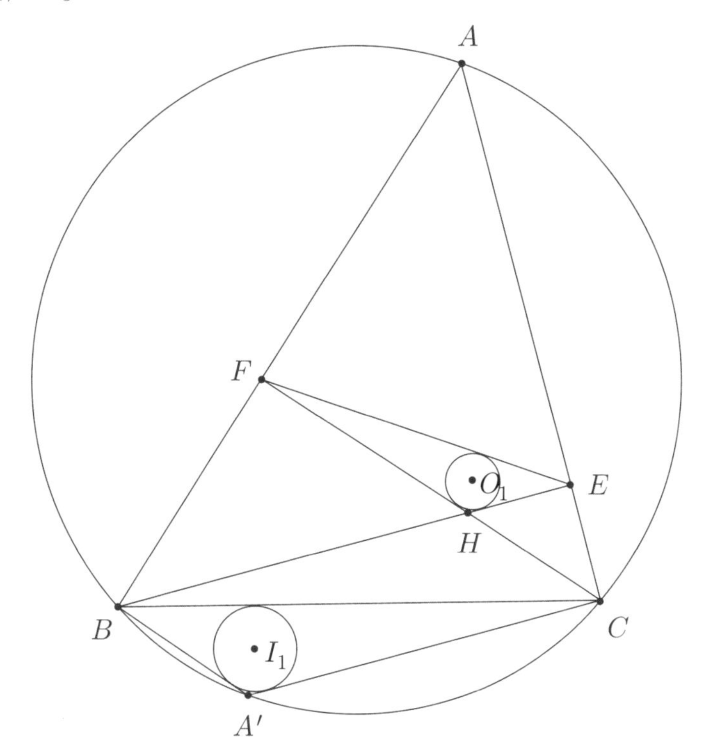

Proof. Let A' be the point diametrically opposite to A on the circumcircle of triangle ABC. Then since quadrilateral $AFHE$ is cyclic (its vertices lie on the circle with diameter AH) we have that

$$\angle A'BC = \angle A'AC = 90° - \angle B = \angle HAF = \angle HEF$$

and similarly

$$\angle A'CB = \angle HFE.$$

Moreover since quadrilateral $BCEF$ is cyclic (its vertices lie on the circle with diameter BC) we have that $\angle AEF = \angle B$ and $\angle AFE = \angle C$ - hence, quadrilaterals $AFHE$ and $ACA'B$ are similar. Now, let I_1 be the incenter of triangle $A'BC$. Since I_1 and O_1 are corresponding points in quadrilaterals $ACA'B$ and $AFHE$ we have that $\angle BAO_1 = \angle FAO_1 = \angle CAI_1$ so line AI_1 is the reflection of line AO_1 over the A-internal angle bisector of triangle ABC. Defining I_2 and I_3 similarly, by **Delta 3.6** it suffices to show that lines AI_1, BI_2, CI_3 concur. But we know that

$$\begin{aligned}
\angle I_1CB &= \frac{\angle A'CB}{2} \\
&= \frac{\angle CFE}{2} \\
&= \frac{\angle HAC}{2} \\
&= \frac{\angle HBC}{2} \\
&= \frac{\angle CFD}{2} \\
&= \frac{\angle A'CB}{2} \\
&= \angle I_2CA,
\end{aligned}$$

and similarly $\angle I_1BC = \angle I_3BA$ and $\angle I_2AC = \angle I_3AB$ so by Jacobi's Theorem we obtain the desired concurrency. □

Jacobi can also be applied in some degenerate cases as well. The following problem is an example:

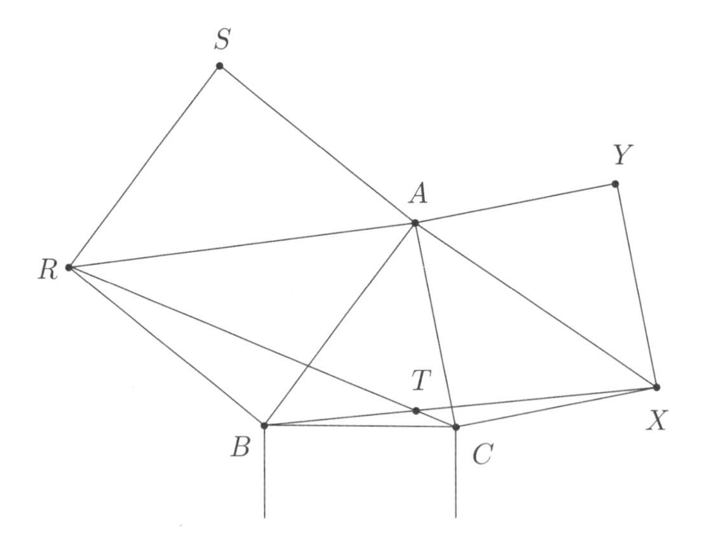

Delta 6.5. Let ABC be a triangle and let $ACXY$ and $ABRS$ be the squares erected on the sides AC and AB that are directed towards the exterior of triangle ABC. Let BX and CR intersect at T. Prove that AT is the A-altitude of triangle ABC.

Proof. Consider the point at infinity A_∞ on the A-altitude of triangle ABC. We have that $\angle RAB = \angle XAC = 45^\circ$ and $\angle RBA = \angle A_\infty BC = 90^\circ$ and $\angle XCA = \angle A_\infty CB = 90^\circ$, so by Jacobi's Theorem, we have that lines BX and CR intersect on the A-altitude of triangle ABC as desired. □

Assigned Problems

Epsilon 6.1. Let ABC be a given triangle. Let d_{A_1}, d_{A_2} be two lines passing through the vertex A, so that they are symmetric to each other with respect to the internal angle bisector of $\angle BAC$ (in other words, the lines d_{A_1}, d_{A_2} are two isogonals with respect to $\angle BAC$). In the same manner, take two isogonals d_{B_1}, d_{B_2} with respect to $\angle ABC$ and two isogonals d_{C_1}, d_{C_2} with respect to $\angle BCA$. Prove that the hexagon determined by these 6 lines has concurrent diagonals.

Epsilon 6.2. Given a triangle ABC with incenter I, let X, Y, Z be the reflections of I into the sidelines BC, CA, and AB, respectively. Prove that the lines AX, BY, CZ are concurrent.

Epsilon 6.3. Let ABC be a triangle with incenter I and let I_1, I_2, I_3 be the incenters of triangles BIC, CIA, AIB respectively. Prove that lines AI_1, BI_2, CI_3 concur.

Epsilon 6.4. (existence of the Kosnita point) Let ABC be a triangle with circumcenter O. Let X, Y, Z be the circumcenters of triangles BOC, COA, AOB respectively. Prove that lines AX, BY, CZ concur.

Epsilon 6.5. Let ABC be a triangle and let D, E, F be points in the plane of triangle ABC such that triangles BCD, CAE, ABF are similar, isosceles, and all don't intersect the interior of triangle ABC. Let H_1 and H_2 be the orthocenters of triangles CAE and ABF respectively. Show that $AD \perp H_1H_2$.

Epsilon 6.6. (Floor van Lamoen) Let A', B', C' be three points in the plane of a triangle ABC such that $\angle B'AC = \angle BAC'$, $\angle C'BA = \angle CBA'$ and $\angle A'CB = \angle ACB'$. Let X, Y, Z be the feet of the perpendiculars from the points A', B', C' to the lines BC, CA, AB. Then, the lines AX, BY, CZ are concurrent.

Epsilon 6.7. (Kiepert Hyperbola) Show that the locus of points described in **Corollary 6.3** is a hyperbola.

Chapter 7

Isogonal Conjugates and Pedal Triangles

We saw that Jacobi dealt with pairs of isogonal lines with respect to the angles of a given triangle ABC. Now, we transition to the more general concept of isogonal conjugates, as they were defined in **Section 3**. We will prove three essential properties that connect them to pedal triangles (which we will define shortly).

Definition. Let ABC be a triangle and let P be a point in its plane. Consider the reflection r_a of the line PA in the internal angle bisector of angle A and similarly define r_b and r_c. Then, the lines r_a, r_b, r_c are concurrent and this concurrency point is called the **isogonal conjugate** of P with respect to triangle ABC.

Definition. Let ABC be a triangle and let P be a point in its plane. Let X, Y, Z be the feet of the perpendiculars from P to lines BC, CA, AB respectively. Then triangle XYZ is the **pedal triangle** of P with respect to triangle ABC.

Now, let us note something completely obvious: when starting with P, to get its isogonal conjugate Q, we took the reflections of the lines PA, PB, PC in their corresponding internal angle bisectors and got three concurrent lines all passing through the isogonal conjugate Q of P; well, if we start with the reflections and take their reflections across the same internal angle bisectors, we obviously get the lines PA, PB, PC back; hence, we can also say that P is the isogonal conjugate of Q. Therefore, it makes sense to talk about pairs of isogonal conjugates with respect to a triangle ABC.

What are some important pairs, you might ask? Well, first, note that the incenter I is clearly its own isogonal conjugate! What about the orthocenter?

Surprisingly enough, it turns out the the orthocenter and circumcenter of a triangle are isogonal conjugates. Indeed, if ABC is our triangle, which we can assume without loss of generality is acute, and H, O are its orthocenter and circumcenter respectively, we have that $\angle BAH = \angle CAO = 90^\circ - \angle B$, and similarly for the other two pairs of angles, which implies the claim. The third pair to remember is the centroid G of triangle ABC and the Symmedian point K, which clearly follows from the definition of K.

Also, keep in mind that the isogonal conjugate of a point P is a point at infinity if and only if P lies on the circumcircle. This is justified by the following mini-lemma.

Delta 7.1. Let P be a point in the plane of a triangle ABC. Prove that the reflections of the lines PA, PB, PC in the corresponding internal angle bisectors are parallel if and only if P lies on the circumcircle of ABC.

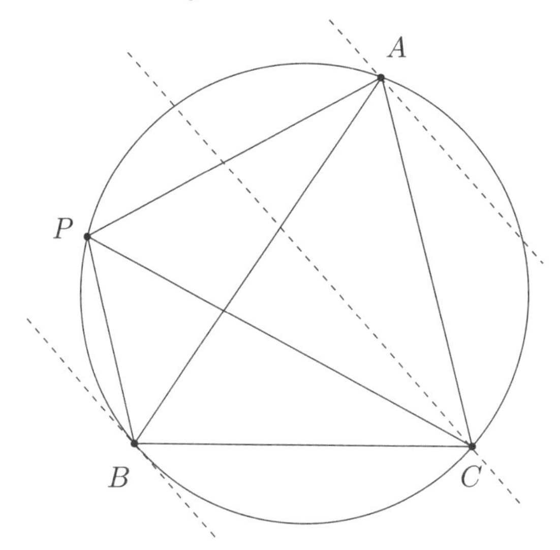

Proof. First, assume that P lies on the circumcircle of ABC. It clearly suffices to show that the reflections of the lines PA and PB in their corresponding internal angle bisectors are parallel since then we can do the same thing for another pair and conclude via the transitivity of parallelism. Thus, if r_a, r_b, r_c denote the reflections, we want to show that $\angle(r_a, AB) = \angle(r_b, AB)$. But this is immediate! We have that $\angle(r_a, AB) = \angle PAC = \angle PBC$, where the latter holds because $PABC$ is cyclic; however, $\angle PBC = \angle(r_b, AB)$ since r_b is the reflection of PB in the $B-$ internal angle bisector of $\angle ABC$. Thus, we conclude that $\angle(r_a, AB) = \angle(r_b, AB)$, as desired.

Conversely, we retrace the angle chasing from above. Now, we know that $\angle(r_a, AB) = \angle(r_b, AB)$ without any assumptions about the position of P and we would like to show that $PABC$ is cyclic, i.e. that $\angle PAC = \angle PBC$. But note that $\angle PAC = \angle(r_a, AB)$ and so $\angle PAC = \angle(r_b, AB) = \angle PBC$, as claimed. This completes the proof.

Let's move to the three properties we mentioned earlier.

Theorem 7.1. Let P be a point in the plane of the triangle ABC and let XYZ be the pedal triangle of P with respect to ABC - i.e. X, Y, Z are the projections of P on the sidelines BC, CA, and AB, respectively. Then, the perpendiculars from the vertices A, B, C to the lines YZ, ZX, and XY, respectively, are concurrent at the isogonal conjugate of P with respect to ABC.

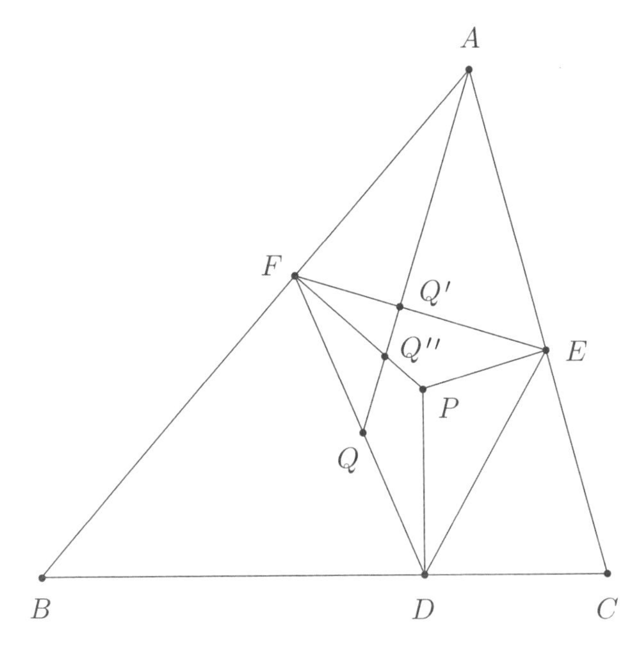

Proof. Simple angle chasing! Let Q be the isogonal conjugate of P with respect to ABC. It suffices to show that $AQ \perp EF$. And indeed this is immediate, since by definition we have that $\angle QAB = \angle PAC$, whereas $\angle PAC = \angle PAE = \angle PFE$, because $AFPE$ is cyclic; hence, the two triangles AFQ'' and $FQ'Q''$ are similar, where Q' and Q'' denote the intersections of AQ with EF and PF respectively. Therefore, $\angle FQ'Q'' = \angle AFQ'' = 90°$, and so $AQ \perp EF$, which proves the claim. □

Corollary 7.1. If E, F are the feet of the altitudes from the vertices B and C of triangle ABC, then the lines AO and EF are perpendicular, where O denotes as usual the circumcenter of ABC.

Corollary 7.2. Let ABC be a triangle with excenters I_a, I_b, I_c. Let D, E, F be the tangency points of the A-excircle with BC, of the B-excircle with CA, and of the C-excircle with AB, respectively. Prove that the lines I_aD, I_bE, I_cF are concurrent at the circumcenter of triangle $I_aI_bI_c$. This is usually called the **Bevan point** of triangle ABC.

Proof. Just note that ABC is the orthic triangle of triangle $I_aI_bI_c$ and apply the converse of **Corollary 7.1**. □

Delta 7.2. Prove that P coincides with its isogonal conjugate with respect to triangle ABC if and only if P is the incenter or one of the excenters.

Proof. Consider the reflection of the line AP over the A-internal angle bisector of triangle ABC. This reflection coincides with AP if and only if AP is parallel or perpendicular to the A-internal angle bisector of triangle ABC, which implies the desired result since similar results hold for lines BP and CP. □

The next theorem is a quick (but careful) application of **Theorem 7.1**.

Theorem 7.2. Let P be a point in the plane of the triangle ABC and let R, S, T be the reflections of the point P in the sidelines BC, CA, and AB, respectively. Then, the isogonal conjugate of P with respect to ABC is the circumcenter of triangle RST.

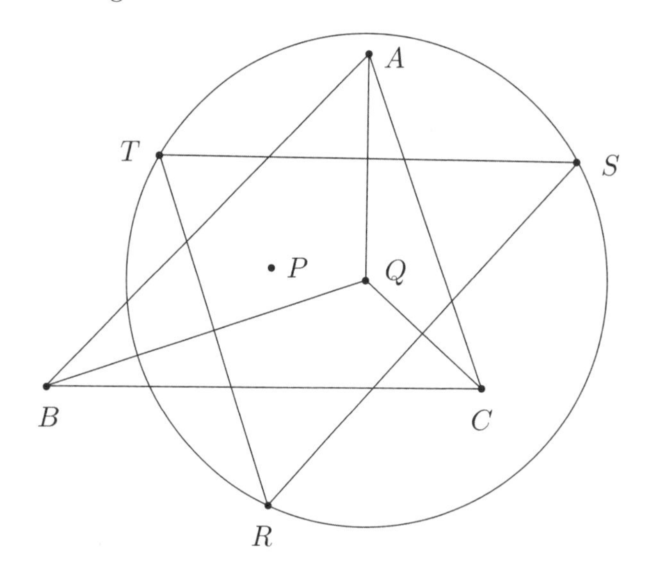

Proof. Let X, Y, Z be the midpoints of segments PR, PS, PT. Obviously, triangle XYZ is the pedal triangle of P with respect to ABC and is homothetic with triangle RST; hence, by **Theorem 7.1**, we immediately get that if Q denotes the isogonal conjugate of P, then $AQ \perp ST$. However, we still need to show that AQ is the perpendicular bisector of ST. But this is clear! Note that S is the reflection of P across AC, so $AP = AS$, and T is the reflection of P across AB, so $AP = AT$; hence $AS = AT$, i.e. triangle AST is isosceles and so A needs to lie on the perpendicular bisector of segment ST as claimed. This completes the proof, since after doing the same thing for BQ and CQ, we can conclude that they are the perpendicular bisectors of segments TR and RS respectively, so they need to meet at the circumcenter of RST as desired. □

Corollary 7.3. Recall that the reflections of the orthocenter into the sidelines of the triangle lie on the circumcircle.

The next theorem is perhaps the most remarkable of the three and its proof is really wonderful.

Theorem 7.3. (The Six Point Circle Theorem) Let P be a point in the plane of the triangle ABC and let Q be its isogonal conjugate with respect to ABC. If DEF denotes the pedal triangle of P and XYZ denotes the pedal triangle of Q (both with respect to ABC, of course), then the points D, E, F, X, Y, Z all lie on the same circle, whose center is the midpoint of the segment PQ.

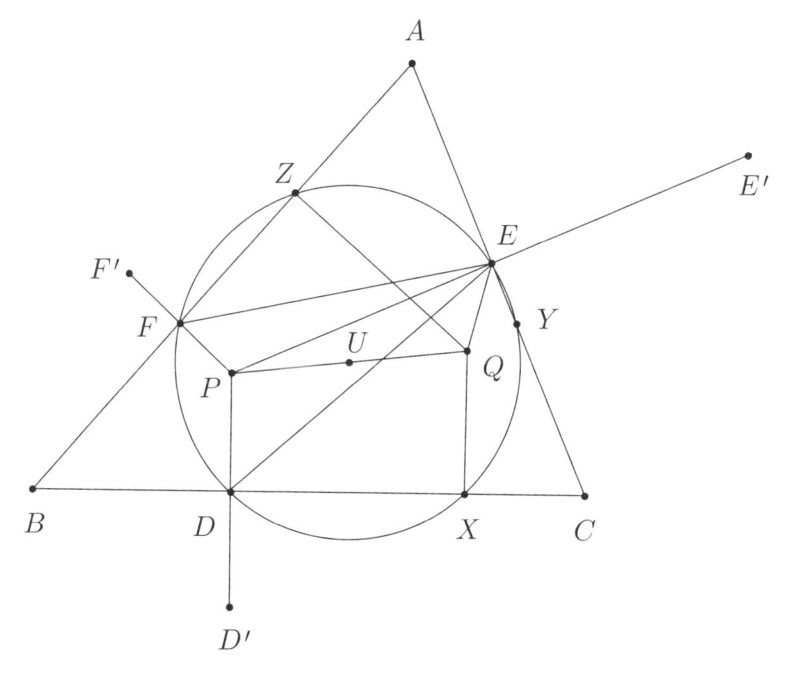

Proof. The proof uses **Theorem 7.2** in a very clever way. More precisely, let D', E', F' be the reflections of P across BC, CA, AB respectively. Also let X', Y', Z' be the reflections of Q across BC, CA, AB respectively. Let U be the midpoint of PQ.

Now, note that the lines UD, UE, UF are the U-midlines in triangles PQD', PQE', PQF'; thus we have that

$$UD = \frac{1}{2}QD', \quad UE = \frac{1}{2}QE', \quad \text{and} \quad UF = \frac{1}{2}QF'.$$

However, **Theorem 7.2** tells us that Q is the circumcenter of triangle $D'E'F'$, so $QD' = QE' = QF'$, which implies that $UD = UE = UF$.

Similarly, we can deduce that $UX = UY = UF$ by looking at triangle $X'Y'Z'$ which has circumcenter P. To finish, we still need to prove that, for example, $UD = UX$. But this is actually immediate, since U is the midpoint of PQ and since PD and QX are both perpendicular to DX! Combining this with what we found above, it follows that $UD = UE = UF = UX = UY = UZ$, so we can conclude that the points D, E, F, X, Y, Z all lie on a circle with center U. This completes the proof. □

Even though it feels like using an atomic bomb, let us note that **Theorem 7.3** gives a very simple justification for the existence of the nine-point circle.

Corollary 7.4. Given a triangle ABC with orthocenter H, the midpoints of its sides, the feet of the altitudes, and the midpoints of the segments HA, HB, HC all lie on the same circle - the **nine-point circle** or **the Euler circle** of triangle ABC. The center of this circle is the midpoint of the segment OH where O is the circumcenter of triangle ABC, and this point is called the **ninepoint center** or **the Euler point** of triangle ABC.

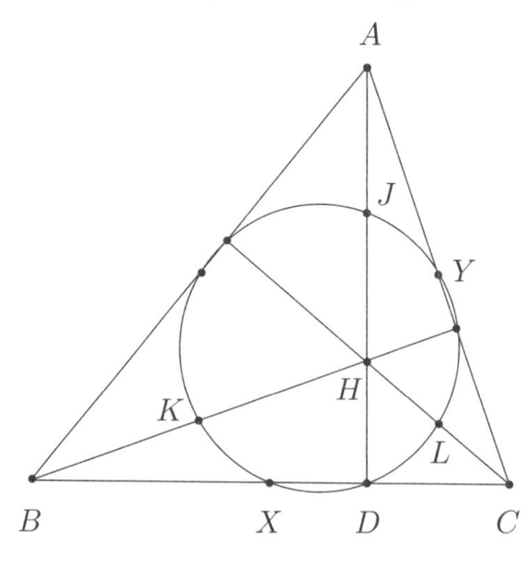

Proof. Of course, we know from **Theorem 7.3** that the midpoints of the sides and the feet of the altitudes are concyclic, all lying on the circle centered at the midpoint of OH (as O and H are isogonal conjugates and the medial and orthic triangles are their pedal triangles), but we still need to show that the midpoints J, K, L of segments HA, HB, HC lie on this circle. But this is just a simple angle chase! Indeed, note that it is sufficient to take the foot of the $A-$altitude, call it D, and the midpoints X, Y of BC and CA, and prove that J, D, X, Y are concyclic. To do this, we would like to prove that $XY \perp YJ$, since we already know that $JD \perp DX$. But JY is the A-midline in triangle AHC so $JY \parallel HC$ and $HC \perp AB$, so $JY \perp AB$, and therefore $JY \perp XY$ (as XY is the $C-$midline in triangle ABC). Doing the same thing for K and L gives us the concyclicity of all nine points. This completes the proof. □

Now, let's see a few applications before the exercises. We begin with a problem from an IMO Shortlist.

Delta 7.3. (IMO Shortlist 1998) Let P be a point in the plane of a triangle ABC and let Q be its isogonal conjugate with respect to ABC. Prove that

$$\frac{AP \cdot AQ}{AB \cdot AC} + \frac{BP \cdot BQ}{BA \cdot BC} + \frac{CP \cdot CQ}{CA \cdot CB} = 1.$$

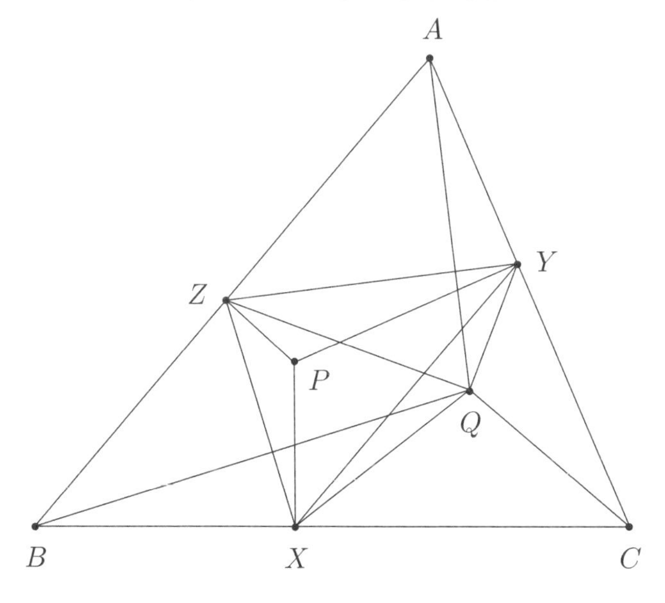

Proof. Let X, Y, Z be the feet of the projections from P onto BC, CA, AB respectively. Then points A, Y, P, Z lie on a circle with diameter AP so by the Extended Law of Sines we have that

$$YZ = AP \sin A$$

Moreover from **Theorem 7.1** we know that $AQ \perp YZ$ which means that

$$[AYQZ] = \frac{AQ \cdot YZ}{2} = \frac{AQ \cdot AP \sin A}{2} = \frac{AP \cdot AQ}{AB \cdot AC}[ABC]$$

where we used the fact that $[ABC] = \frac{AB \cdot AC \sin A}{2}$. Finding similar expressions for $[BZQX]$ and $[CXQY]$ and summing then yields the desired result, since $[AYQZ] + [BZQX] + [CXQY] = [ABC]$. □

The next problem appeared in Gazeta Matematica, and also has a very surprising proof.

Delta 7.4. (Gazeta Matematica) Let ABC be a triangle and let P be a point in its interior with pedal triangle DEF. Suppose that that the lines DE and DF are perpendicular. Prove that if Q is the isogonal conjugate of P with respect to triangle ABC then Q is the orthocenter of triangle AEF.

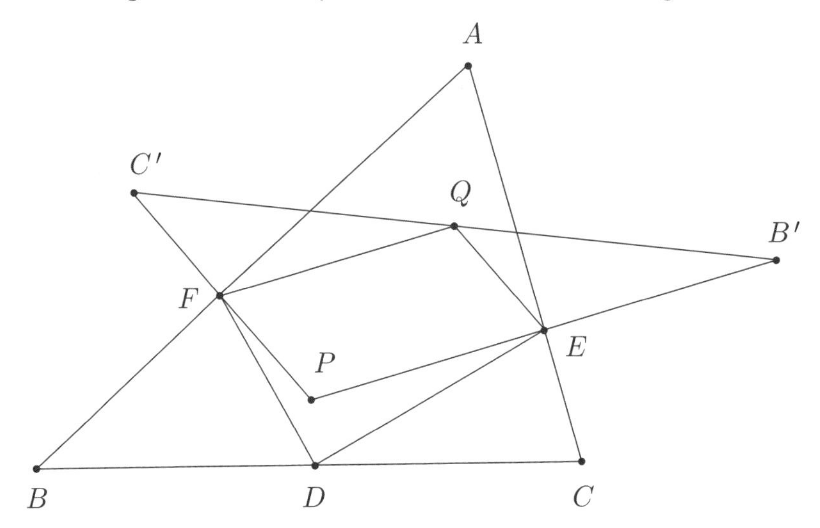

Proof. Let the reflections of P over BC, CA, AB be A', B', C' respectively. Since $DE \perp DF$ we also have that $A'B' \perp A'C'$ so the circumcenter of triangle $A'B'C'$ is the midpoint of $B'C'$. Therefore by **Theorem 7.2** we have that Q is the midpoint of $B'C'$. Since QE is the B'-midline of triangle $PB'C'$ we have that $QE \parallel PC'$ so $QE \perp AF$. Similarly $QF \perp AE$ and so Q is the orthocenter of triangle AEF as desired. □

Delta 7.5. Let P and Q be two isogonal conjugates with respect to a triangle ABC. Then, the reflection of the line AP with respect to the internal angle bisector of $\angle BPC$ and the reflection of the line AQ with respect to internal angle bisector of $\angle BQC$ are symmetric to each other with respect to the sideline BC.

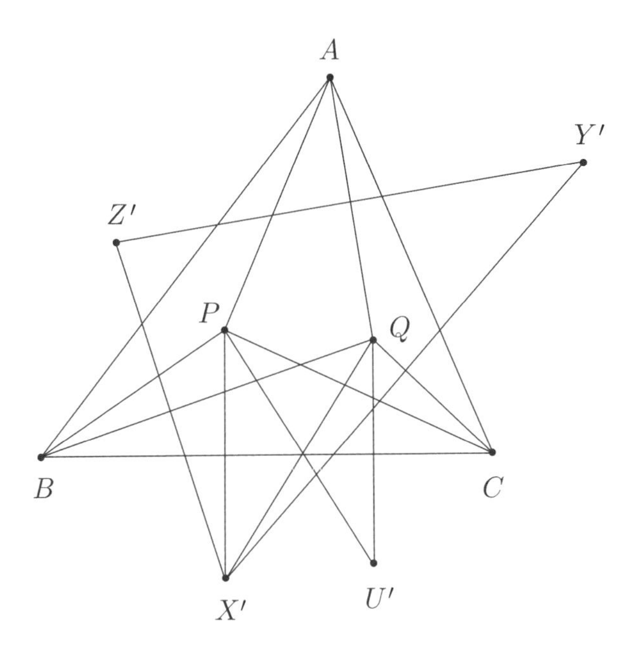

Proof. Let X', Y', Z' be the reflections of the point P across the lines BC, CA, AB, and let U' be the reflection of the point Q in the line BC. Then, the point Q is, in turn, the reflection of U' in the line BC. And since the points Q and X' are the reflections of U' and P in the line BC, the line QX' is the reflection of the line $U'P$ in the line BC. This means that the lines PU' and QX' are symmetric to each other with respect to the line BC.

However, by **Theorem 7.2**, the point Q is the circumcenter of triangle $X'Y'Z'$. Hence, we get that $\angle QX'Z' = 90° - \angle Z'Y'X'$. In other words, $\angle(QX', Z'X') = 90° - \angle(Y'Z', X'Y')$. Furthermore, **Theorem 7.2** also tells us that the lines AQ, BQ, CQ are the perpendicular bisectors of the segments $Y'Z'$, $Z'X'$, $X'Y'$, thus $\angle(Y'Z', AQ) = 90°$, $\angle(BQ, Z'X') = 90°$ and $\angle(X'Y', CQ) = 90°$. Hence, we immediately get that $\angle(BQ, QX') = \angle(CQ, AQ)$. Thus, the line QX' is the reflection of the line AQ with respect to the internal angle bisector of $\angle BQC$. Similarly, the line PU' is the isogonal conjugate of AP with respect to $\angle BPC$. And since we know that the lines PU' and QX' are symmetric to each other with respect to BC, the conclusion follows. This completes the proof. □

The preceding **Delta 7.5** is a crucial lemma in the proof of the following beautiful result by Hatzipolakis. The proof is after [DG] and it is due to Ehrmann.

Delta 7.6. (Hatzipolakis/Ehrmann) Let P be a point in the plane of a triangle ABC. The lines AP, BP, CP intersect the lines BC, CA, AB at the points

A', B', C'. Let Q be the isogonal conjugate of the point P with respect to triangle ABC. Then, the reflections of the lines AQ, BQ, CQ in the lines $B'C'$, $C'A'$, $A'B'$ are concurrent.

Proof. Let A_1, B_1, C_1, P_1 be the isogonal conjugates of the points A, B, C, P with respect to triangle $A'B'C'$. Since P_1 is the isogonal conjugate of P with respect to triangle $A'B'C'$, the line $A'P_1$ is the reflection of the line $A'P$ with respect to the internal angle bisector of angle $\angle C'A'B'$. Since A_1 is the isogonal conjugate of A with respect to $A'B'C'$, the line $A'A_1$ is the isogonal of the line $A'A$ with respect to the angle $C'A'B'$. But since the lines $A'P$ and $A'A$ coincide, their isogonals with respect to the angle $C'A'B'$ must also coincide; i.e., the lines $A'P_1$ and $A'A_1$ coincide. Hence, the points A', A_1, P_1 are collinear, and similarly we get that the points B', B_1, P_1 are collinear, and the points C', C_1, P_1 are collinear.

Since B_1 is the isogonal conjugate of B with respect to triangle $A'B'C'$, the line $A'B_1$ is the isogonal of the line $A'B$ with respect to $\angle C'A'B'$. Also, since C_1 is the isogonal conjugate of C with respect to triangle $A'B'C'$, the line $A'C_1$ is the isogonal of the line $A'C$ with respect to $\angle C'A'B'$. But again, since the lines $A'B$ and $A'C$ coincide, their isogonal with respect to $\angle C'A'B'$ need to coincide, so $A'B_1$ and $A'C_1$ coincide. Thus, we got that the points A', B_1, C_1 are collinear, and similarly, the points B', C_1, A_1 are collinear, and the points C', A_1, B_1 are collinear.

Now, since the point Q is the isogonal conjugate of the point P with respect to triangle ABC, the line AQ is the isogonal of the line AP with respect to the angle $\angle CAB$. However, A and A_1 are isogonal conjugates with respect to triangle $A'B'C'$, thus the result from **Delta 7.5** yields that the isogonal of $A'A$ with respect to $\angle B'AC'$ and the isogonal of $A'A_1$ with respect to $\angle B'A_1C'$ are symmetric to each other with respect to $B'C'$. Moreover, the isogonal of the line $A'A$ with respect to $B'AC'$ is the isogonal of the line AP with respect to CAB (since the line $A'A$ is the line $A'P$, and the angle $\angle B'AC'$ is the angle $\angle CAB$), and this is the line AQ. Further, the isogonal of the line $A'A_1$ with respect to the angle $\angle B'A_1C'$ is the isogonal of the line A_1P_1 with respect to the angle $C_1A_1B_1$ (since the line $A'A_1$ is the line A_1P_1 and the angle $\angle B'A_1C'$ is the angle $\angle C_1A_1B_1$). Thus, we got that the line AQ and the isogonal of the line A_1P_1 with respect to $\angle C_1A_1B_1$ are symmetric to each other with respect to the line $B'C'$. In other words, the reflection of AQ in the line $B'C'$ is the isogonal of the line A_1P_1 with respect to $\angle C_1A_1B_1$. We can say analogous statements about the reflections of the lines BQ and CQ across $C'A'$ and $A'B'$; thus, we conclude that the reflections of the lines AQ, BQ, CQ in the lines $B'C'$, $C'A'$, $A'B'$ are concurrent, as these lines are

the isogonals of the concurrent lines A_1P_1, B_1P_1, C_1P_1 with respect to the angles of triangle $A_1B_1C_1$, so they need to meet at the isogonal conjugate of the point P_1 with respect to triangle $A_1B_1C_1$. This completes the proof. □

We proceed with an unexpected computational result.

Theorem 7.4. (Euler's Pedal Triangle Theorem) Let P be a point in the plane of the triangle ABC. If $A_1B_1C_1$ is the pedal triangle of P with respect to ABC, then

$$\frac{K_{A_1B_1C_1}}{K_{ABC}} = \frac{|R^2 - OP^2|}{4R^2},$$

where K_{DEF} denotes the area of triangle DEF for any triangle DEF and R is the circumradius of triangle ABC.

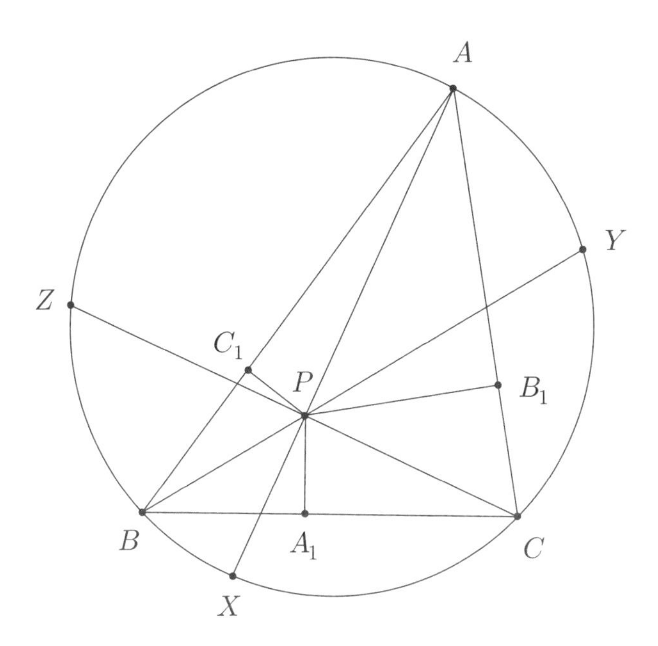

Proof. We need a preliminary result that will provide a lot of insight into what's going on.

Claim. Let X, Y, Z be the second intersections of the lines AP, BP, CP with the circumcircle of ABC. Then, triangles $A_1B_1C_1$ and XYZ are similar.

Indeed, we can write

$$\begin{aligned}\angle B_1A_1C_1 &= \angle PA_1B_1 + \angle PA_1C_1\\ &= \angle PCB_1 + \angle PBC_1\\ &= \angle YBA + \angle ZCA\\ &= \angle YXA + \angle ZXA\\ &= \angle YXZ.\end{aligned}$$

And similarly, we can show that $\angle B_1C_1A_1 = \angle YZX$ and $\angle A_1B_1C_1 = \angle XYZ$, so the claim is indeed true.

Returning to the problem, we start computing the ratio $\frac{K_{A_1B_1C_1}}{K_{ABC}}$ using the well-known formula $EF \cdot FD \cdot DE = 4R \cdot K_{DEF}$ for any triangle DEF. Consequently, we write

$$\frac{K_{A_1B_1C_1}}{K_{ABC}} = \frac{B_1C_1}{BC} \cdot \frac{C_1A_1}{CA} \cdot \frac{A_1B_1}{AB} \cdot \frac{R}{R_{A_1B_1C_1}}.$$

where $R_{A_1B_1C_1}$ is the circumradius of triangle $A_1B_1C_1$. However, triangle $A_1B_1C_1$ and the circumcevian triangle XYZ from the claim are similar so

$$\frac{R}{R_{A_1B_1C_1}} = \frac{YZ}{B_1C_1}.$$

Hence, we get that

$$\frac{K_{A_1B_1C_1}}{K_{ABC}} = \frac{YZ}{BC} \cdot \frac{C_1A_1}{CA} \cdot \frac{A_1B_1}{AB}.$$

On the other hand, the quadrilateral $BCYZ$ is cyclic, so triangles PBC and PZY are similar, and so $\frac{PB}{PZ} = \frac{PC}{PY} = \frac{BC}{YZ}$; thus keeping in mind that $C_1A_1 = PB\sin B$ and $A_1B_1 = PC\sin C$ (which follows from the Law of Sines in triangle BA_1C_1 and CA_1B_1), we can write

$$\begin{aligned}\frac{K_{A_1B_1C_1}}{K_{ABC}} &= \frac{YZ}{BC} \cdot \frac{C_1A_1}{CA} \cdot \frac{A_1B_1}{AB}\\ &= \frac{PZ}{PB} \cdot \frac{PB\sin B}{CA} \cdot \frac{PC\sin C}{AB}\\ &= \frac{PZ}{PB} \cdot \frac{PB}{2R} \cdot \frac{PC}{2R}\\ &= \frac{PC \cdot PZ}{4R^2}\\ &= \frac{|R^2 - OP^2|}{4R^2},\end{aligned}$$

where the last equality holds since $PC \cdot PZ$ is precisely the power of P with respect to the circumcircle of ABC. This completes the proof. $\square$

//Do remember that the sidelengths of the pedal triangle $A_1B_1C_1$ of P with respect to ABC are given by the nice formulas $B_1C_1 = PA \sin A$, $C_1A_1 = PB \sin B$, $A_1B_1 = PC \sin C$.

We end the section with a difficult problem from the 2014 International Math Olympiad.

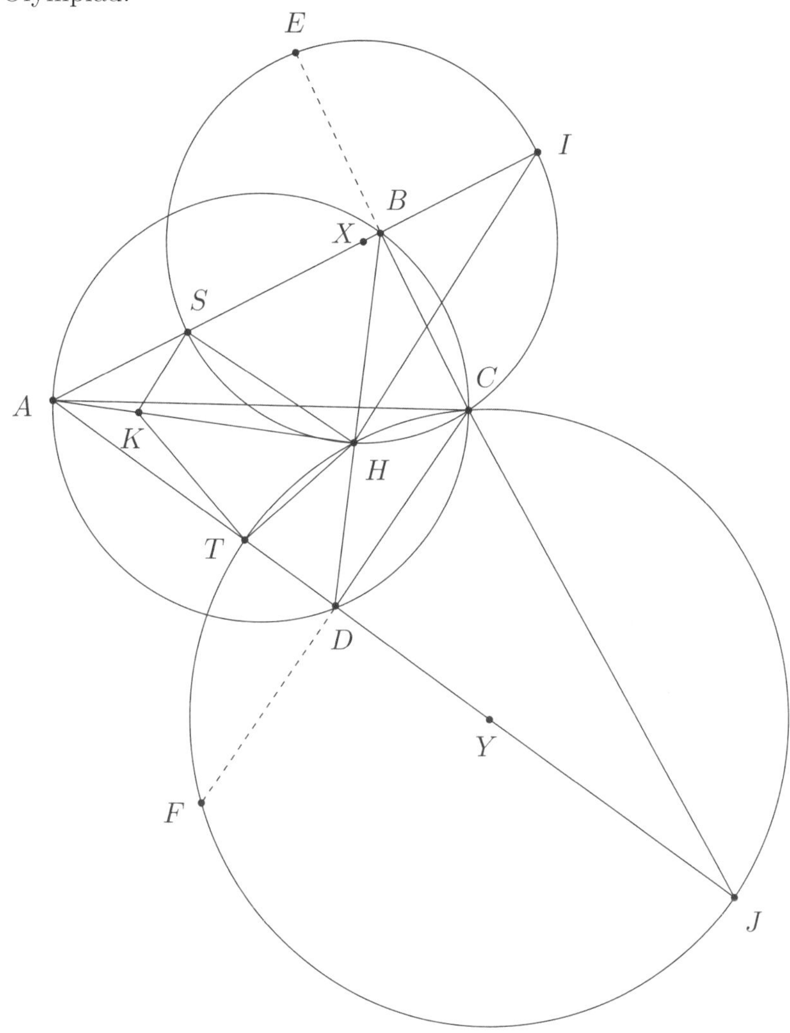

Delta 7.7. (IMO 2014) Convex quadrilateral $ABCD$ has $\angle ABC = \angle CDA = 90°$. Point H is the foot of the perpendicular from A to BD. Points S and T lie on sides AB and AD, respectively, such that H lies inside triangle SCT

and

$$\angle CHS - \angle CSB = 90^\circ, \quad \angle THC - \angle DTC = 90^\circ.$$

Prove that line BD is tangent to the circumcircle of triangle TSH.

Proof. Reflect C over lines AB and AD to obtain points E and F respectively. Then quadrilaterals $ESHC$ and $FTHC$ are easily found to be cyclic - let their circumcenters be X and Y respectively. Clearly it suffices to show that the circumcenter of triangle TSH lies on line AH. Let the perpendicular through S to SH intersect AH at K and let the perpendicular through T to TH intersect AH at K'. It clearly suffices to show that $K = K'$.

Let I be the point on AB such that $HI \parallel KS$ and let J be the point on AD such that $HJ \parallel K'T$. Then since we have that $\frac{AS}{AI} = \frac{AK}{AH}$ and $\frac{AT}{AJ} = \frac{AK'}{AH}$ it suffices to show that $\frac{AS}{AI} = \frac{AT}{AJ}$. Since X and Y are the midpoints of segments SI and TJ respectively, this is equivalent to $ST \parallel XY$ which itself is equivalent to $CH \perp ST$.

This means we want to prove that $\angle STH + (180^\circ - \angle THC) = 90^\circ$. But since $FTHC$ is cyclic, we know that $180^\circ - \angle THC = \angle TFC = \angle TCD$ and since $\angle TCD + \angle CTD = 180^\circ - \angle TDC = 90^\circ$ it really suffices to show that $\angle STH = \angle CTD$.

Now, we know from the introduction that lines AH and AC are isogonal with respect to angle $\angle BAC$ so it suffices to show that points C and H are isogonal conjugates with respect to triangle AST. Let H' be the isogonal conjugate of C with respect to triangle AST. Then we have that

$$\begin{aligned}\angle SCT &= 180^\circ - \angle STC - \angle TSC \\ &= 180^\circ - (180^\circ - \angle H'SA) - (180^\circ - \angle H'TA) \\ &= 180^\circ - \angle A - \angle SH'T\end{aligned}$$

Now also note that

$$\begin{aligned}\angle SCT &= 180^\circ - \angle DCJ - \angle SCB - \angle TCD \\ &= 180^\circ - \angle A - \angle SEC - \angle TFC \\ &= 180^\circ - \angle A - (180^\circ - \angle SHC) - (180^\circ - \angle THC) \\ &= 180^\circ - \angle A - \angle SHT\end{aligned}$$

so since H' must lie on line CH we have that $H = H'$. This completes the proof. □

//Remember the fact that if P and Q are isogonal conjugates with respect to a triangle ABC then $\angle BPC + \angle BQC = 180^\circ + \angle A$ if P and Q are inside triangle ABC and $\angle BPC + \angle BQC = 180^\circ - \angle A$ otherwise (this was used in the final step of the proof).

Assigned Problems

Epsilon 7.1. Let P, Q be isogonal conjugates with respect to a triangle ABC. Prove that $AP \sin BPC = AQ \sin BQC$.

Epsilon 7.2. Let ABC be a triangle and let Γ be a circle that meets the line BC at A_1, A_2, the line CA at B_1 and B_2, and the line C_1 and C_2. Let Ω_1, Ω_2, Ω_3 be the circles with diameters A_1A_2, B_1B_2, and C_1C_2, respectively. Prove that the radical center of these three circles is the isogonal conjugate of the center of Γ with respect to triangle ABC.

Epsilon 7.3. (Generalization of ELMO 2014) Let P_1, P_2 be isogonal conjugates with respect to triangle ABC. Point Q_1 is on the circumcircle of triangle BCP_1 such that points P_1 and Q_1 are diametrically opposite, and Q_2 is constructed similarly. Prove that Q_1, Q_2 are also isogonal conjugates with respect to triangle ABC.

Epsilon 7.4. Let Γ be an ellipse inscribed in a triangle ABC with foci P and Q. Prove that P, Q are isogonal conjugates with respect to triangle ABC.

Epsilon 7.5. (Romania JBMO TST 2009) Let $ABCD$ be a quadrilateral. The diagonals AC and BD are perpendicular at point O. The perpendiculars from O on the sides of the quadrilateral meet AB, BC, CD, DA at M, N, P, Q, respectively, and meet again CD, DA, AB, BC at M', N', P', Q', respectively. Then, the points M, N, P, Q, M', N', P', Q' are all concyclic. (Note: this is just an angle chase, but is an amazingly useful lemma for the next problem).

Epsilon 7.6. (Mathematical Reflections) Let $ABCD$ be a quadrilateral and let $P = AC \cap BD$, $E = AD \cap BC$, and $F = AB \cap CD$. Denote by $\text{isog}_{XYZ}(P)$ the isogonal conjugate of P with respect to triangle XYZ. Prove that $\text{isog}_{ABE}(P) = \text{isog}_{CDE}(P) = \text{isog}_{ADF}(P) = \text{isog}_{BCF}(P)$ provided that AC and BD are perpendicular. (Hint: use the previous problem).

Epsilon 7.7. Let ABC be an acute triangle. Denote by A_1, B_1, C_1 the projections of the centroid G onto the sides BC, CA, and AB, respectively. Prove that

$$\frac{2}{9} \le \frac{K_{A_1B_1C_1}}{K_{ABC}} \le \frac{1}{4},$$

where $K_{\mathcal{P}}$ denotes the area of the convex polygon $\mathcal{P}$.

Epsilon 7.8. Let P and Q be isogonal conjugates with respect to a triangle ABC and let their pedal triangles with respect to triangle ABC be $P_aP_bP_c$ and $Q_aQ_bQ_c$ respectively. Let $X = P_bP_c \cap Q_bQ_c$. Prove that $AX \perp PQ$.

Epsilon 7.9. (IMO 2004) In a convex quadrilateral $ABCD$, the diagonal BD bisects neither the angle ABC nor the angle CDA. The point P lies inside $ABCD$ and satisfies

$$\angle PBC = \angle DBA \quad \text{and} \quad \angle PDC = \angle BDA.$$

Prove that $ABCD$ is a cyclic quadrilateral if and only if $AP = CP$

Epsilon 7.10. (USAMO 2011) Let P be a given point inside quadrilateral $ABCD$. Points Q_1 and Q_2 are located within $ABCD$ such that

$$\angle Q_1BC = \angle ABP, \ \angle Q_1CB = \angle DCP, \ \angle Q_2AD = \angle BAP, \ \angle Q_2DA = \angle CDP.$$

Prove that $\overline{Q_1Q_2} || \overline{AB}$ if and only if $\overline{Q_1Q_2} || \overline{CD}$.

Chapter 8

Simson and Steiner

The heart of this section is represented by the following very famous result.

Theorem 8.1. (Simson) Let ABC be a triangle and let P be a point in its plane. If X, Y, Z are the projections of P on the sidelines of BC, CA, and AB, respectively, then the points X, Y, Z are collinear if and only if P is on the circumcircle of ABC.

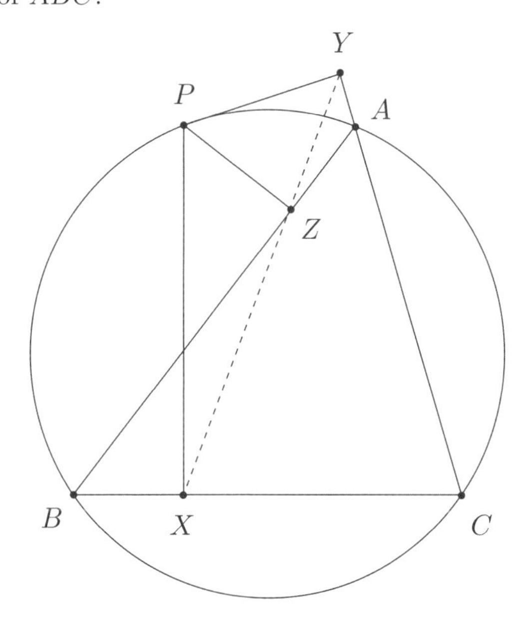

When P is on the circumcircle, the line determined by the projections X, Y, Z (or the degenerate pedal triangle XYZ) is called the **Simson line** of P with respect to ABC. The degeneracy of the pedal triangle in this case was also foreseen in the previous section. We saw in **Delta 7.1** that the points that have isogonal conjugates at infinity are precisely the points lying on the circumcircle, and **Theorem 7.3** told us that this happens if and only if the pedal triangle is degenerate. We, however, give the classic angle-chasing proof below.

Proof. First, assume that P is on the circumcircle of ABC; in order to prove that X, Y, Z are collinear, we would like to show that $\angle XYC = \angle ZYA$. And this is a simple angle chase! Note that $XYPC$ is cyclic, so $\angle XYC = \angle XPC = 90^\circ - \angle PCX = 90^\circ - \angle PCB = 90^\circ - \angle PAZ$ (the latter equality holds because $ABCP$ is cyclic). But $90^\circ - \angle PAZ = \angle APZ = \angle ZYA$; thus, we get that $\angle XYC = \angle ZYA$, as desired. Thus, the points X, Y, Z are indeed collinear.

Conversely, we now know that $\angle XYC = \angle ZYA$, so going backwards via the same angle chasing, we get that $\angle PAZ = \angle PCB$, which implies that $ABCP$ is cyclic, i.e. P needs to lie on the circumcircle of ABC. □

The above theorem by itself represents a very important tool when doing Olympiad geometry. Let's see it in use in a few examples.

Delta 8.1. Let E, F be the tangency points of the incircle of triangle ABC with the sides AC, and AB, respectively. Let Γ be an arbitrary circle passing through the vertex A and the incenter I and denote by X and Y the intersections of this circle with the sides AC and AB. Prove that the midpoint of XY lies on the line EF.

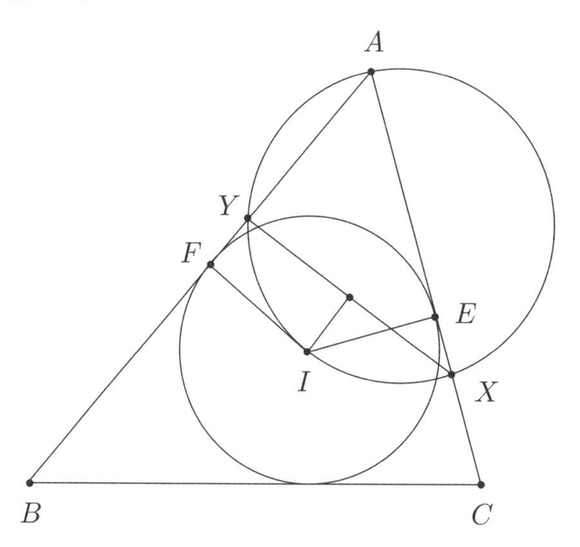

Proof. Note that E and F are the projections of the incenter I on the sides AY and AX of triangle AXY; thus, if we manage to show that the projection of I on the side XY is precisely the midpoint of XY, then we arrive at the conclusion via Simson's theorem as I is on the circumcircle of AXY. And this is indeed the case; since AI is the A–internal angle bisector of triangle AXY and thus I is the midpoint of the arc XY of the circumcircle of ABC not containing the vertex A, we have that $IX = IY$. This completes the proof. □

Delta 8.2. Suppose DEF is the orthic triangle of triangle ABC and let M be the midpoint of BC. Let the feet of the perpendiculars from M to the lines DE and DF be Y and Z, respectively. Prove that YZ is parallel to AD.

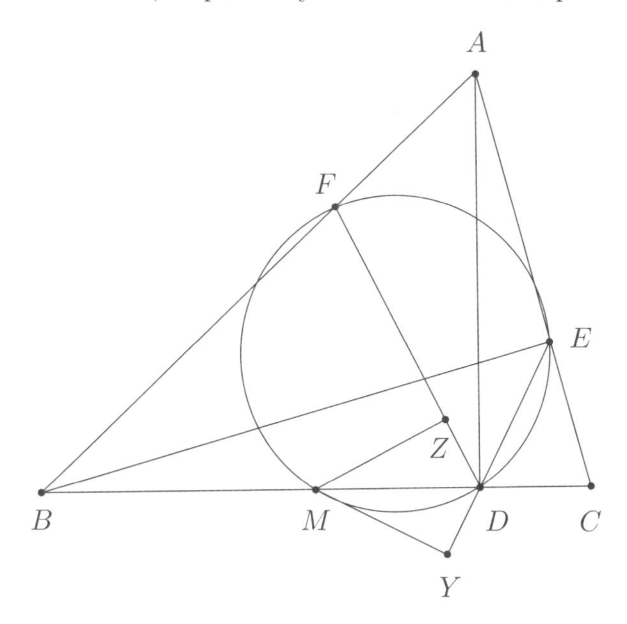

Proof. Recall that M, D, E, F are concyclic, as they lie on the nine-point circle of triangle ABC. Therefore line YZ is actually the Simson line of M with respect to triangle DEF. Now, we proceed with a simple angle chase. As we saw before in **Delta 3.7**, line AD is the D-internal angle bisector of triangle DEF. And $\angle DZY = \angle DMY = 90° - \angle MDY = 90° - \angle EDC = \angle ADE$, since $DYMZ$ is cyclic. Thus, we conclude that $\angle DZA = \angle ADF$, so the lines AD and YZ are indeed parallel, as claimed. This completes the proof. □

Delta 8.3. Let X, Y, Z be the midpoints of the arcs BC, CA, AB of triangle ABC containing the vertices of the triangle. Prove that the Simson lines of X, Y, Z with respect to ABC are concurrent.

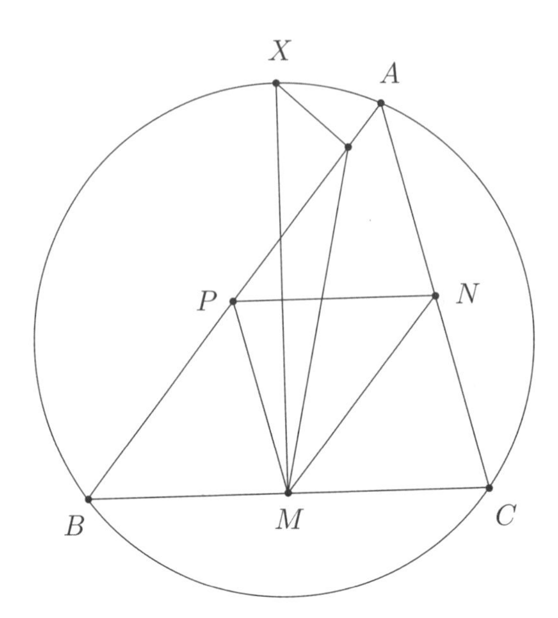

Proof. Let M, N, P be the midpoints of sides BC, CA, AB respectively. We prove that s_X, the Simson line of X with respect to triangle ABC, is the M-internal angle bisector of triangle MNP. To see this, we look at the angle $\angle(s_X, MN)$, and note that

$$\angle(s_X, MN) = \angle(s_X, AB) = \angle A - \angle(s_X, AC) = \angle A - \angle MXC = \frac{1}{2}\angle A.$$

Hence, s_X bisects $\angle NMP$ (as the angle at M in triangle MNP is equal with $\angle A$). This proves that the Simson lines of X, Y, Z with respect to triangle ABC are concurrent at the incenter of triangle MNP. This completes the proof. □

We continue with an *extremely* important result that should be added to our toolbox of lemmas.

Theorem 8.2. The Simson line of P lying on the circumcircle of triangle ABC passes through the midpoint of the segment PH, where H denotes the orthocenter of ABC.

This is a rather difficult lemma to prove synthetically; however, the following idea from Honsberger [18] is quite beautiful.

Proof. The idea is to take the reflection A' of the orthocenter H across BC. We know that it must lie on the circumcircle. Furthermore, let PA' meet BC at E, let D be the foot of the altitude from A, and let X, Y be the

projections of P on the sidelines BC and CA respectively. Since triangle HEA' is isosceles, we have that $\angle HEB = \angle A'EB$. But on the other hand, we have that $\angle YXB = \angle YPC$, as $YXCP$ is cyclic; hence $\angle YXB = 90^\circ - \angle PCA = 90^\circ - \angle A'AP = \angle A'EB$ (as $A'CPA$ is cyclic).

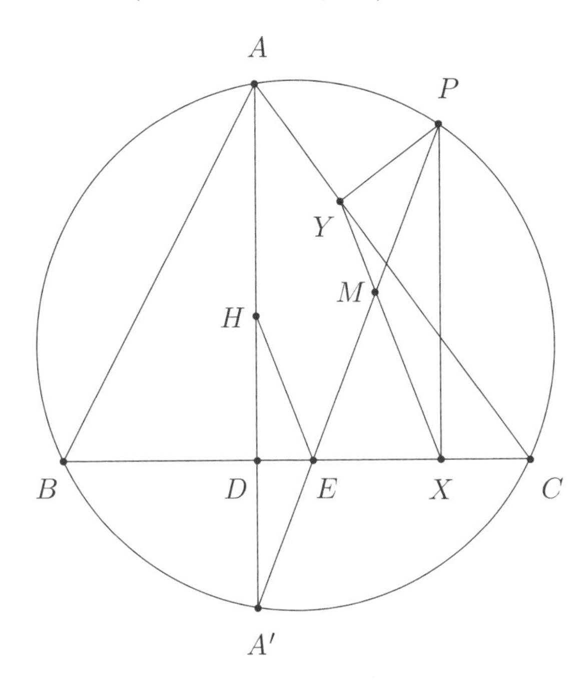

We conclude that $\angle HEB = \angle YXB$, so the Simson line XY is parallel to HE. However, triangle PEX has a right angle at X and has median XM where M is the midpoint of segment PE so we know that $XM = ME$ and hence XM is also parallel to HE. Thus, the Simson line XY coincides with the P-midline of triangle PEH, and so it bisects the segment PH, as desired. This completes the proof. □

Delta 8.4. Let $ABCD$ be a cyclic quadrilateral and let a, b, c, d be the Simson lines of the points A, B, C, D with respect to the triangles BCD, CDA, DAB, and ABC, respectively. Prove that a, b, c, d are concurrent.

Proof. Let H_a, H_b, H_c, H_d denote the orthocenters of triangles BCD, CDA, DAB, and ABC respectively, and let R be the circumradius of $ABCD$. Since lines AH_b and BH_a are both perpendicular to CD we have that they are parallel. Moreover, we know that $AH_b = 2R|\cos DAC|$ and $BH_a = 2R|\cos DBC|$ but since quadrilateral $ABCD$ is cyclic this means that $AH_b = BH_a$ - in other words, quadrilateral AH_bH_aB is a parallelogram. Therefore the midpoints of segments AH_a and BH_b coincide. Similarly we find that the midpoints of segments BH_b and CH_c coincide and that the midpoints of segments CH_c and DH_d coincide and so all four midpoints coincide. Now according to **Theorem 8.2** lines a, b, c, d each pass through the midpoints of segments AH_a, BH_b, CH_c, DH_d respectively but since these midpoints coincide we have that lines a, b, c, d concur as desired. □

A very strong generalization is possible for arbitrary convex quadrilaterals.

Delta 8.5. Let $ABCD$ be a convex quadrilateral and let $\mathcal{A}$ be the circumcircle of the pedal triangle of A with respect to triangle BCD. Similarly, define $\mathcal{B}$, $\mathcal{C}$, and $\mathcal{D}$. Prove that the circles $\mathcal{A}$, $\mathcal{B}$, $\mathcal{C}$, $\mathcal{D}$ all pass through a common point.

Obviously, when $ABCD$ is cyclic, the pedal circles $\mathcal{A}$, $\mathcal{B}$, $\mathcal{C}$, $\mathcal{D}$ become the Simson lines a, b, c, d from **Delta 8.4** because of **Theorem 8.1**. The proof for the general statement about pedal circles is however slightly more involved and we won't cover it here. But come back to it and think about it after doing everything else.

Finally, let's talk about Steiner. He was the one to prove **Theorem 8.2** in the first place. However, the way he stated the result was as follows.

Corollary 8.1. Let ABC be a triangle and let P be a point on its circumcircle. Then the reflections X, Y, Z of P across the sidelines of ABC are collinear and the line they determine passes through the orthocenter H of ABC.

Proof. Just consider the homothety with center P and ratio $\frac{1}{2}$. The points X, Y, Z are mapped into the vertices X', Y', Z' of the pedal triangle of P with respect to ABC and H is mapped to H', the midpoint of PH. By **Theorem 8.2**, the line determined by the points X', Y', Z', H' is actually the Simson line of P with respect to ABC, so the collinearity of X, Y, Z, H follows.

Accordingly, the line determined by the reflections X, Y, Z of P is usually called the **Steiner line of P with respect to triangle ABC**. Let's see some examples.

Delta 8.6. (Sam Korsky, ELMO Shortlist 2015) Let ω be the circumcircle of a triangle ABC and let P be a point in the interior of this triangle. Assume P is not the orthocenter of triangle ABC. Let D, E, F be the second intersections of lines AP, BP, CP respectively with ω. Let X, Y, Z be the reflections of P over lines BC, CA, AB respectively. Prove that the circumcircles of triangles PDX, PEY, PFZ concur at a point on ω.

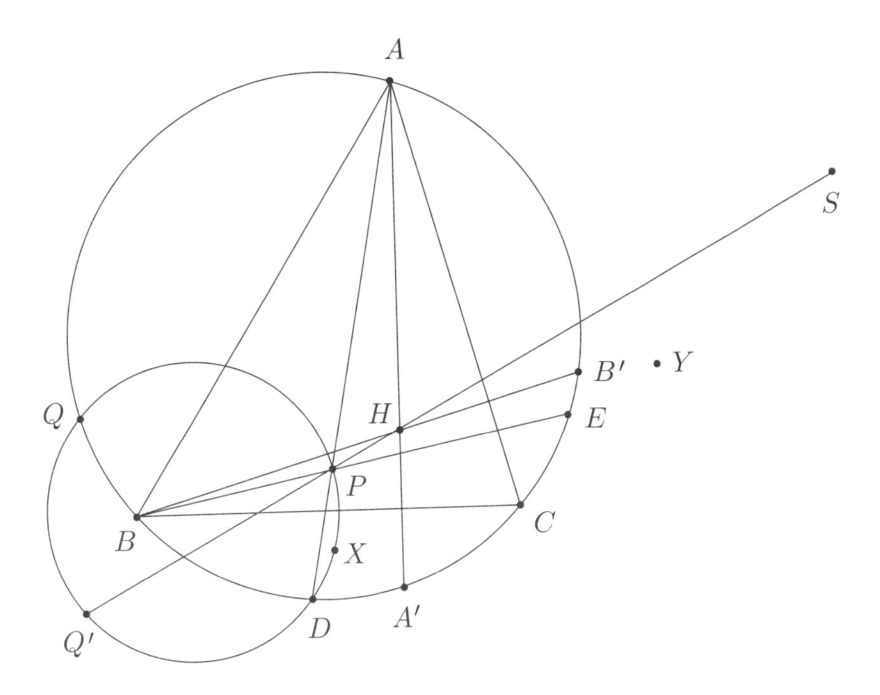

Proof. Let H be the orthocenter of triangle ABC and let the circumcircle of triangle PDX intersect ω again at Q. We have $\angle HAD = \angle XPD = \angle XQD$, so if AH intersects ω at A' then A', Q, X are collinear. Let the reflection of Q over BC be Q' - we know that Q' lies on HP. It suffices to show Q lies on the circumcircle of triangle PEY. If S is the reflection of Q over AC by **Corollary 8.1** we have that S lies on HP. Reflecting back over AC we get, if BH intersects ω at B', that Q lies on $B'Y$. So then we have $\angle PYQ = \angle BB'Q = \angle BEQ = \angle PEQ$, so Q lies on the circumcircle of triangle PYE as desired. This completes the proof. □

Delta 8.7. (Mongolian TST 2004) Let O be the circumcenter of the acute-angled triangle ABC, and let M be a point on the circumcircle of triangle ABC. Let X, Y, and Z be the projections of M onto OA, OB, and OC, respectively. Prove that the incenter of triangle XYZ lies on the Simson line of M with respect to triangle ABC.

Proof. Assume without loss of generality that M lies on minor arc AC and let $m(a)$ denote the clockwise measure of an arc a around the circumcircle of triangle ABC. Since X, Y and Z lie on a circle with diameter OM, the homothety centered at M with ratio 2 sends X, Y and Z to points X', Y' and Z' on the circumcircle of triangle ABC. Note that lines OA, OB and OC are the perpendicular bisectors of segments MX', MY' and MZ', respectively. Therefore $m(MA) = m(AX')$, $m(MB) = m(BY')$ and $m(MC) = m(CZ')$.

Now let P, Q and R denote the midpoints of arcs $Y'Z'$, $X'Z'$ and $X'Y'$ on the circumcircle of triangle ABC not containing X', Y' and Z', respectively. Angle chasing with the given directed arc lengths yields that PM is perpendicular to BC, QM is perpendicular to AC and RM is perpendicular to AB. Further angle chasing yields that triangles ABC and PQR are congruent and of opposite orientation.

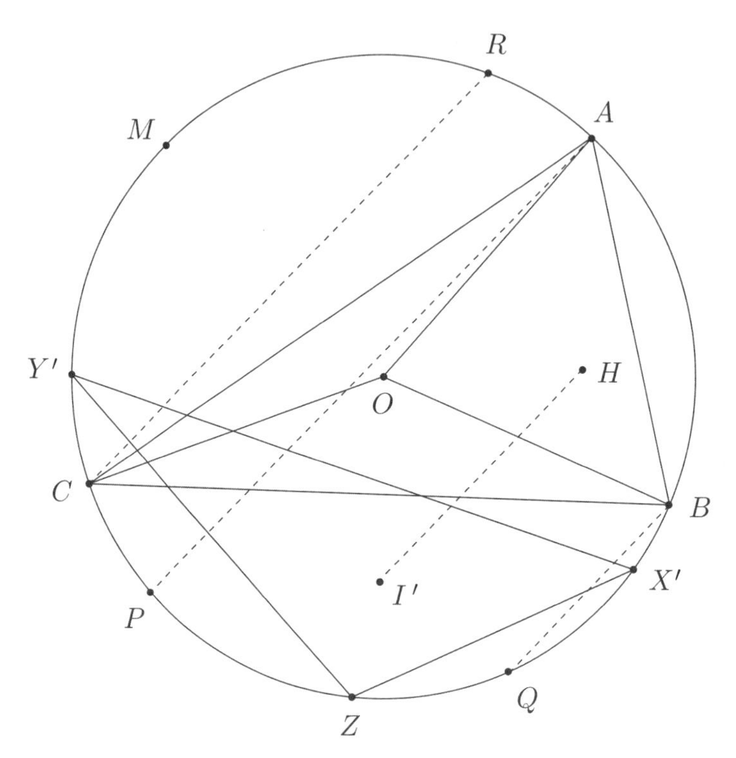

This implies that they are reflections in a line perpendicular to AP, BQ and CR. Angle chasing with cyclic quadrilaterals implies that the Simson line of M with respect to triangle ABC is parallel to AP, BQ and CR. But, **Corollary 8.1** tells us that the homothety centered at M with ratio 2 sends the Simson line of M with respect to triangle ABC to a line ℓ containing the orthocenter H of triangle ABC. Thus, if I' denotes the orthocenter of triangle PQR, then H and I' are reflections of one another in a line perpendicular to AP, BQ and CR, which implies that HI' is parallel to AP, BQ and CR. Therefore, I' lies on line ℓ. However, since P, Q and R denote the midpoints of arcs $Y'Z'$, $X'Z'$ and $X'Y'$ not containing X', Y' and Z', respectively, I' is the incenter of $X'Y'Z'$. The obvious homothety then completes the proof. □

//Make sure you do the angle chasing we playfully omitted.

We also have the following follow-up Lemma by Collings.

Theorem 8.3. (Anti-Steiner Points) Starting this time with a line ℓ passing through the orthocenter H of triangle ABC and taking its reflections r_a, r_b, r_c across the sidelines BC, CA, AB, we have that lines r_a, r_b, r_c are concurrent on the circumcircle of triangle ABC.

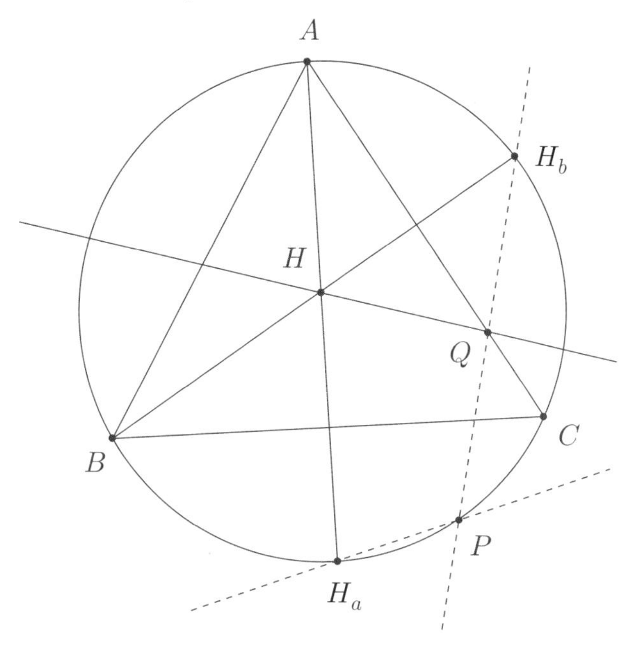

Proof. Let $P = r_a \cap r_b$ and let H_a, H_b be the reflections of H across sides BC, CA respectively. It clearly suffices to show that P lies on the circumcircle of triangle ABC, and since we know H_a and H_b lie on this circle it suffices to show that quadrilateral AH_aPH_b is cyclic. Now let $Q = \ell \cap CA$. It's clear that $\angle AH_aP = \angle H_aHQ$ and $\angle AH_bP = \angle AHQ$ so $\angle AH_aP + \angle AH_bP = \angle H_aHQ + \angle AHQ = 180°$ which implies the desired cyclicity and completes the proof. □

So, the concurrency point of the reflections r_a, r_b, r_c of ℓ is known as the **Anti-Steiner point** of ℓ with respect to triangle ABC. Notice that this tells us that for example the reflections of the Euler line of ABC into the sidelines are concurrent. This point is called the Euler reflection point, as one can imagine why. Nonetheless, we won't dwell much on this; the interested reader can find some very beautiful applications in the Bibliography. Now, let's see what else we can find in this configuration!

Delta 8.8. (The Droz-Farny Line Theorem) Let ℓ_1, ℓ_2 be two perpendicular lines passing through the orthocenter H of triangle ABC and intersecting the sidelines BC, CA, AB at X_1, Y_1, Z_1 and X_2, Y_2, Z_2, respectively. Then, the midpoints of the segments X_1X_2, Y_1Y_2, Z_1Z_2 are collinear.

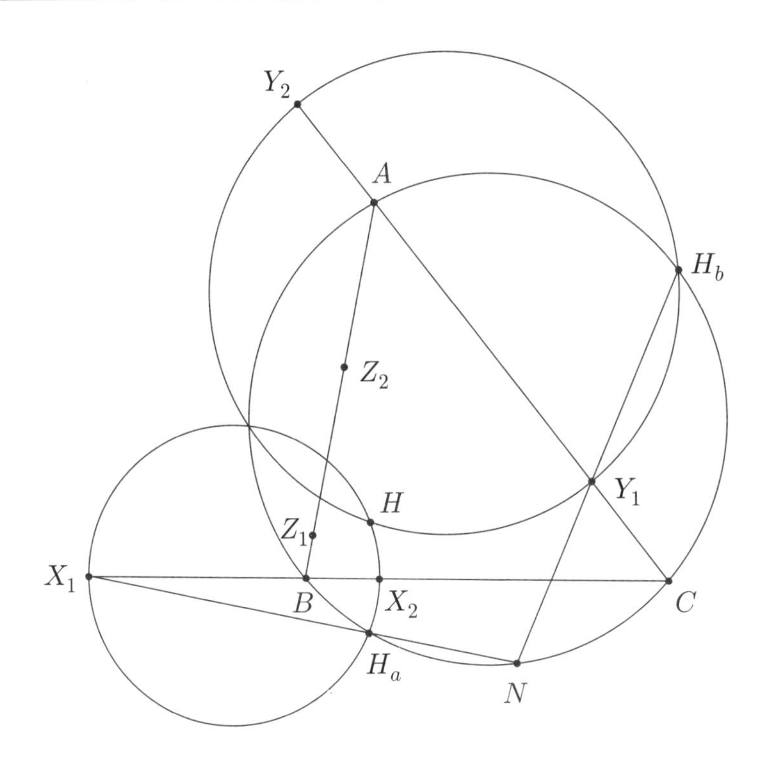

Proof. We begin with a claim commonly referred to as **Miquel's Pivot Theorem**

Claim Let D, E, F be points on sides BC, CA, AB respectively of a triangle ABC. Then the circumcircles of triangles AEF, BFD, CDE concur at a point.

Proof. This is a simple angle chase. Let the circumcircles of triangles BFD and CDE intersect again at point P. Then $\angle EPF = 360^\circ - \angle FPD - \angle DPE = 360^\circ - (180^\circ - \angle B) - (180^\circ - \angle C) = 180^\circ - \angle A$ so quadrilateral $AEPF$ is cyclic. This completes the proof.

Returning to the problem, let H be the orthocenter of triangle ABC and k, k_a, k_b, k_c be the circumcircles of triangles $ABC, HX_1X_2, HY_1Y_2, HZ_1Z_2$ respectively. Let M_a, M_b, M_c be the centers of circles k_a, k_bk_c respectively. Also let H_a, H_b, H_c be the reflections of H across sides BC, CA, AB respectively. Now since $l_1 \perp l_2$ we have that X_1X_2 is a diameter of k_a which implies that H_a lies on k_a. Similarly H_b and H_c lie on k_b and k_c respectively. Moreover we know that H_a, H_b, H_c all lie on k. Now, from **Theorem 8.3** we know that lines H_aX, H_bY, H_cZ concur at a point N on k. And by Miquel's Pivot Theorem on triangle NX_1Y_1 with points H, H_b, H_a we have that circles k_a, k_b, k concur. Similarly we find that circles k, k_c, k_a concur and that circles

k, k_a, k_b concur and so we have that circles k_a, k_b, k_c concur at a point other than H. This means that these three circles are coaxial and so their centers M_a, M_b, M_c are collinear. But these centers are precisely the midpoints of segments X_1X_2, Y_1Y_2, Z_1Z_2 respectively, hence the proof is complete. □

We continue with a related lemma.

Theorem 8.4. If P and Q are two points lying on the circumcircle of ABC, then the (acute) angle between their two Simson lines is half of the angle $\angle POQ$ where O is the circumcenter of triangle ABC.

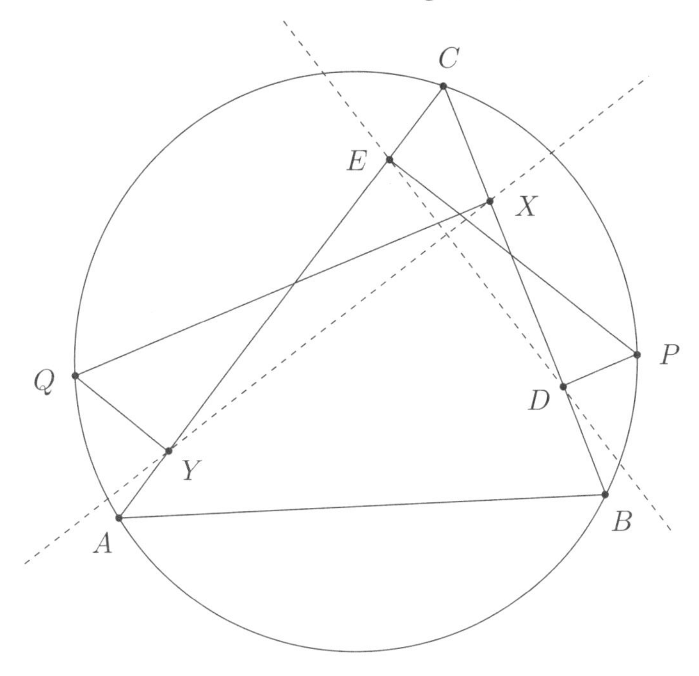

Proof. Without loss of generality, assume that P lies on the arc BC not containing the vertex A and that Q lies on the arc CA not containing B. Let D, E be the projections of P on the sidelines BC, CA and X, Y the projections of Q on BC and CA, respectively. Note that

$$\begin{aligned}
\angle(s_P, s_Q) &= 180^\circ - \angle(s_P, AC) - \angle(AC, s_Q) \\
&= 180^\circ - \angle DEA - \angle XYC \\
&= 180^\circ - \angle DPC - \angle XQC \\
&= 180^\circ - (90^\circ - \angle PCB) - (90^\circ - \angle QCB) \\
&= \angle PCB + \angle QCB \\
&= \angle PCQ \\
&= \frac{1}{2}\angle POQ.
\end{aligned}$$

This completes the proof. □

Corollary 8.2. Let P and P' be two antipodal points on the circumcircle of triangle ABC. Then, their Simson lines are perpendicular and the intersection point of their Simson lines lies on the nine-point circle of triangle ABC.

Proof. The fact that the two Simson lines are perpendicular follows immediately from **Theorem 8.4**. Now, to see why their intersection point must lie on the nine-point circle of ABC, we use **Theorem 8.2**. The Simson line s_P of P with respect to ABC must pass through the midpoint X of the segment HP, whereas the Simson line $s_{P'}$ of P' must pass through the midpoint Y of HP'; but PP' is a diameter in the circumcircle of triangle ABC, hence, XY is a diameter in the nine-point circle of triangle ABC since XY is the H-midline in triangle HPP' and $HN = NO$, where N is the nine-point center of triangle ABC. Thus, because the two Simson lines s_P and $s_{P'}$ meet at a right angle, it follows that their intersection point lies on the circle with diameter PP', which is precisely the nine-point circle. This completes the proof. □

Delta 8.9. Let A_1, A_2, A_3, A_4, and A_5 be five concyclic points. For $1 \leq i < j \leq 5$, let $X_{i,j}$ be the intersection of the Simson lines of A_i and A_j with respect to the triangle formed by the other three points. Show that all ten such points $X_{i,j}$ are concyclic.

Proof. Given four concyclic points A, B, C, D call the concurrency point X of the Simson lines in **Delta 8.4** the **anticenter** of $ABCD$. We prove the following preliminary result.

Claim. Given 5 points A_1, A_2, A_3, A_4, A_5 on the circle (O), denote by H_1, H_2, H_3, H_4, H_5 the anticenters of $A_2A_3A_4A_5$, $A_1A_3A_4A_5$, $A_1A_2A_4A_5$, $A_1A_2A_3A_5$, $A_1A_2A_3A_4$, respectively. Then, the points H_1, H_2, H_3, H_4, H_5 are concyclic.

We proceed using vectors. Assume without loss of generality the circumcircle of $A_1A_2A_3A_4A_5$ is the unit circle. Then it is well known that the orthocenter of triangle $A_1A_2A_3$ has vector coordinate

$$\overrightarrow{OA_1} + \overrightarrow{OA_2} + \overrightarrow{OA_3}$$

so the anticenter of quadrilateral $A_1A_2A_3A_4$ has vector coordinate

$$\frac{\overrightarrow{OA_1} + \overrightarrow{OA_2} + \overrightarrow{OA_3} + \overrightarrow{OA_4}}{2}$$

and we can obtain similar expressions for the other four anticenters. It's clear now that each of these anticenters lies on a circle with radius $\frac{1}{2}$ and center

$$\frac{\overrightarrow{OA_1} + \overrightarrow{OA_2} + \overrightarrow{OA_3} + \overrightarrow{OA_4} + \overrightarrow{OA_5}}{2}$$

Returning to the problem, we let H_1, H_2, H_3, H_4, H_5 be the anticenters of triangles $A_iA_jA_k$ $(1 \le i < j < k \le 5)$. By the Claim above, the points H_1, H_2, H_3, H_4, H_5 are concyclic. On the other hand, by **Delta 8.4**, the Simson lines d and l of A_3, A_1 with respect to triangles $A_2A_4A_5$, $A_2A_4A_5$ pass through H_1, H_3 respectively. Thus, if we let F be the intersection of d and l, **Theorem 8.4** yields

$$\angle H_1FH_3 = \angle A_1A_4A_3 = \angle H_1H_4H_3,$$

and so F lies on the circle determined by H_1, H_2, H_3, H_4, H_5. Similarly, all the other nine intersections of Simson lines to consider lie on this circle. This completes the proof! □

We conclude the section with a strong generalization of Simson's theorem.

Let ABC be a triangle and let ℓ be a line intersecting lines BC, CA, AB at points D, E, F respectively. Let the line at E perpendicular to CA intersect the line at F perpendicular to AB at point A_1 and define points B_1, C_1 similarly. Then triangle $A_1B_1C_1$ is called the **paralogic triangle of triangle** ABC **with respect to** ℓ. Let's find some properties of this configuration!

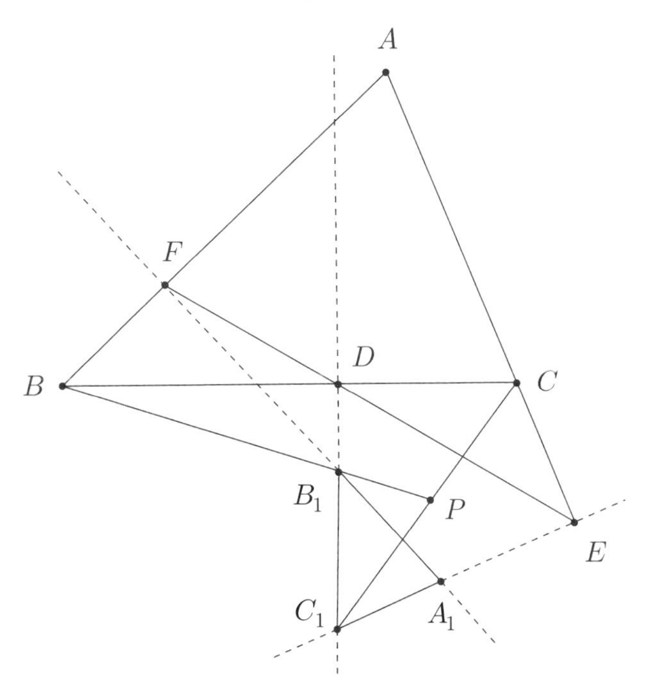

A quick angle chase shows that triangle $A_1B_1C_1$ is similar to triangle ABC, and it's easy to see that these two triangles are perspective with perspectrix ℓ. This means that by Desargues' Theorem, lines AA_1, BB_1, CC_1 concur at a point P, the perspector of the two triangles. We now proceed with a generalization of **Theorem 8.1**.

Theorem 8.5. P lies on the circumcircle of triangle ABC and the circumcircle of triangle $A_1B_1C_1$.

Proof. This is just an angle chase. Note that

$$\begin{aligned}\angle BPC &= \angle PC_1B_1 + \angle PB_1C_1\\ &= \angle AEF + \angle BB_1D\\ &= \angle AEF + \angle AFE\\ &= 180^\circ - \angle A,\end{aligned}$$

where we used the fact that quadrilaterals CDC_1E and BFB_1D are cyclic.

This implies that P lies on the circumcircle of triangle ABC and we can similarly find that it lies on the circumcircle of triangle $A_1B_1C_1$. In fact, with some more effort we could have shown that the circumcircles of triangles ABC and $A_1B_1C_1$ are orthogonal as well. The proof is complete. □

Now we give a surprising generalization of **Theorem 8.2**.

Theorem 8.6. (Sondat's Theorem) Let H, H_1 be the orthocenters of triangles $ABC, A_1B_1C_1$ respectively. Then line ℓ bisects the segment HH_1.

Proof. Let X, Y, Z be the midpoints of segments AA_1, BB_1, CC_1 respectively and let O be the circumcenter of triangle ABC. We begin with a claim that is essentially Miquel's Pivot Theorem for quadrilaterals.

Claim. The circumcircles of triangles ABC, AEF, BFD, CDE concur at a point M. Moreover, points M, O, X, Y, Z are concyclic.

The first part of the claim is proven by two applications of Miquel's Pivot Theorem, and the second part is proven by noting that X, Y, Z are the circumcenters of triangles AEF, BFD, CDE respectively and then performing a simple angle chase. We will provide more detailed proofs later in the book.

Returning to the problem, let H' be the midpoint of segment HH_1. Since triangles ABC and $A_1B_1C_1$ are similar, a quick application of vectors yields that triangle XYZ is similar to triangle ABC and that H' is the orthocenter of triangle XYZ.

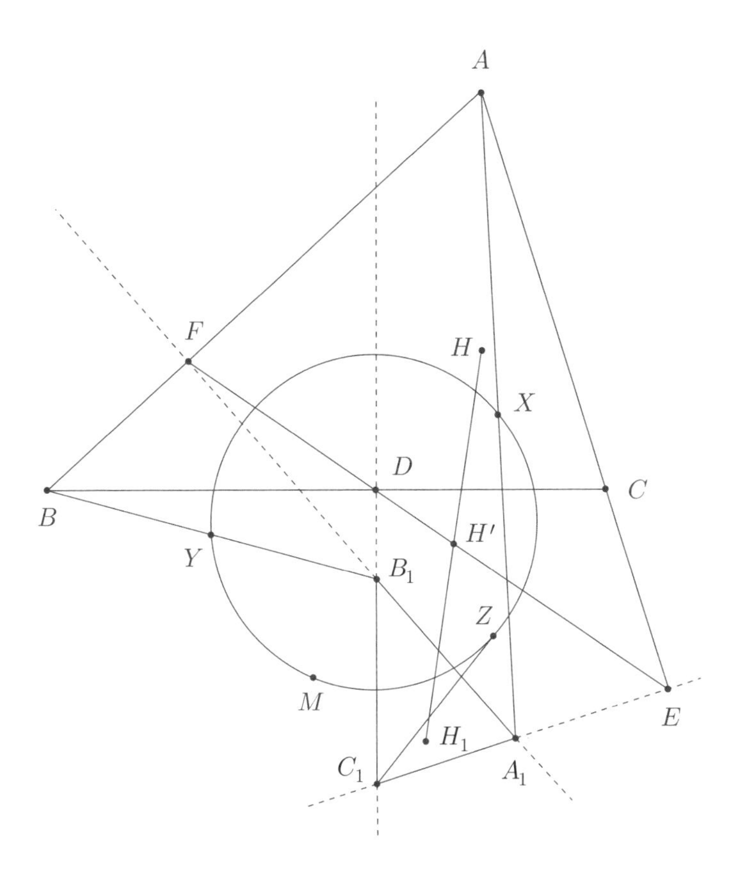

Now D, E, F are the reflections of M over lines YZ, ZX, XY respectively since X, Y, Z are the centers of the circumcircles of triangles AEF, BFD, CDE respectively and F, E, D are the pairwise intersections of these circles. Therefore, since the Claim guarantees that M lies on the circumcircle of triangle XYZ, we have that ℓ is the Steiner line of M with respect to triangle XYZ so ℓ passes through H'. This completes the proof. $\square$

Assigned Problems

Epsilon 8.1. Let ABC be an isosceles triangle where $AC = BC$, and let M be the midpoint of the side AB. Furthermore, let Γ be the circle with center C with radius less than CM; from A and B draw the tangents to Γ and denote by P, Q two intersections of these tangents such that PQ does not intersect the segment CM. Prove that P, Q, and M are collinear.

Epsilon 8.2. (USAMO 2010) Let $AXYZB$ be a convex pentagon inscribed in a semicircle of diameter AB. Denote by P, Q, R, S the feet of the perpendiculars from Y onto lines AX, BX, AZ, BZ, respectively. Prove that the acute angle formed by lines PQ and RS is half the size of $\angle XOZ$, where O is the midpoint of segment AB.

Epsilon 8.3. (IMO 2003) Let $ABCD$ be a cyclic quadrilateral. Let P, Q, R be the feet of the perpendiculars from D to the lines BC, CA, AB, respectively. Show that $PQ = QR$ if and only if the bisectors of $\angle ABC$ and $\angle ADC$ are concurrent with AC.

Epsilon 8.4. (Romania TST 1999) Let ABC be a triangle with orthocenter H, circumcenter O and circumcenter R. Let D, E, F be the reflections of the vertices A, B, C across the opposite sides. Prove that they are collinear if and only if $OH = 2R$.

Epsilon 8.5. Let A, B, C, P, Q, and R be six concyclic points. Show that if the Simson lines of P, Q, and R with respect to triangle ABC are concurrent, then the Simson lines of A, B, and C with respect to triangle PQR are concurrent. Furthermore, show that the points of concurrency are the same.

Epsilon 8.6. Let ABC be a triangle and let D, E, F be the tangency points of the incircle with the sidelines BC, CA, AB. Also, let the incircle intersect the segments AI, BI, CI at points M, N, P, respectively. Prove that the Simson lines of any point lying on the incircle with respect to triangles DEF and MNP are perpendicular.

Epsilon 8.7. (Romania TST 2009) Prove that the circumcircle of a triangle contains exactly 3 points whose Simson lines are tangent to the triangle's nine-point circle and these points are the vertices of an equilateral triangle

Epsilon 8.8. Let ABC be a triangle and let P, Q be two points lying on its circumcircle. Prove that their Simson lines meet on the A-altitude of triangle ABC if and only if $PQ||BC$.

Epsilon 8.9. (The Parry Reflection Point) Suppose triangle ABC has circumcenter O and orthocenter H. Parallel lines α, β, γ are drawn through the vertices A, B, C, respectively. Let α', β', γ' be the reflections of α, β, γ over the sides BC, CA, AB, respectively. Then, these reflections are concurrent if and only if α, β, γ are parallel to the Euler line OH of triangle ABC. In this case, their point of concurrency P is the reflection of O over the Euler reflection point (the anti-Steiner point of the Euler line).

Epsilon 8.10. (Sharygin 2010) Let ABC be a triangle. From A, B, C draw pairwise parallel lines d_a, d_b, d_c respectively. Let l_a, l_b, l_c be the reflections of d_a, d_b, d_c through BC, CA, AB respectively. If XYZ is the triangle formed by lines l_a, l_b, l_c, find the locus of incenters of triangles XYZ.

Chapter 9

Symmedians

We will now discuss the Symmedian point, which was briefly mentioned in **Section 7**. We will prove a lot of beautiful properties of symmedians that will help us solve many Olympiad problems and establish a lot of connections between the triangle centers. Before beginning this journey, make sure you take another look at the Ratio Lemma from **Section 3** because we will use it a lot.

We start with a well-known result about isogonal lines in general.

Theorem 9.1. (Steiner's Theorem) If D is a point on the sideline BC of triangle ABC, and if the reflection of the line AD in the internal angle bisector of the angle A intersects the line BC at a point E, then

$$\frac{DB}{DC} \cdot \frac{EB}{EC} = \frac{AB^2}{AC^2}.$$

Proof. From the Ratio Lemma, we write

$$\frac{DB}{DC} = \frac{AB}{AC} \cdot \frac{\sin DAB}{\sin DAC} \text{ and } \frac{EB}{EC} = \frac{AB}{AC} \cdot \frac{\sin EAB}{\sin EAC}.$$

Thus, keeping in mind that $\angle DAB = \angle EAC$ and $\angle DAC = \angle EAB$, by multiplying, we obtain that

$$\frac{DB}{DC} \cdot \frac{EB}{EC} = \frac{AB^2}{AC^2},$$

as claimed. □

//The converse of Steiner's theorem also holds; the reader is encouraged to fill in the details.

Corollary 9.1. In a triangle ABC with X on the side BC, we have that

$$\frac{XB}{XC} = \frac{AB^2}{AC^2}$$

if and only if AX is the A-symmedian of triangle ABC.

This represents what is perhaps the most important characterization of the A-symmedian of the triangle and all the lemmas that we shall see here will use this.

Theorem 9.2. Let the tangents at vertices B and C of triangle ABC to the circumcircle of triangle ABC meet at a point X. Prove that the line AX is the A-symmedian of triangle ABC.

This one is one of the most beautiful results you'll ever see about symmedians. Make sure you remember it. We will give three proofs, each emphasizing a different technique.

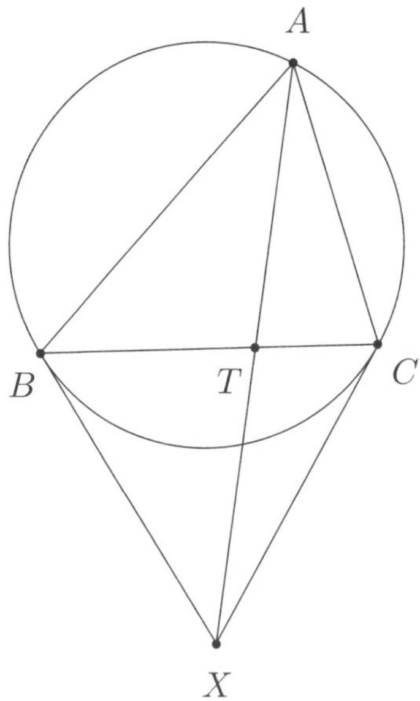

First Proof. Let T be the intersection of AX with the side BC. Since this point lies in the interior of the segment BC, we notice that because of **Corollary 9.1** it is enough to show that $\frac{TB}{TC} = \frac{AB^2}{AC^2}$. This is where the Ratio Lemma comes in - we have that

$$\frac{TB}{TC} = \frac{XB}{XC} \cdot \frac{\sin TXB}{\sin TXC}$$

but $XB = XC$ as they are both tangents from the same point to the circumcircle of ABC; hence $\frac{TB}{TC} = \frac{\sin TXB}{\sin TXC}$.

Now, we apply the Law of Sines twice, in triangles XAB and XAC. We get that

$$\frac{AB}{\sin TXB} = \frac{AX}{\sin XBA} = \frac{AX}{\sin (B + XBC)}$$

and

$$\frac{AC}{\sin TXC} = \frac{AX}{\sin XCA} = \frac{AX}{\sin (C + XCB)}.$$

But $\angle XBC = \angle XCB = \angle A$, since the lines XB and XC are both tangent to the circumcircle of ABC. Hence, it follows that

$$\frac{AB}{\sin TXB} = \frac{AX}{\sin C} \text{ and } \frac{AC}{\sin TXC} = \frac{AX}{\sin B}.$$

Therefore, by dividing the two relations, we conclude that

$$\begin{aligned} \frac{TB}{TC} &= \frac{\sin TXB}{\sin TXC} \\ &= \frac{AB}{AC} \cdot \frac{\sin C}{\sin B} \\ &= \frac{AB^2}{AC^2}, \end{aligned}$$

where the last equality holds because of the Law of Sines applied in triangle ABC. This completes the proof. □

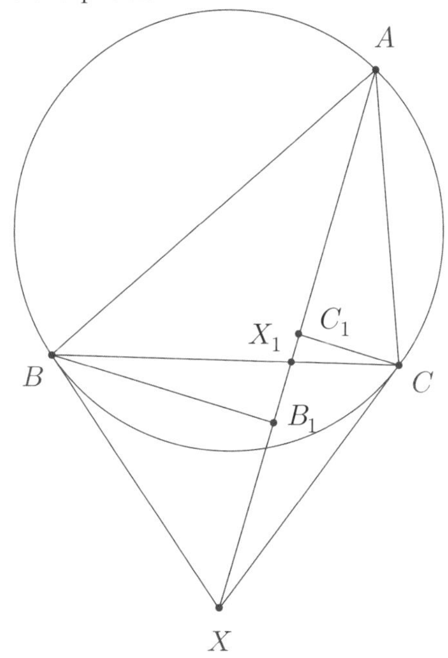

Second Proof. This is also rather computational as it involves sines, but it gives you a different way of looking at things. Denote by X_1 the intersection of AX with the side BC. Let B_1, C_1 be the orthogonal projections of the vertices B, C on the line AX, respectively. We have that

$$\frac{X_1B}{X_1C} = \frac{[ABX_1]}{[AX_1C]} = \frac{BB_1}{CC_1} = \frac{[ABX]}{[XCA]} = \frac{AB \cdot BX \cdot \sin ABX}{CA \cdot XC \cdot \sin XCA}.$$

Furthermore, we know that

$$\sin ABX = \sin(ABC + CAB) = \sin BCA$$

and

$$\sin XCA = \sin(BCA + CAB) = \sin ABC.$$

Hence, since $BX = XC$,

$$\frac{X_1B}{X_1C} = \frac{AB}{CA} \cdot \frac{\sin B}{\sin C} = \left(\frac{AB}{CA}\right)^2.$$

We thus conclude as above that in this case, the line AX is the A-symmedian of triangle ABC. □

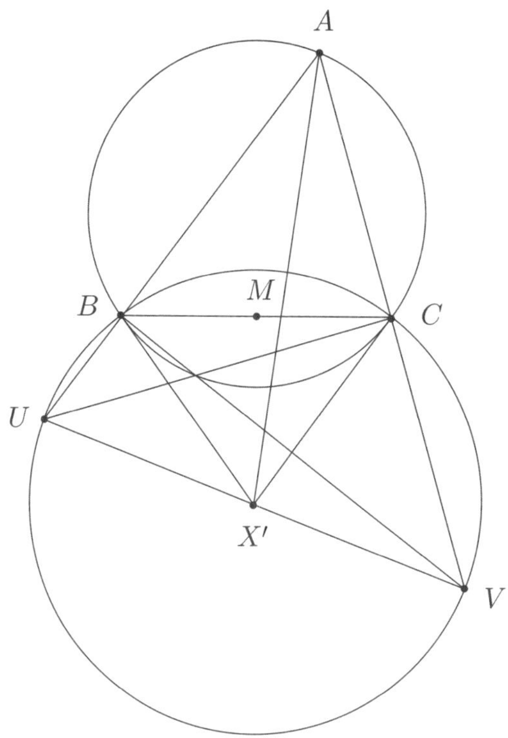

Third Proof. Time for a more clever proof. Since line BX is the tangent to the circumcircle of triangle ABC at the point B, we have as above that

$\measuredangle(BX, BC) = \measuredangle BAC$. In other words, $\measuredangle XBC = \measuredangle BAC$. Let the perpendicular to the line CA at the point C meet the line AB at U, and let the perpendicular to the line AB at the point B meet the line CA at V. Then, since $\measuredangle UBV = 90°$ and $\measuredangle UCV = 90°$, the points B and C lie on the circle with diameter UV. The center of this circle is the midpoint of the segment UV; denote this midpoint by X'. Then, $\measuredangle X'BC = 90° - \measuredangle CUB$. Since the lines CU and CA are perpendicular,

$$\measuredangle X'BC = 90° - \measuredangle CUB = \measuredangle(CU, CA) - \measuredangle(CU, AB)$$

$$= \measuredangle(AB, CA) = \measuredangle BAC = \measuredangle XBC.$$

Thus, the point X' lies on the line BX. Similarly, the point X' lies on the line CX. But the lines BX and CX have only one point in common, namely the point X. Thus, the point X' coincides with the point X. As we know that the point X' is the midpoint of segment UV, we conclude that the point X is the midpoint of the segment UV. Now, since the points B and C lie on the circle with diameter UV, the angles $\measuredangle BUV$ and $\measuredangle BCV$ are congruent, which can be written as $\measuredangle AUV =$ - $\measuredangle ACB$. Similarly, $\measuredangle AVU = -\measuredangle ABC$. Hence, the triangles ABC and AVU are similar and oppositely oriented. Now, if M is the midpoint of the segment BC, then the points M and X are corresponding points in these oppositely oriented similar triangles ABC and AVU (in fact, they are the midpoints of the corresponding sides BC and VU of these triangles). Since corresponding points in oppositely oriented similar triangles form oppositely equal angles, we thus conclude that $\measuredangle BAM = -\measuredangle VAX$, which can be written as $\measuredangle(AB,\ AM) = -\measuredangle(CA,\ AX)$. Now, if ω is the angle bisector of the angle CAB, then $\measuredangle(AB,\ \omega) = -\measuredangle(CA,\ \omega)$. Thus,

$$\measuredangle(\omega,\ AM) = \measuredangle(AB,\ AM) - \measuredangle(AB,\ \omega) = (-\measuredangle(CA,\ AX)) - (-\measuredangle(CA,\ \omega))$$

$$= \measuredangle(CA,\ \omega) - \measuredangle(CA,\ AX) = \measuredangle(AX,\ \omega).$$

Hence, since the line AM is the A-median of triangle ABC (as the point M is the midpoint of the side BC), and the line ω is the angle bisector of the angle CAB, we thus see that the line AX is the reflection of the A-median of triangle ABC in the angle bisector of the angle CAB. In other words, the line AX is the A-symmedian of triangle ABC. □

We immediately see a nice consequence!

Corollary 9.2. If D, E, F denote the tangency points of the incircle with the sides BC, CA, AB of triangle ABC, then the symmedian point of triangle DEF is the Gergonne point of triangle ABC - i.e. the intersection point of the lines AD, BE and CF.

We emphasize this with the following nice application from Mathematical Reflections:

Delta 9.1. (Mathematical Reflections) Let D, E, F be the points of tangency of the incircle of a triangle ABC with its sides BC, CA, AB, respectively. Then, the triangle ABC is equilateral if and only if the centroids of DEF and ABC are isogonal with respect to triangle DEF.

Proof. It is clear that if ABC is equilateral the centroids of triangles ABC and DEF coincide with the incenter of triangle DEF, and thus they are isogonal. Conversely, by **Corollary 9.2**, the lines AD, BE, CF are the symmedians of the vertices D, E, F in triangle DEF, the symmedian point of DEF coincides with the Gergonne point of ABC. The symmedian point of DEF is isogonal with the centroid of DEF, thus we conclude that the centroid of ABC coincides with the symmedian point of DEF, and hereby, the Gergonne point and the centroid of triangle ABC coincide. Thus, triangle ABC is equilateral as desired. □

As a remark, the problem remains true when the isogonality becomes with respect to triangle ABC. We however don't know a nice proof, so we leave it as an exercise for the reader.

We now proceed with a corollary of the second proof: an identity establishing the length of the segment AX.

Delta 9.2. Let the tangents to the circumcircle of a triangle ABC at the vertices B and C intersect each other at a point X, and let M be the midpoint of the side BC of triangle ABC. Then, $AM = AX \cdot |\cos A|$.

Proof. In the third proof of **Theorem 9.2**, we showed that the points M and X are corresponding points in the similar triangles ABC and AVU. Corresponding points in similar triangles form similar triangles themselves; thus, the triangles ABM and AVX are similar, so $\frac{AM}{AX} = \frac{AB}{AV}$. But in the right-angled triangle ABV, we have

$$\frac{AB}{AV} = |\cos \angle BAV| = |\cos A|,$$

and thus, $\frac{AM}{AX} = |\cos A|$. Thus, $AM = AX \cdot |\cos A|$, which completes the proof. □

Now, time for another characterization of the symmedian!

Theorem 9.3. The A-symmedian is the locus of the midpoints of the antiparallels to BC bounded by the lines AB and AC in triangle ABC.

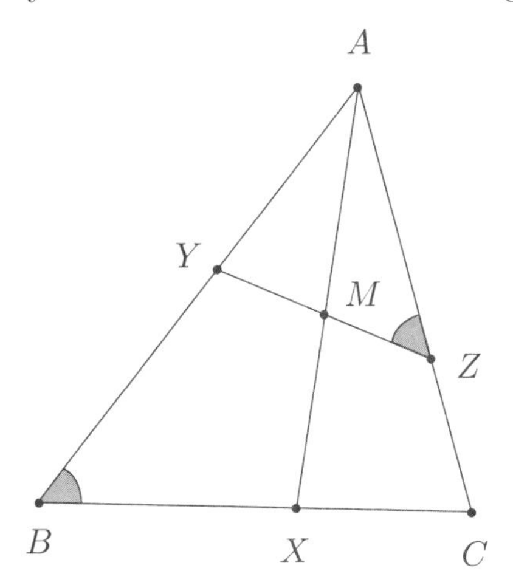

Proof. Let YZ be an antiparallel to the line BC with Y on AB and Z on AC and let M be the midpoint of YZ. It suffices to show that AM is the A-symmedian. Let X be the intersection of AM with BC and let's try to prove that $\frac{XB}{XC} = \frac{AB^2}{AC^2}$.

We use the Ratio Lemma. More precisely, we have that

$$\frac{XB}{XC} = \frac{AB}{AC} \cdot \frac{\sin XAB}{\sin XAC} = \frac{AB}{AC} \cdot \frac{\sin MAY}{\sin MAZ}.$$

And from the way we wrote the angles $\angle XAB$ and $\angle XAC$ in the last term, we already know what's the next step. The Ratio Lemma applied again, only this time in triangle AYZ, gives us

$$1 = \frac{MY}{MZ} = \frac{AY}{AZ} \cdot \frac{\sin MAY}{\sin MAZ},$$

hence

$$\frac{\sin MAY}{\sin MAZ} = \frac{AZ}{AY} = \frac{AB}{AC},$$

where the last equality holds because of the similarity of triangles ABC and AZY. Thus, we conclude that

$$\frac{XB}{XC} = \frac{AB^2}{AC^2},$$

which completes the proof. □

A very nice consequence of this fact is the following beautiful result due to Lemoine - remember, Lemoine did a lot with symmedians!

Delta 9.3. Let K be the symmedian point of triangle ABC and let x, y, z be the antiparallels drawn through K to the lines BC, CA, and AB, respectively. Prove that the six points determined by x, y, z on the sides of ABC all lie on a same circle.

//This circle is known in literature as the **First Lemoine Circle**.

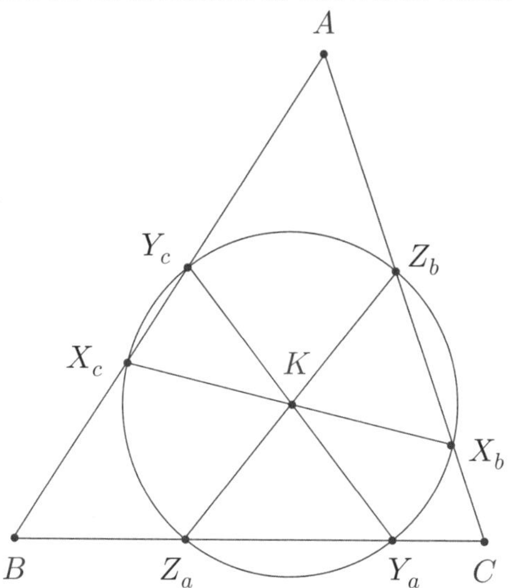

Proof. Let X_b, X_c be the intersections of x with CA, AB, respectively. Similarly, let Y_c, Y_a be the intersections of y with AB, BC, and Z_a, Z_b the intersections of z with BC, CA. By **Theorem 9.3**, we know that $KX_b = KX_c$, $KY_c = KY_a$, $KZ_a = KZ_b$. Moreover, since y, z are antiparallels, we have that $\angle KZ_aY_a = \angle KY_aZ_a = \angle A$, thus triangle KY_aZ_a is isosceles, i.e. $KY_a = KZ_a$. Hence, $KY_a = KZ_a = KY_c = KZ_b$. Moreover, we can do the same thing for triangles KX_bZ_b, KY_cX_c to argue that they are isosceles, so we also have that $KX_b = KZ_b$ and $KY_c = KX_c$. Therefore, we conclude that

$$KZ_a = KY_a = KX_b = KZ_b = KY_c = KX_c,$$

so we get that all six points X_b, X_c, Y_c, Y_a, Z_a, Z_b lie on a same circle that is centered at K. This completes the proof. □

Of course, given this name, you expect to have a second Lemoine circle. Indeed, this is the case!

Delta 9.4. (The Second Lemoine Circle) Let K be the symmedian point of the triangle ABC and let x, y, z this time be the *parallels* drawn through K to BC, CA, and AB, respectively. Prove that the six points determined by x, y, z on the sides of ABC all lie on a same circle.

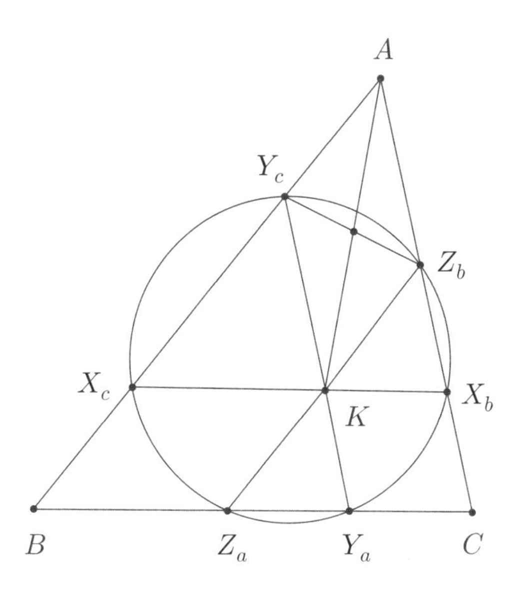

Proof. Let the line x meet AC and AB at X_b and X_c, y meet BC, BA at Y_a, Y_c, and z meet CA, CB at Z_b, Z_a. First, note that AY_cKZ_b is a parallelogram, thus the line AK bisects the segment Y_cZ_b. However, AK is the $A-$symmedian of triangle ABC; hence the line supporting the segment Y_cZ_b needs to be antiparallel to BC, according to **Theorem 9.3**. Thus, $\angle AZ_bY_c = \angle B = \angle Y_cX_cX_b$; hence, we get that Y_c, X_c, X_b, Z_b all need to lie on a same circle, say Γ_1. Similarly, the points Y_c, X_c, Z_a, Y_a need to lie on a same circle Γ_2, and the points Z_a, Y_a, X_b, Z_b need to lie on a same circle Γ_3. However, these three circles need to be the same, for otherwise, their pairwise radical axes are not concurrent (since they are just the sidelines of the triangle!) and that's impossible. Thus, $\Gamma_1 = \Gamma_2 = \Gamma_3$ and so all six points X_b, X_c, Y_c, Y_a, Z_a, Z_b are concyclic. This completes the proof. □

Furthermore, the proof of **Delta 9.3** gives us the following beautiful result.

Delta 9.5. Let ABC be a triangle and let M be the midpoint of BC and X be the midpoint of the A-altitude of ABC. Prove that the symmedian point of ABC lies on the line MX.

Proof. The locus of the centers of the rectangles inscribed in triangle ABC and having one side on BC is precisely the line MX! Why? Well, in the first place, this locus is a line. The reason is as follows: Take a rectangle $X_1X_2Y_1Z_1$ inscribed in ABC with X_1, X_2 on BC. Erect the perpendiculars to BC at the vertices B and C and intersect these perpendiculars with the lines AX_1, AX_2 at two points X_1', X_2'. Then the rectangle $X_1X_2Y_1Z_1$ is the image of the rectangle $BCX_2'X_1'$ under a homothety with center A; hence, since the locus of the centers of the rectangles $BCX_2'X_1'$ is the perpendicular bisector of BC (and thus a line), it follows that the locus of the centers of rectangles $X_1X_2Y_1Z_1$ is also a line (the image of the perpendicular bisector under a certain homothety!) - fill in the details yourselves.

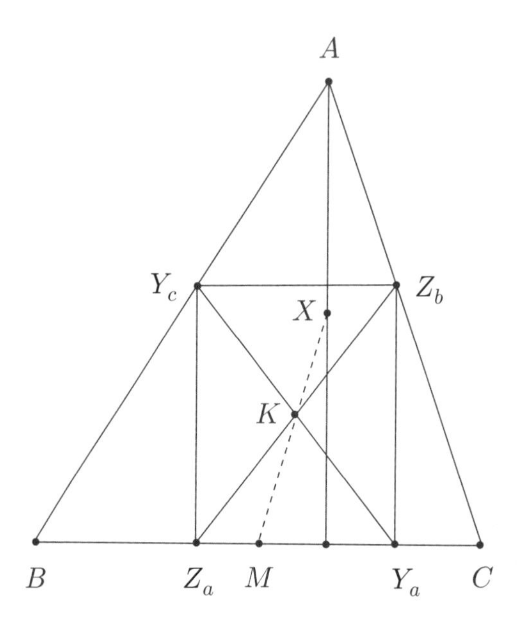

Now, it is clear that the midpoint of BC and the midpoint of the A-altitude belong to this line, since they are the centers of the two degenerate rectangles inscribed in ABC with one side on BC; hence the locus is precisely the line MX. Now, why does K lie on this locus? Well, because K is the center of a rectangle inscribed in ABC which has one side on BC! Indeed, recall from the proof of **Delta 9.3** that $KZ_b = KY_c = KZ_a = KY_a$, so $Z_aY_aZ_bY_c$ is a rectangle inscribed in ABC with Z_aY_a on BC with center K. □

//This remarkable result about the symmedian point lies at the heart of one of the authors' favorite concurrency in triangle geometry. The statement goes like this and it is only directed towards the die-hards who found everything trivial up until this point. Unfortunately, no simple proofs are known, so do treat this carefully.

Delta 9.6. Let ABC be a triangle with incenter I and circumcenter O. Let X, Y, Z be the midpoints of the segments IA, IB, IC and let K_a, K_b, K_c be the symmedian points of triangles IBC, ICA, IAB. Prove that the lines XK_a, YK_b, ZK_c are concurrent on the line OI.

We now give our last characterization that we want to emphasize. It is as important and useful as the previous ones so make sure you remember it as well!

Theorem 9.4. Let ABC be a triangle and let X be a point on the side BC. Obviously, for any point P on the line AX, we have that

$$\frac{\delta(P, AB)}{\delta(P, AC)} = \frac{\delta(X, AB)}{\delta(X, AC)}.$$

In other words, the ratio of the distances from P to the sides is independent of the point P chosen on AX.

Now, the claim is the following. For any point P on AX, we have that

$$\frac{\delta(P, AB)}{\delta(P, AC)} = \frac{\delta(X, AB)}{\delta(X, AC)} = \frac{AB}{AC}$$

if and only if AX is the A-symmedian of triangle ABC.

Proof. The proof is very easy. Recall that AX is the A-symmedian if and only if

$$\frac{XB}{XC} = \frac{AB^2}{AC^2}.$$

Hence, by now using the Ratio Lemma, we get that AX is the A-symmedian and only if

$$\frac{\sin XAB}{\sin XAC} = \frac{AB}{AC}.$$

But for any point P on AX, we have that

$$\frac{\delta(P, AB)}{\delta(P, AC)} = \frac{\delta(X, AB)}{\delta(X, AC)} = \frac{\sin XAB}{\sin XAC}.$$

Hence, we immediately obtain the conclusion that

$$\frac{\delta(P, AB)}{\delta(P, AC)} = \frac{\delta(X, AB)}{\delta(X, AC)} = \frac{AB}{AC}$$

if and only if AX is the A-symmedian of triangle ABC. □

While pretty immediate, this can prove to be incredibly useful and we shall use it to show a very important corollary: **The Lemoine Pedal Triangle Theorem**.

Delta 9.7. (Lemoine Pedal Triangle Theorem) The symmedian point K of triangle ABC is the only point in the plane of ABC which is the centroid of its own pedal triangle.

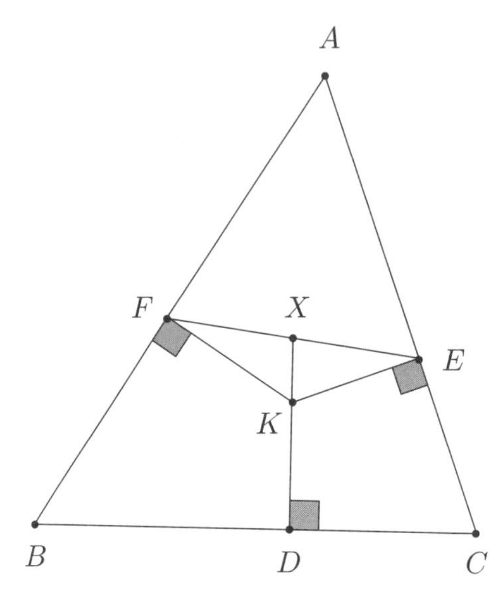

Proof. The idea is to use **Theorem 9.4**, as already mentioned. For the direct implication, let D, E, F be the projections of K on the sides BC, CA, AB and take X to be the intersection of DK with EF. We would like to show that X is the midpoint of EF, since after that we could just repeat the argument for EY and FZ and conclude that K is the centroid of DEF. By the Ratio Lemma, we know that

$$\frac{XE}{XF} = \frac{KE}{KF} \cdot \frac{\sin XKE}{\sin XKF}.$$

However, K obviously lies on the A-symmedian, thus by **Theorem 9.4**,

$$\frac{KE}{KF} = \frac{\delta(K, AC)}{\delta(K, AB)} = \frac{AC}{AB}.$$

Furthermore, $\angle XKE = \angle C$ and $\angle XKF = \angle B$ since the quadrilaterals $KDCE$ and $KFBD$ are cyclic; thus, we conclude that

$$\begin{aligned} \frac{XE}{XF} &= \frac{AC}{AB} \cdot \frac{\sin C}{\sin B} \\ &= \frac{AC}{AB} \cdot \frac{AB}{AC} \\ &= 1. \end{aligned}$$

This proves that X is the midpoint of EF and settles the direct implication. As for the converse, things are essentially similar. Now, we know that K is a point having projections D, E, F so that

$$1 = \frac{XE}{XF} = \frac{KE}{KF} \cdot \frac{\sin XKE}{\sin XKF}.$$

The equalities $\angle XKE = \angle C$ and $\angle XKF = \angle B$ hold because of cyclic quadrilaterals $KDCE$ and $KFBD$; thus, we immediately get that

$$\frac{KE}{KF} = \frac{AB}{AC}.$$

Hence, by **Theorem 9.4**, we conclude that K needs to lie on the A-symmedian, and similarly it must lie on the B and C-symmedians; thus we get that K is the symmedian point of the triangle. This completes the proof. □

Now, let's finally get to some Olympiad problems!

Delta 9.8. (Poland NMO 2000) Let ABC be a triangle with $AC = BC$, and P a point inside the triangle such that $\angle PAB = \angle PBC$. If M is the midpoint of AB, then show that $\angle APM + \angle BPC = 180°$.

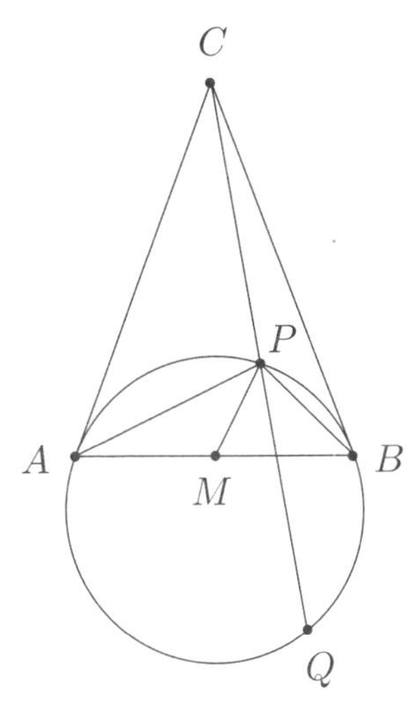

Proof. Let ω be the circumcircle of triangle PAB. Since $\angle PAB = \angle PBC$ we have that BC is tangent to ω and since $AC = BC$ we have that AC is tangent to ω as well. Let line CP intersect ω again at Q. Then from **Theorem 9.2** we have that line QP is the Q-symmedian in triangle PAB and so $\angle APM + \angle BPC = \angle BPQ + \angle BPC = 180°$ as desired. □

Delta 9.9. (BMO 2009) Let MN be a line parallel to the side BC of a triangle ABC, with M on the side AB and N on the side AC. The lines BN and CM meet at point P. The circumcircles of triangles BMP and CNP meet at two distinct points P and Q. Prove that $\angle BAQ = \angle CAP$.

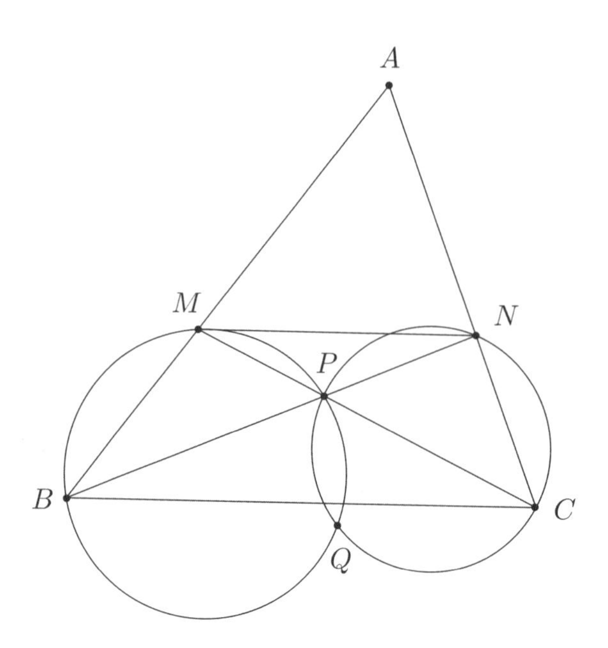

Proof. First of all, it's clear that AP is the A-median (remember **Delta 3.3**). Therefore it suffices to show that AQ is the A-symmedian of triangle ABC. Now since quadrilaterals $QBMP$ and $QCNP$ are cyclic we have that

$$\angle QBM = \angle QPC = \angle QNC$$

and

$$\angle QMB = \angle QPB = \angle QCN$$

so triangles QBM and QNC are similar. Therefore since $MN \parallel BC$ we have

$$\frac{\delta(Q, AB)}{\delta(Q, CA)} = \frac{BM}{CN} = \frac{AB}{CA}$$

so by **Theorem 9.4** we have that AQ is the A-symmedian of triangle ABC as desired. □

Delta 9.10. (APMO 2012) Let ABC be an acute triangle. Denote by D the foot of the perpendicular line drawn from the point A to the side BC, by M the midpoint of BC, and by H the orthocenter of ABC. Let E be the point of intersection of the circumcircle Γ of the triangle ABC and the half line MH, and F be the point of intersection (other than E) of the line ED and the circle Γ. Prove that

$$\frac{BF}{CF} = \frac{AB}{AC}$$

must hold.

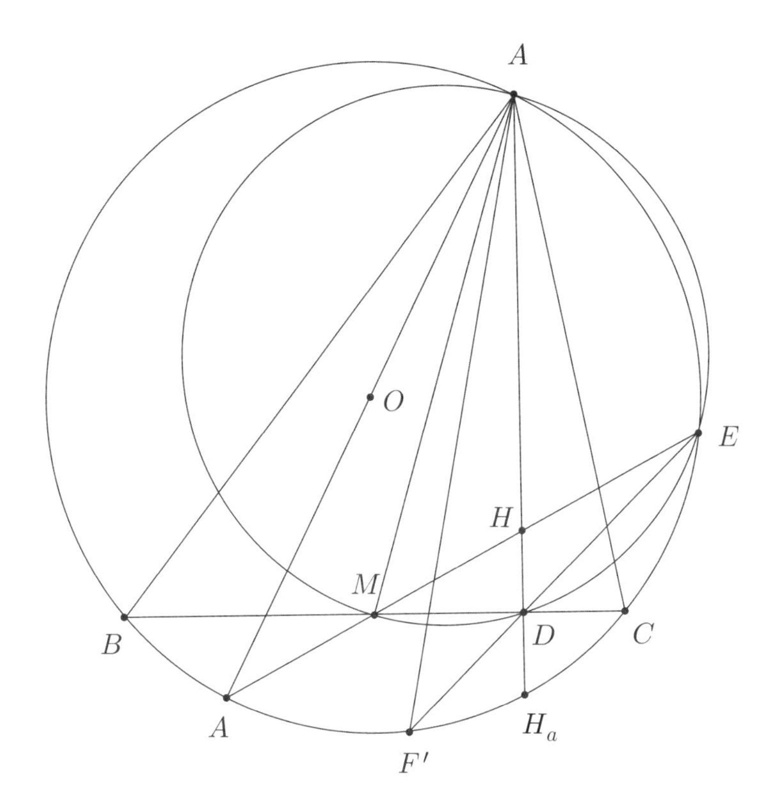

Proof. Let O and ω be the circumcenter and circumcircle respectively of triangle ABC. Let A' be the reflection of H over M and let H_a be the reflection of H over D. We know that H_a lies on the ω and since $OM \parallel AH$ and $AH = 2OM$ we have that OM is a midline of triangle $AA'H$. Therefore A' is the antipode of A with respect to ω. Since points A, E, H_a, A' all lie on ω we have that $HA \cdot HH_a = HE \cdot HA'$ and so

$$HM \cdot HE = \frac{1}{2}HA' \cdot HE = \frac{1}{2}HH_a \cdot HA = HD \cdot HA$$

so quadrilateral $AMDE$ is cyclic. Therefore

$$\angle HAM = \angle DAM = \angle FEA' = \angle FAA' = \angle FAO$$

and since AO and AH are isogonal conjugates with respect to angle A in triangle ABC we know that AM and AF are also isogonal. Therefore AF is the A-symmedian of triangle ABC. Now let the tangents to ω at B and C intersect at X. From **Theorem 9.2** we know that points A, F, X are collinear so line FA is the F-symmedian of triangle FBC. Letting Z be the intersection of AF and BC we apply **Corollary 9.1** twice and see that

$$\frac{BF^2}{CF^2} = \frac{BZ}{CZ} = \frac{AB^2}{AC^2}$$

which completes the proof. □

Delta 9.11. (Cosmin Pohoata, Romania TST 2014) Let the tangents to the circumcircle of a triangle ABC at the vertices B and C intersect each other at X. Consider the circle ω centered at X with radius XB, and let M be the point of intersection of the internal angle bisector of angle $\angle BAC$ with ω such that M lies in the interior of triangle ABC. If O is the circumcenter of triangle ABC, denote by P the intersection of OM with the sideline BC, and let E, F be the orthogonal projections of M on lines CA, AB, respectively. Prove that the lines AP, BE and CF are concurrent.

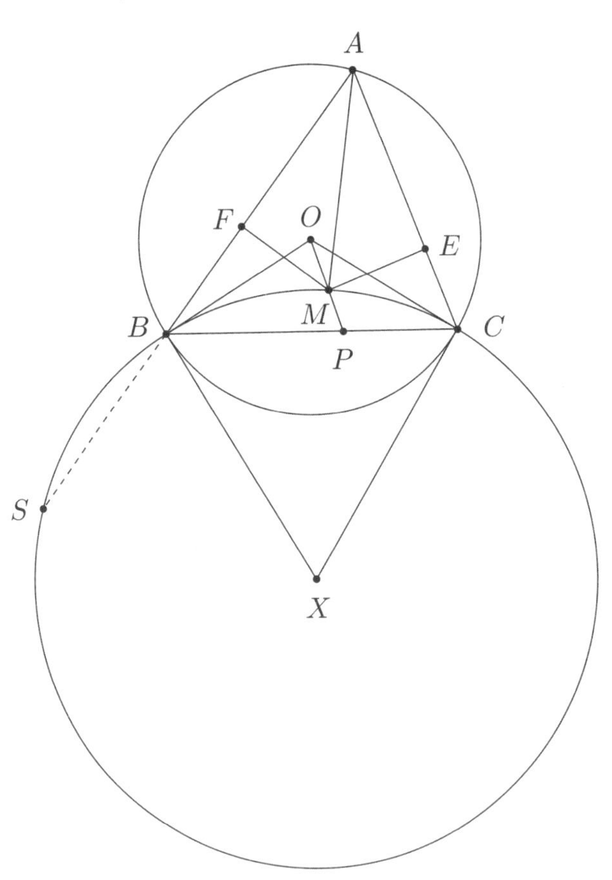

Proof. Let S be the second intersection of line AB with ω. We have that $\angle BSC = 90° - \angle A$, which means that $CS \perp CA$, and thus $CS \parallel ME$. This yields $\angle MBF = \angle MCS = \angle CME$; therefore, triangles MBF and CME are similar. This yields

$$\frac{MB}{MC} = \frac{MF}{CE} = \frac{BF}{ME},$$

and so, since $ME = MF$, we have

$$\left(\frac{MB}{MC}\right)^2 = \frac{MF}{CE} \cdot \frac{BF}{ME} = \frac{BF}{CE}.$$

On other hand, since X is the circumcenter of triangle MBC and since $\angle XBO = \angle XCO = 90°$, by **Theorem 9.2** we have that the line OM is the M-symmedian of triangle MBC; consequently, it follows that

$$\frac{PB}{PC} = \frac{MB^2}{MC^2}.$$

Therefore, since $AE = AF$, we have

$$\frac{PB}{PC} \cdot \frac{EC}{EA} \cdot \frac{FA}{FB} = 1,$$

which, by Ceva's Theorem, means that the lines AP, BE, CF are concurrent as desired. □

Assigned Problems

Epsilon 9.1. Let be ABC be a right triangle with $\angle A = 90°$. Points M, N are on AB, AC, and P, Q on BC so that $MNPQ$ is a rectangle. Furthermore, let BN intersect MQ at E and CM intersect NP at F. Prove that $\angle EAB = \angle FAC$.

Epsilon 9.2. In the cyclic pentagon $ABCDE$ we have $AC \| DE$ and $\angle AMB = \angle BMC$, where M is the midpoint of BD. Show that the line BE bisects segment AC.

Epsilon 9.3. Given triangle ABC, two circles are drawn tangent to BC at B and C, respectively, and both pass through A. These two circles intersect again at a point D. Prove that the reflection of D over BC lies on the symmedian of ABC from A.

Epsilon 9.4. (IMO Shortlist 2003) Three distinct points A, B, C are fixed on a line in this order. Let Γ be a circle passing through A and C whose center does not lie on the line AC. Denote by P the intersection of the tangents to Γ at A and C. Suppose Γ meets the segment PB at Q. Prove that the intersection of the bisector of $\angle AQC$ and the line AC does not depend on the choice of Γ.

Epsilon 9.5. (Vietnam TST 2001) In the plane, two circles intersect at A and B, and a common tangent intersects the circles at P and Q. Let the tangents at P and Q to the circumcircle of triangle APQ intersect at P and Q. Let the tangents at P and Q to the circumcircle of triangle APQ intersect at S, and let H be the reflection of B across the line PQ. Prove that the points A, S, H are collinear.

Epsilon 9.6. Let ABC be a triangle and let $ACUV$ and $ABST$ be the squares erected on the sides which are directed towards the exterior of the triangle. Let X be the circumcenter of triangle ATV. Prove that AX is the A-symmedian of triangle ABC.

Epsilon 9.7. (USAMO 2008) Let ABC be an acute, scalene triangle, and let M, N, and P be the midpoints of segments BC, CA, and AB, respectively. Let the perpendicular bisectors of segments AB and AC intersect ray AM in points D and E respectively, and let lines BD and CE intersect in point F, inside of triangle ABC. Prove that points A, N, F, and P all lie on one circle.

Epsilon 9.8. (USA TST 2007) Let $\mathcal{O}$ be the circumcircle of a given triangle ABC. The tangents to $\mathcal{O}$ at the vertices B and C meet at a point T. Consider

S the intersection of the line through A perpendicular to AT with the sideline BC. Denote by B_1 and C_1 the points on the line ST, for which $B_1T = BT = C_1T$ and such that C_1 lies between the points B_1 and S. Then, the triangles ABC and AB_1C_1 are directly similar.

Epsilon 9.9. Use the Lemoine Pedal Triangle Theorem to give a second proof of **Delta 9.5**.

Epsilon 9.10. (ELMO Shortlist 2014) Let $ABCD$ be a quadrilateral inscribed in circle ω. Define $E = AA \cap CD$, $F = AA \cap BC$, $G = BE \cap \omega$, $H = BE \cap AD$, $I = DF \cap \omega$, and $J = DF \cap AB$. Prove that GI, HJ, and the B-symmedian of triangle ABC are concurrent.

Chapter 10

Harmonic Divisions

Definition. Let A, B, C, D be four points on a line. Then the **cross-ratio** $(A, B; C, D)$ of these four points is defined as

$$(A, B; C, D) = \frac{CA}{CB} : \frac{DA}{DB}$$

where we use directed lengths.

Definition. Let A, B, C, D be points lying on a circle. The **cross-ratio** $(A, B; C, D)$ of these four points is defined as

$$(A, B; C, D) = \pm\frac{CA}{CB} : \frac{DA}{DB}$$

where we take the $+$ if segments AB and CD do not intersect and take the $-$ otherwise.

Definition. Let A, B, C, D be four points on a line in this order. If $(A, C; B, D) = -1$ then $(A, C; B, D)$ is called a **harmonic division** or a **harmonic bundle** (or is simply described as being **harmonic**)

Definition. Let A, B, C, D be four points lying on a circle in this order. If $(A, C; B, D) = -1$ then the quadrilateral $ABCD$ is called a **harmonic quadrilateral**. In other words, a cyclic quadrilateral $ABCD$ is harmonic if and only if $AB \cdot CD = DA \cdot BC$.

Corollary 10.1. Notice that we have the very nice implication: if $(A, C; B, D)$ and $(A, C; B, D')$ are both harmonic, then $D = D'$, and a similar result holds if the points A, B, C, D, D' are concyclic.

So, let's see some properties of these harmonic divisions! We will cover four basic lemmas in this material - they will be more than enough to solve a really significant chunk of interesting problems! But before we get to that, we begin with two easy properties about harmonic divisions and midpoints that we leave as exercises to the reader.

Delta 10.1. Prove that $(A, C; B, D)$ is harmonic if and only if

$$MB \cdot MD = MA^2,$$

where M is the midpoint of the segment AC. (Hint: imagine that the line determined by A, B, C, D, M is the real axis and associate real numbers a, b, c, d to A, B, C, D; what is the number associated to M?)

Delta 10.2. Let M be the midpoint of a segment AB. Prove that $(A, B; M, P)$ is harmonic if and only if P is the point at infinity on line AB.

Now, we talk about the four major lemmas. The following result demonstrates the easiest way to generate harmonic divisions.

Theorem 10.1. In a triangle ABC, consider three points X, Y, Z on the interior of sides BC, CA, and AB, respectively. If X' is the point of intersection of the line YZ with the extended side BC (suppose C lies between B and X'), then $(B, C; X, X')$ is a harmonic division if and only if the cevians AX, BY, CZ are concurrent.

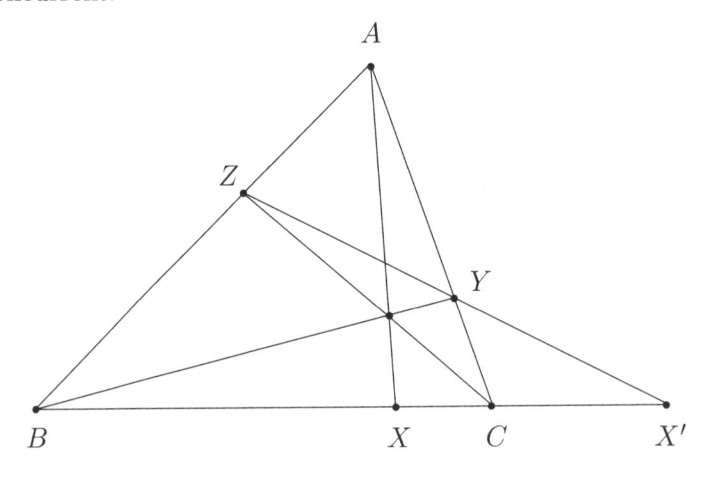

Proof. Since the points Y, Z, X' are collinear, by Menelaus' Theorem, we have that

$$\frac{X'B}{X'C} \cdot \frac{YC}{YA} \cdot \frac{ZA}{ZB} = -1.$$

Now by Ceva's Theorem, the lines AX, BY, CZ are concurrent if and only if

$$\frac{XB}{XC} \cdot \frac{YC}{YA} \cdot \frac{ZA}{ZB} = 1,$$

hence by dividing the two expressions we deduce that this happens if and only if $\frac{XB}{XC} = \frac{X'B}{X'C}$ or $(B, C; X, X') = -1$, i.e. if and only if $(B, C; X, X')$ is a harmonic division. □

The following result also demonstrates an easy way to generate harmonic quadrilaterals.

Theorem 10.2. Let A, B, C be three points lying on a circle ω. Let the tangents at A and C to ω intersect at a point P and let the line PB intersect ω again at D. Then $ABCD$ is a harmonic quadrilateral.

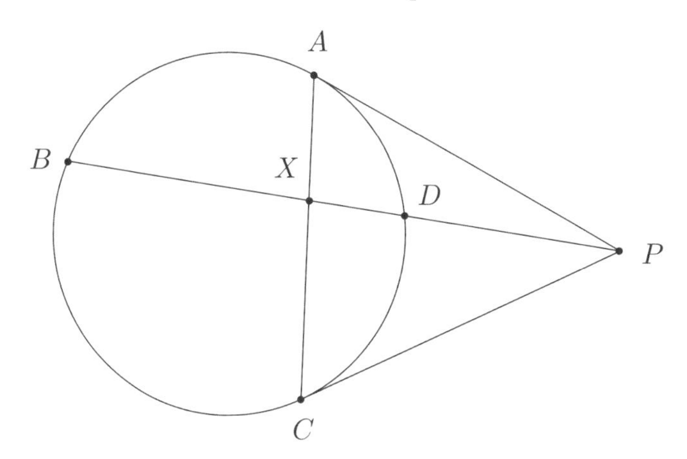

Proof. Let $X = AC \cap BD$. Then we know from the last section that BD is the B-symmedian of triangle ABC and the D-symmedian of triangle ADC so we have that

$$\frac{BA^2}{BC^2} = \frac{AX}{CX} = \frac{DA^2}{DC^2}.$$

In other words, that $(A, C; B, D) = -1$ as desired. □

The converse of **Theorem 10.2** holds as well. This gives a very useful criterion for harmonic quadrilaterals - a cyclic quadrilateral; $ABCD$ is harmonic if and only if AC is a symmedian in triangles BAD and BCD and BD is a symmedian in triangles ABC and ADC.

The next result puts the "projective" in projective geometry.

Theorem 10.3. Let A, B, C, D be four points lying in this order on a line d, and let P be a point not lying on this line. Take another line d' and consider the intersections A', B', C', D' of the lines PA, PB, PC, and PD, respectively, with d'. Then $(A, C; B, D) = (A', C'; B', D')$.

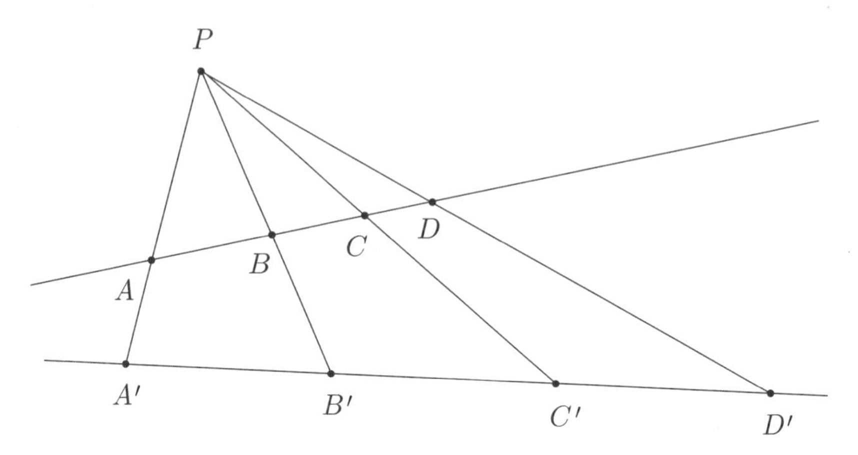

Proof. How can we evaluate these ratios? The Ratio Lemma! Let $x = \angle APB, y = \angle BPC, z = \angle CPD$; we have that

$$\frac{BA}{BC} = \frac{PA}{PC} \cdot \frac{\sin x}{\sin y} \text{ and } \frac{DA}{DC} = \frac{PA}{PC} \cdot \frac{\sin(x+y+z)}{\sin z}.$$

Thus,

$$(A, C; B, D) = -\frac{\sin x \cdot \sin z}{\sin y \cdot \sin(x+y+z)}$$

and now since we can do the same thing for $(A', C'; B', D')$ the result is clear. $\square$

This configuration can be denoted by $P(A, C; B, D)$, and is called a **pencil**. We are taking "perspective" at point P and "projecting" the bundle $(A, C; B, D)$ to the bundle $(A', C'; B', D')$. This can be written as $(A, C; B, D) \stackrel{P}{=} (A', C'; B', D')$. **Theorem 10.3** can be restated as follows; projections preserve cross-ratios. Unsurprisingly, we can do more than project from lines to lines - we can project from lines to circles and vice-versa, as long as the point we are taking perspective at lies on the circle! We leave the proof again as an exercise to the reader.

Delta 10.3. Let A, B, C, D be four points lying in this order on a circle ω and let P be a point also lying on ω. Let lines PA, PB, PC, PD intersect a line d at points A', B', C', D' respectively. Prove that $(A, C; B, D) = (A', C'; B', D')$.

The fourth and final lemma gives us a nice way to deal with perpendicularities and angle bisectors.

Theorem 10.4. Let A, B, C, D be four points lying in this order on a line d. If X is a point not lying on this line, then if two of the following three propositions are true, then the third is also true:

(a) $(A, C; B, D)$ is harmonic.

(b) XB is the internal angle bisector of $\angle AXC$.

(c) $XB \perp XD$.

Proof. First, note that if (a) and (b) are true, then (c) is obviously also true, since we have that XB and XD are the internal and external angle bisectors of angle $\angle AXC$ and we know that they are perpendicular. Likewise, if (b) and (c) are true, then by the Angle-Bisector Theorem, we have that $(A, C; B, D)$ is harmonic, so (a) holds. The "trickier" part is deducing (b) from (a) and (c), and as we will see in a few examples, this is the implication that is very interesting to spot in crowded configurations - it will often help you get the right idea about the solution if not solve the problem directly! But let's first deal with the proof. Again, label the angles $x = \angle AXB$, $y = \angle BXC$, $z = \angle CXD$. From (c) we know that $y + z = 90°$. Looking at the proof of **Theorem 10.3** we know that

$$(A, C; B, D) = -\frac{\sin x \cdot \sin z}{\sin y \cdot \sin (x + y + z)} = -\frac{\sin x \cdot \cos y}{\sin y \cdot \cos x} = -\frac{\tan x}{\tan y}$$

and since the tangent function is monotonic on the interval $(0, 90°)$ we have that $(A, C; B, D) = -1$ only if $x = y$ as desired. □

Now, let's see some applications.

Delta 10.4. (IMO 1995 Shortlist) Let ABC be a triangle, and let D, E, F be the points of tangency of the incircle of triangle ABC with the sides BC, CA, and AB, respectively. Let X be in the interior of ABC such that the incircle of XBC touches XB, XC, and BC at Z, Y, and D, respectively. Prove that $EFZY$ is cyclic.

Proof. Let T be the intersection of BC with EF. Because of the concurrency of the lines AD, BE, CF at the Gergonne point of triangle ABC, we deduce that the bundle $(B, C; D, T)$ is harmonic by **Theorem 10.1**.

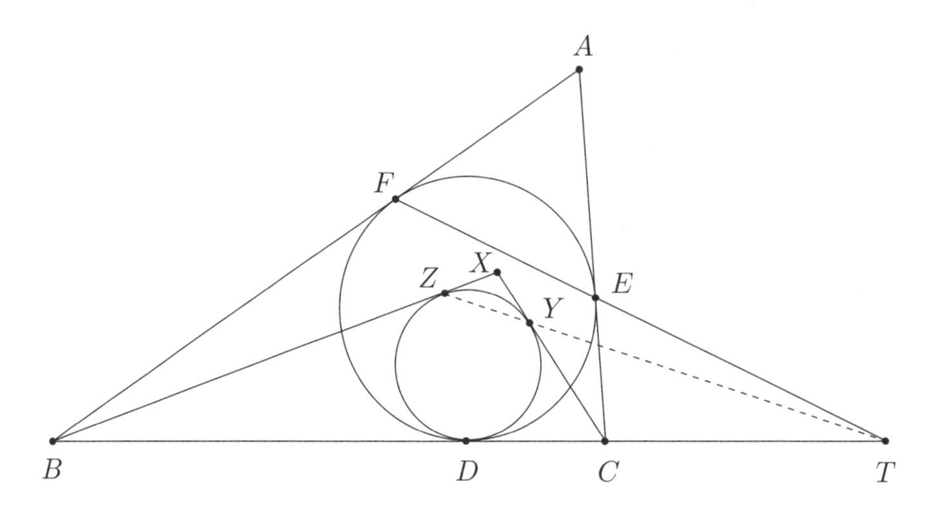

Similarly, the lines XD, BY and CZ are concurrent at the Gergonne point of triangle XBC, so $(B, C; D, T')$ is also harmonic. Hence, by **Corollary 10.1**, we get that $T = T'$ and thus T lies on YZ.

Now expressing the power of point T with respect to the incircles of ABC and XBC we get that $TD^2 = TE \cdot TF$ and $TD^2 = TZ \cdot TY$; so $TE \cdot TF = TZ \cdot TY$, which means that the quadrilateral $EFZY$ is cyclic, as desired. □

Delta 10.5. (Chinese TST 2002) Let $ABCD$ be a convex quadrilateral for which we label the intersections $E = AB \cap CD$, $F = AD \cap BC$, $P = AC \cap BD$. Let O the foot of the perpendicular from P to the line EF. Prove that $\angle BOC = \angle AOD$.

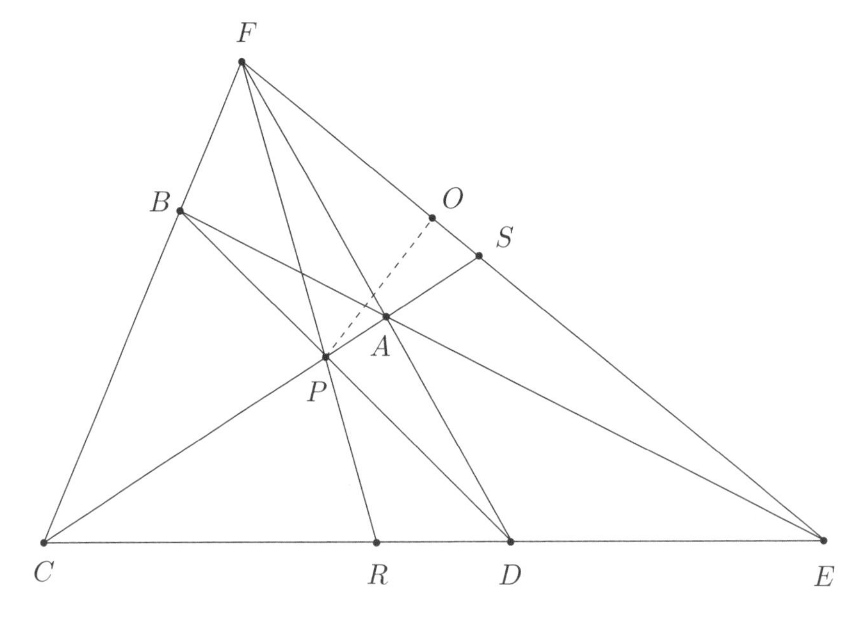

Proof. Let $R = FP \cap CD$ and $S = AC \cap EF$. Then since lines FR, CA, DB concur at P by **Theorem 10.1** we have that $(C, D; R, E)$ is harmonic. Also, note that $(C, A; P, S) \stackrel{F}{=} (C, D; R, E)$ so $(C, A; P, S)$ is harmonic as well. And since $OS \perp OP$ by **Theorem 10.4** we find that OP bisects angle $\angle AOC$. Similarly we can show that OP bisects $\angle BOD$, and the combination of these two bisections clearly implies the desired result. $\square$

Deduce the following nice consequence on your own.

Delta 10.6. (Romanian TST 2008) Let $ABCD$ be a convex quadrilateral and let $O \in AC \cap BD$, $P \in AB \cap CD$, $Q \in BC \cap DA$. If R is the orthogonal projection of O on the line PQ prove that the orthogonal projections of R on the sidelines of $ABCD$ are concyclic. (Hint: use the result in **Delta 10.5** and angle chase).

Let's tackle something even more involved!

Delta 10.7. Let ABC be a right triangle with $\angle A = 90°$, and let D be a point lying on the side AC. Denote by E the reflection of A over line BD, and by F the intersection of CE with the perpendicular through D to the line BC. Prove that AF, DE and BC are concurrent.

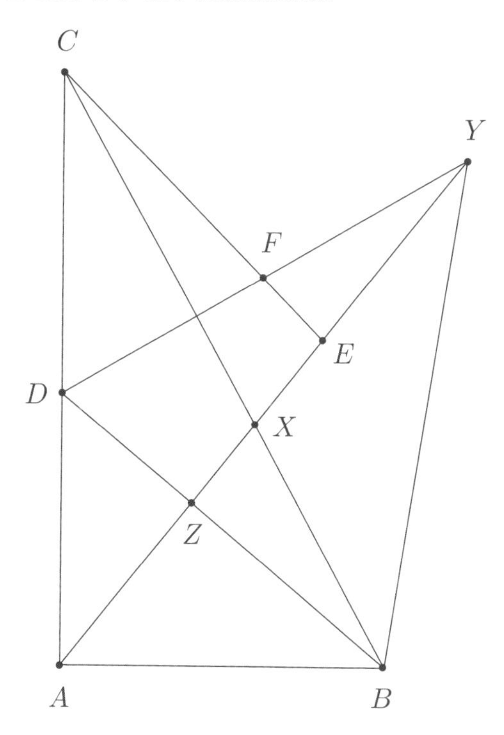

Proof. Let $X = BC \cap AE$ and $Y = DF \cap AE$ and $Z = BD \cap AE$. Then by **Theorem 10.1** lines AF, DE, CX concur if and only if $(A, E; X, Y)$ is harmonic. But since Z is the midpoint of AE by **Delta 10.1** it suffices to show that $ZX \cdot ZY = ZA^2$. Now, note that $XY \perp BD$ and $BX \perp DY$ so X is the orthocenter of triangle BDY. This means that $ZX \cdot ZY = ZD \cdot ZB$ but since triangle ABD has a right angle at A and since Z is the foot of the A-altitude of this triangle we know that $ZD \cdot ZB = ZA^2$. This completes the proof. □

Delta 10.8. Let ω be the incircle of triangle ABC and let D, E, F be the points of tangency of ω with sides BC, CA, AB, respectively. Let M be the second intersection of AD with ω, N the second intersection of line DF with the circumcircle of triangle CDM, and G the intersection of lines CN and AB. Prove that $CD = 3FG$.

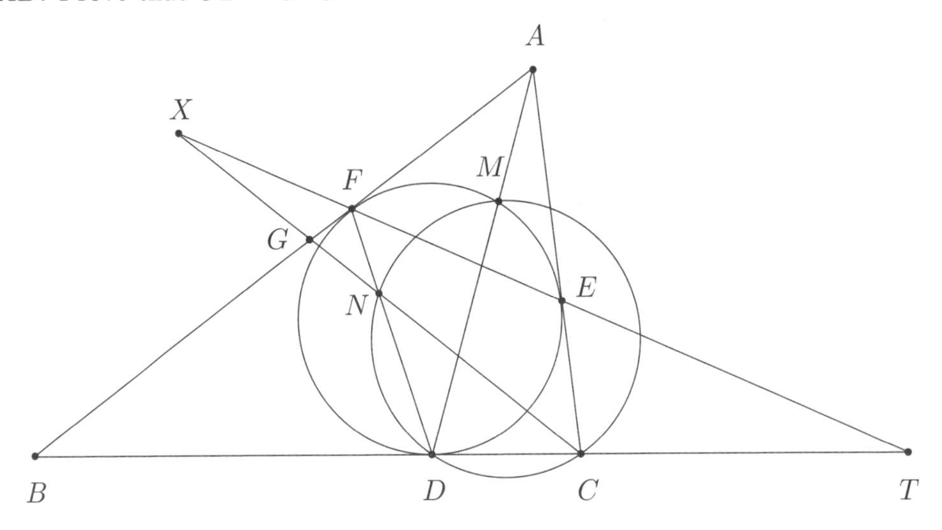

Proof. Let $X = EF \cap CG$ and $T = EF \cap BC$. By **Theorem 10.1**, since lines AD, BE, CF concur at the Gergonne point of triangle ABC, we have that $(B, C; D, T)$ is a harmonic division. But since $(G, C; N, X) \stackrel{F}{=} (B, C; D, T)$ we have that $(G, C; N, X)$ is harmonic as well.

On the other hand, by Menelaus' Theorem applied in triangle BCG for the collinear points D, N, F, we see that in order to show that $CD = 3GF$ it suffices to prove that $CN = 3NG$. However, $(G, C; N, X)$ is harmonic, so $\frac{NC}{NG} = \frac{XC}{XG}$, and therefore it is enough to prove that N is the midpoint of segment CX.

Now, observe that $\angle MEX = \angle MDF = \angle MCX$, and therefore the quadrilateral $MECX$ is cyclic, which implies that $\angle MXC = \angle MEA =$

$\angle ADE$ and $\angle MCX = \angle ADF$. Furthermore, $\angle CMN = \angle FDB$ and $\angle XMN = \angle XMC - \angle CMN = \angle CEF - \angle FDB = \angle EDC$.

Applying the Ratio Lemma and using these equalities in

$$\frac{NX}{NC} = \frac{MX}{MC} \cdot \frac{\sin XMN}{\sin CMN} = \frac{\sin MCX}{\sin MXC} \cdot \frac{\sin XMN}{\sin CMN}$$

we obtain that

$$NC = NX \text{ if and only if } \frac{\sin FDA}{\sin EDA} = \frac{\sin BDF}{\sin CDE}.$$

However, DA is the D-symmedian of triangle DEF, so

$$\frac{\sin FDA}{\sin EDA} = \frac{FD}{ED} = \frac{\sin DEF}{\sin DFE} = \frac{\sin BDF}{\sin CDE}.$$

Therefore, N is the midpoint of segment CX, as claimed. This completes the proof. □

Delta 10.9. (ELMO Shortlist 2015) Let CA, CB be the tangent segments from a point C to a circle ω. Let X be the reflection of A over B, and let the circumcircle ω' of CBX intersect ω again at D. If CD intersects ω again at E, prove that EX is tangent to ω'

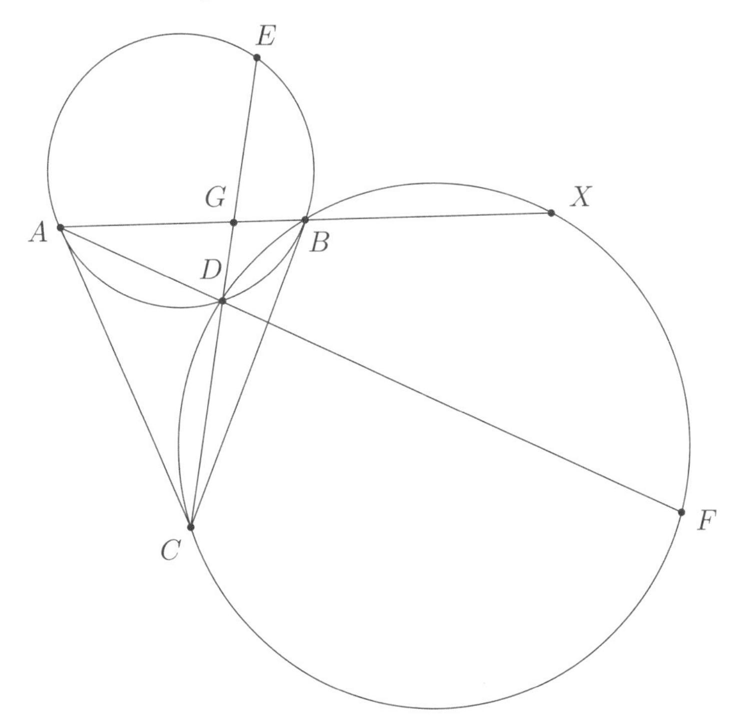

Proof. First note that $\angle ECX = \angle DBA = \angle CEA$ which implies that $EA \parallel CX$. Now let F be the second intersection of line AD with ω'. We have that $\angle AFC = \angle DBC = \angle XAF$ so $FC \parallel AX$. Therefore by **Delta 10.2** we have that the pencil $F(X, A; B, C)$ is harmonic and projecting onto ω' this means that quadrilateral $CDBX$ is harmonic. Let $G = AB \cap ED$. From **Theorem 10.2** we know that quadrilateral $ADBE$ is harmonic and since $(C, G; D, E) \stackrel{A}{=} (A, B; D, E)$ we have that $(C, G; D, E)$ is harmonic. Now, let line EX intersect ω' again at X'. We have $(C, B; D, X') \stackrel{X}{=} (C, G; D, E) = -1$ so quadrilateral $CDBX'$ is harmonic. But we proved earlier than quadrilateral $CDBX$ is harmonic so we must have $X = X'$ and thus EX is tangent to ω' as desired. □

We proceed with a famous lemma that is surprisingly difficult to prove (unless one knows about cross-ratios or Haruki's Lemma, that is...).

Theorem 10.5. (Butterfly Theorem) Let AB, CD, XY be three chords of a circle ω that are all concurrent at a point M. Assume without loss of generality that points A and C are on the same side of line XY and let $R = AD \cap XY$ and $S = BC \cap XY$. Then if M is the midpoint of XY, it is also the midpoint of RS.

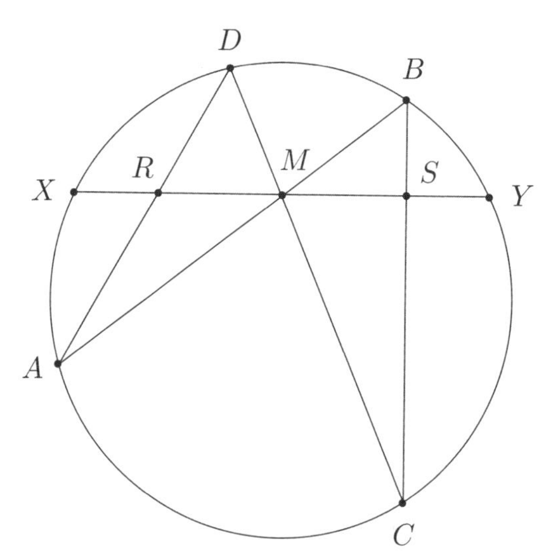

Proof. Note that $(X, Y; M, R) \stackrel{A}{=} (X, Y; B, D) \stackrel{C}{=} (X, Y; S, M)$ which since $MX = MY$ implies that

$$\frac{RX}{RY} = \frac{SY}{SX}$$

which immediately yields that M is the midpoint of RS as desired. □

Delta 10.10. Let M be the midpoint of the side BC of a given triangle ABC. Denote by D, E the intersections of the circle ω with diameter AM with the sides AB, AC, respectively. The tangents at D, E of the circle ω intersect each other at P. Prove that $PB = PC$.

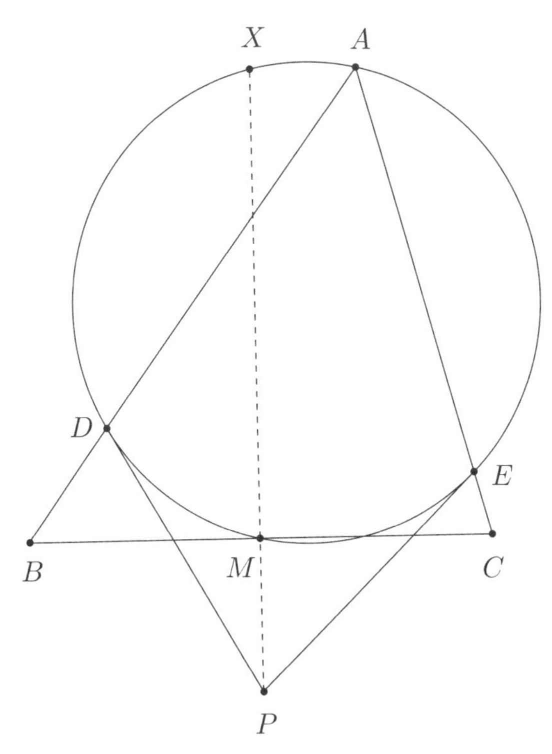

Proof. Let the perpendicular through M to BC intersect ω again at X. Since $\angle AXM = 90^\circ$ we have that $AX \parallel BC$. Now, let A_∞ be the point at infinity on line BC. By **Delta 10.2** we have that $(B, C; M, A_\infty)$ is harmonic, so since $(D, E; M, X) \stackrel{A}{=} (B, C; M, A_\infty)$ we have that quadrilateral $DXEM$ is harmonic. Therefore by **Theorem 10.2** P lies on line XM, which is precisely the perpendicular bisector of segment BC. Hence $PB = PC$ as desired. $\square$

Delta 10.11. (IMO Shortlist 2002) The incircle Ω of the acute-angled triangle ABC is tangent to its side BC at a point K. Let AD be an altitude of triangle ABC, and let M be the midpoint of the segment AD. If N is the common point of the circle Ω and the line KM (distinct from K), then prove that line NK bisects angle $\angle BNC$.

Proof. Let J be the second intersection of line AK with Ω and let K' be the antipode of K with respect to Ω. Let R and S be the tangency points of

Ω with lines AB and AC respectively and let $X = RS \cap BC$. Let A_∞ be the point at infinity on line AD. By **Delta 10.2** we have that $(A, D; M, A_\infty)$ is harmonic so since $(J, K; N, K') \stackrel{K}{=} (A, D; M, A_\infty)$ we have that quadrilateral $KK'JN$ is harmonic. Moreover since the tangents from R and S to Ω intersect at A and since A, J, K are collinear this implies that quadrilateral $KRJS$ is harmonic as well. Since BC is the tangent from K to Ω this means that X is on the tangent from J to Ω.

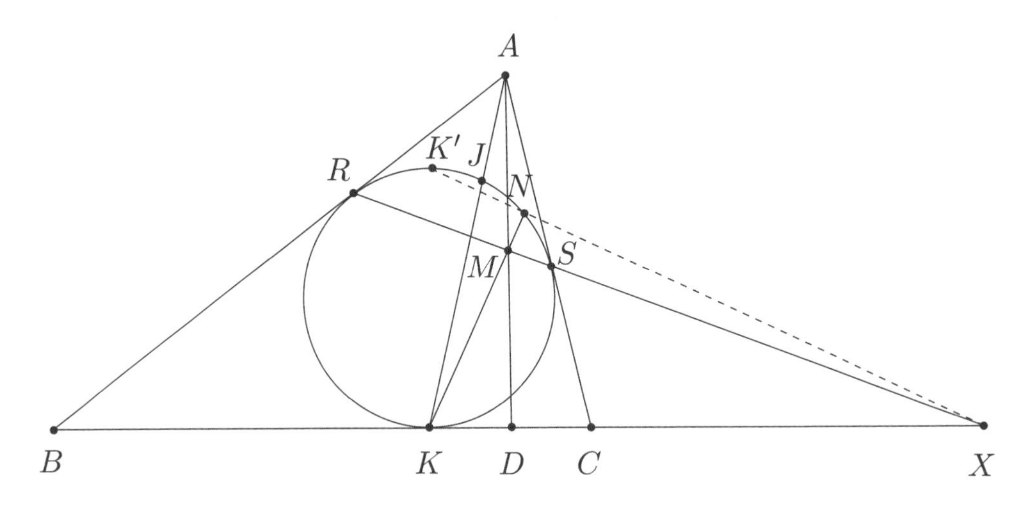

Therefore X, N, K' are collinear, so $\angle XNK = 180° - \angle K'NK = 90°$. Now since lines AK, BS, CR concur at the Gergonne point of triangle ABC, by **Theorem 10.1** we have that $(B, C; K, X)$ is harmonic. An application of **Theorem 10.4** then implies the desired result. □

We finish the section by scaling the G8 summit!

Delta 10.12. (IMO Shortlist 2004 G8) Given a cyclic quadrilateral $ABCD$, let M be the midpoint of the side CD, and let N be a point on the circumcircle of triangle ABM. Assume that the point N is different from the point M and satisfies $\frac{AN}{BN} = \frac{AM}{BM}$. Prove that the points E, F, N are collinear, where $E = AC \cap BD$ and $F = BC \cap DA$.

Proof. Let line CM intersect the circumcircle of triangle ABM again at P, and let $G = AB \cap CD$. We have that

$$MP \cdot MG = MG^2 - GP \cdot GM = MG^2 - GA \cdot GB = MG^2 - GC \cdot GD = MC^2$$

so by **Delta 10.1** we know that $(C, D; P, G)$ is harmonic.

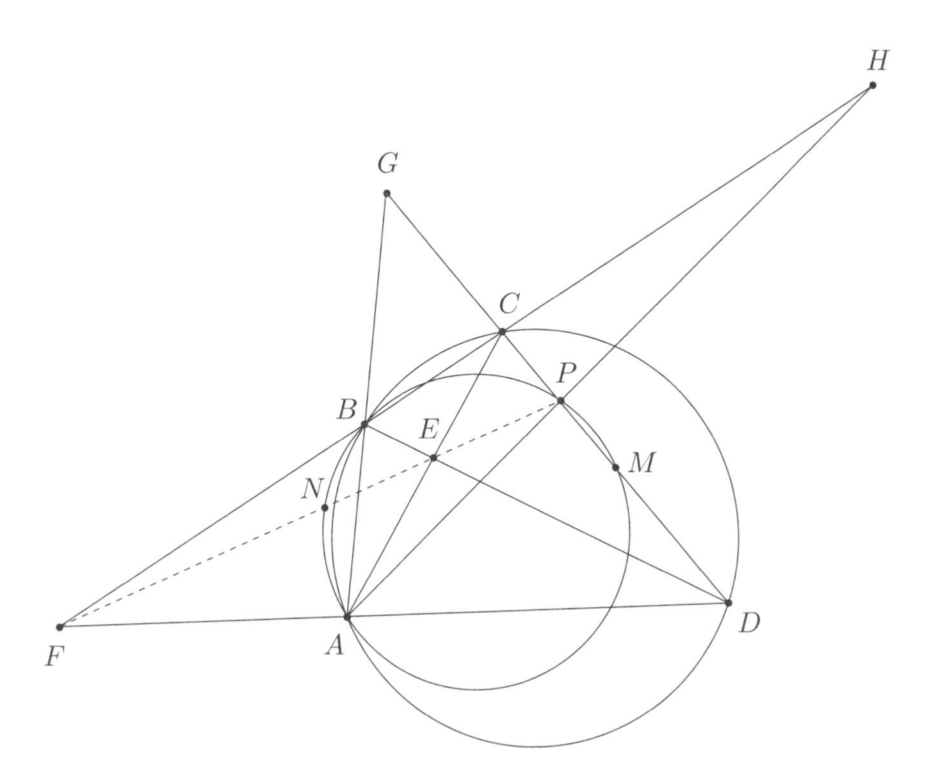

By **Theorem 10.1**, we must have that FP, CA, DB concur so P lies on line EF. Now let $H = AP \cap BC$. Since lines FP, CA, DB concur at E by **Theorem 10.1** again we have that $(C, F; B, H)$ is harmonic. Now let line FP intersect the circumcircle of triangle ABM again at N'. We have that $(M, N'; B, A) \stackrel{P}{=} (C, F; B, H) = -1$ so quadrilateral $AMBN'$ is harmonic. But by definition quadrilateral $AMBN$ is also harmonic, and so we must have $N = N'$. This completes the proof. $\square$

Assigned Problems

Epsilon 10.1. Let ABC be a triangle with orthocenter H and let D, E, F be the feet of the altitudes lying on the sides BC, CA, and AB, respectively. Let T be the intersection of the lines EF and BC. Prove that the line TH is perpendicular to the A-median of triangle ABC.

Epsilon 10.2. In triangle ABC let M be the midpoint of side AB; let X be the second intersection of BC with the circumcircle of triangle AMC and let ω be the circle which is tangent to AC at C and passes through X. Furthermore, let XM intersect ω at Z and AC at Y (so that A lies between C and Y). Prove that the lines AX, BY, CZ are concurrent.

Epsilon 10.3. (IMO 2012) Given triangle ABC the point J is the center of the excircle opposite the vertex A. This excircle is tangent to the side BC at M, and to the lines AB and AC at K and L, respectively. The lines LM and BJ meet at F, and the lines KM and CJ meet at G. Let S be the point of intersection of the lines AF and BC, and let T be the point of intersection of the lines AG and BC. Prove that M is the midpoint of ST.

Epsilon 10.4. (IMO 2003) Let $ABCD$ be a cyclic quadrilateral. Let P, Q, R be the feet of the perpendiculars from D to the lines BC, CA, AB, respectively. Show that $PQ = QR$ if and only if quadrilateral $ABCD$ is harmonic.

Epsilon 10.5. (Sharygin 2013) Let D be the foot of the B-internal angle bisector of triangle ABC. Points I_a, I_c are the incenters of triangles ABD, CBD respectively. The line I_aI_c meets AC in point Q. Prove that $\angle DBQ = 90^\circ$

Epsilon 10.6. (USA TST 2011) In an acute scalene triangle ABC, points D, E, F lie on sides BC, CA, AB, respectively, such that $AD \perp BC, BE \perp CA, CF \perp AB$. Altitudes AD, BE, CF meet at orthocenter H. Points P and Q lie on segment EF such that $AP \perp EF$ and $HQ \perp EF$. Lines DP and QH intersect at point R. Compute HQ/HR

Epsilon 10.7. Let ω be a circle with center O and A a point outside it. Denote by B, C the points where the tangents from A to ω meet the circle, D the point on ω for which O lies on line AD, X the foot of the perpendicular from B to CD, Y the midpoint of segment BX, and Z the second intersection of DY with ω. Prove that $ZA \perp ZC$.

Epsilon 10.8. (APMO 2013) Let $ABCD$ be a quadrilateral inscribed in a circle ω, and let P be a point on the extension of AC such that PB and PD are tangent to ω. The tangent at C intersects PD at Q and the line AD at R. Let E be the second point of intersection between AQ and ω. Prove that B, E, R are collinear.

Epsilon 10.9. (Sharygin 2013) Let AD be a bisector of triangle ABC. Points M and N are projections of B and C respectively on AD. The circle with diameter MN intersects BC at points X and Y. Prove that $\angle BAX = \angle CAY$.

Epsilon 10.10. (ELMO Shortlist 2014) Let $AB = AC$ in $\triangle ABC$, and let D be a point on segment AB. The tangent at D to the circumcircle ω of triangle BCD hits AC at E. The other tangent from E to ω touches it at F, and $G = BF \cap CD$, $H = AG \cap BC$. Prove that $BH = 2HC$.

Chapter 11

Appendix A: Some Generalizations of Blanchet's Theorem

This is a rather unusual section, in the sense that it should be thought of as a strict appendix to Chapter 10. As a consequence, we won't be including a list of proposed problems at the end. We start the section with the statement of Blanchet's Theorem.

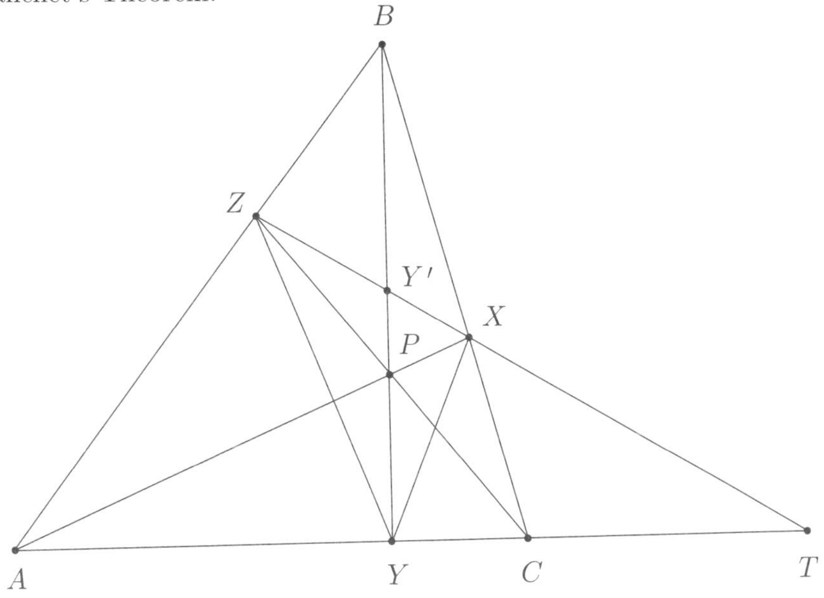

Theorem 11.1. (Blanchet's Theorem) Let Y be the foot of the B-altitude of triangle ABC. Let P be a point on the line BY. Let the lines CP and AP intersect the lines AB and BC at points Z and X respectively. Then, the line

BY bisects the angle $\angle XYZ$.

Proof. Let $Y' = BY \cap ZX$ and let $T = CA \cap ZX$. Since lines AX, BY, CZ concur we know that $(A, C; Y, T)$ is harmonic and since $(Z, X; Y', T) \stackrel{B}{=} (A, C; Y, T)$ we have that $(Z, X; Y', T)$ is harmonic as well. But since $YY' \perp YT$ we must have that YY' bisects angle $\angle ZYX$ as desired. This completes the proof. $\square$

We proceed with a generalization appearing in Engel's problem-solving classic [1] as Problem 88 in section 12.3.2:

Theorem 11.2. Let P be a point in the plane of a triangle ABC. Let the lines AP, BP, CP intersect the lines BC, CA, AB at the points A', B', C' respectively. Let M be the projection of A' on the line $B'C'$. Then, the line MA' bisects the angle $\angle BMC$.

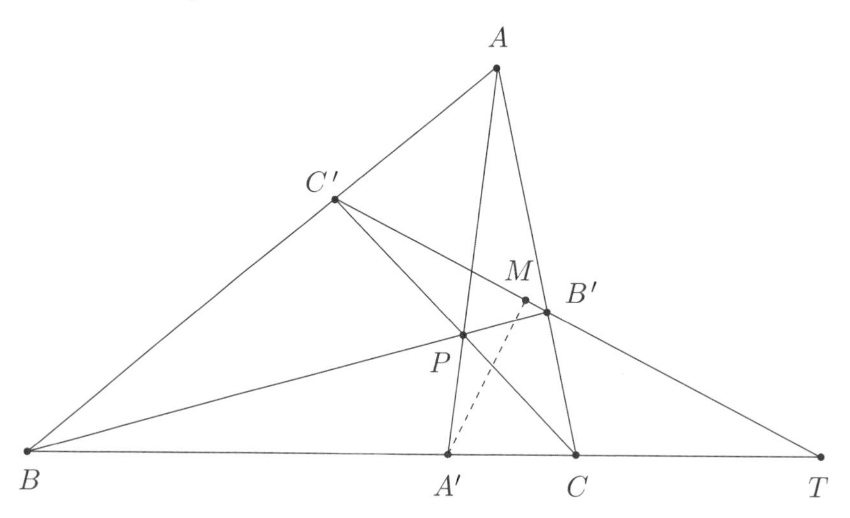

Proof. Let $T = BC \cap B'C'$. Since lines AA', BB', CC' concur at P we have that $(B, C; A', T)$ is harmonic. Then since $MA' \perp MT$ we have that MA' bisects angle $\angle BMC$ as desired. This completes the proof. $\square$

Engel [1] treats the above lemma as a plain geometry exercise. We will see that it is actually a fact worth remembering, having many powerful consequences. Next, we show a related result by Jean-Pierre Ehrmann from [7]:

Delta 11.1. Let P be a point in the plane of a triangle ABC. Let the lines AP, BP, CP intersect the lines BC, CA, AB at the points A', B', C' respectively. Denote by X the orthogonal projection of the point A' on the line $B'C'$. Denote by X' the reflection of the point P in the line $B'C'$. Then, the points A, X and X' are collinear.

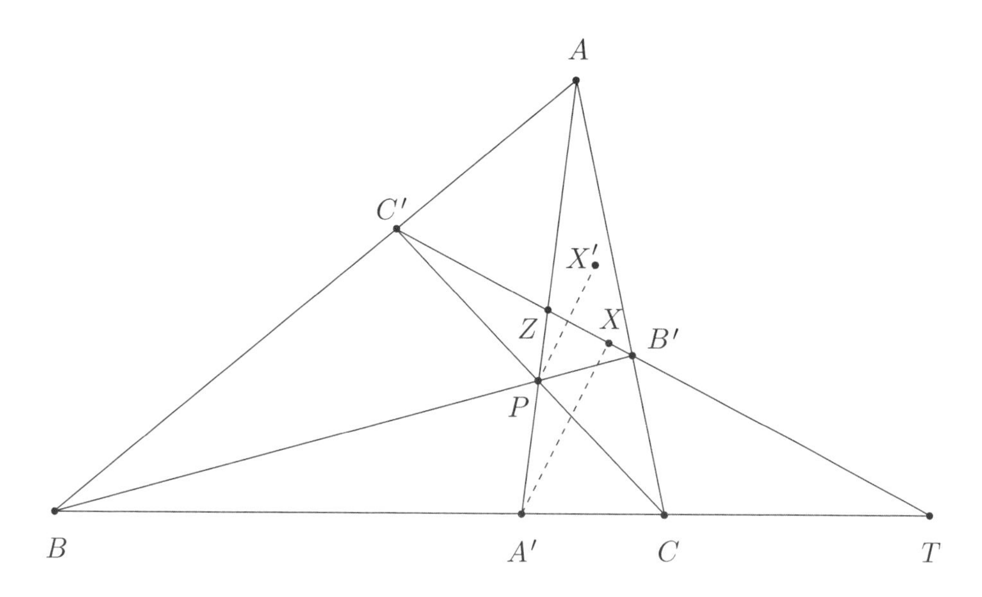

Proof. Let $T = BC \cap B'C'$ and $Z = AA' \cap B'C'$. Since lines AA', BB', CC' concur at P we have that $(B, C; A', T)$ is harmonic. Then since $(A, P; A', Z) \stackrel{C'}{=} (B, C; A', T)$ we have that $(A, P; A', Z)$ is harmonic as well. But since $XZ \perp XA'$ we have that XZ bisects angle $\angle AXP$. Moreover, since X' is the reflection of P about XZ it's clear that XZ also bisects angle $\angle X'XP$ and so A, X, X' must be collinear as desired. □

In [7], the result from **Delta 11.1** is just a preliminary result for the following fact:

Delta 11.2. Let P be a point in the plane of a triangle ABC. Let the lines AP, BP, CP intersect the lines BC, CA, AB at the points A', B', C' respectively. Let X, Y, Z be the feet of the altitudes of triangle $A'B'C'$ issuing from A', B', C', respectively. Let X', Y', Z' be the reflections of the point P in the lines $B'C'$, $C'A'$, $A'B'$ respectively. Then, the lines AX', BY', CZ' concur.

Proof. From the result in **Delta 11.1** it suffices to show that lines AX, BY, CZ concur. However, we know that lines AA', BB', CC' concur at P and lines $A'X, B'Y, C'Z$ concur at the orthocenter of triangle $A'B'C'$ so by the Cevian Nest Theorem (**Delta 3.11**) we have that lines AX, BY, CZ concur as desired. □

The next result is a kind of "**Delta 11.1** stretched by a factor of $\frac{1}{2}$":

Delta 11.3. Let P be a point in the plane of a triangle ABC. Let the lines AP, BP, CP intersect the lines BC, CA, AB at the points A', B', C' respectively. Let X_1 be the orthogonal projection of P on the line $B'C'$. Let X be the projection of A' on the line $B'C'$. Let X_2 be the projection of the point P on the line $A'X$. Let A_1 be the midpoint of the segment AP. Then, the points A_1, X_1 and X_2 are collinear.

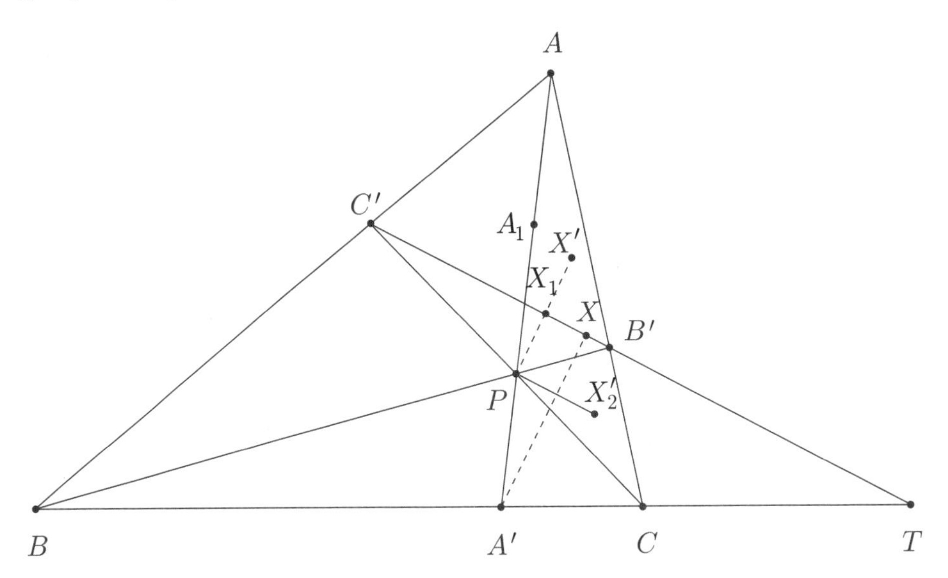

Proof. Consider the homothety centered at P with ratio 2. This homothety takes A_1 to A, X_1 to the reflection of P over $B'C'$, and X_2 to the reflection of P over $A'X$. Let X' be the reflection of P over $B'C'$ and let X_2' be the reflection of P over $A'X$. Assume without loss of generality that X is on the same side of AA' as B'. We know from the proof of **Delta 11.1** that XC' is the angle bisector of angles $\angle AXP$ and $\angle X'XP$. Hence, since $\angle X_2'XA = \angle X_2'XP + \angle AXP = 2\angle C'XP + 2\angle A'XP = 180°$, we have that points A, X, X', X_2' are all collinear. This completes the proof. □

Delta 11.3 generalizes a well-forgotten result from the early 19th Century - a result that appeared in [2], with a reference to "W. Dixon Rangeley, Gentleman's Diary, 1822, p. 47".

Delta 11.4. Let ABC be a triangle. Let the incircle of triangle ABC have center I and touch side BC at X. Let S be the foot of the A-altitude of triangle ABC. Let Q be the projection of I on line AS. Also, let P be the midpoint of the arc BC not containing A on the circumcircle of triangle ABC. Then, points P, X, and Q are collinear.

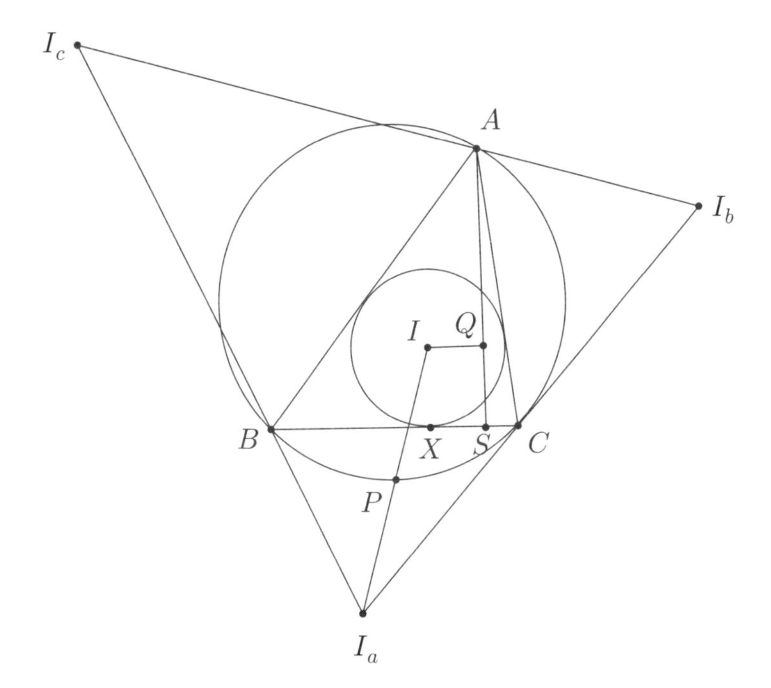

Proof. Let I_a, I_b, I_c be the AB, C-excenters of triangle ABC respectively. Note that

$$\begin{aligned}\angle PI_aB &= 180^\circ - \angle ABI_a - \angle BAP \\ &= 180^\circ - \left(90 + \frac{\angle B}{2}\right) - \frac{\angle A}{2} \\ &= \frac{\angle C}{2}\end{aligned}$$

and

$$\begin{aligned}\angle PBI_a &= \angle CBI_a - \angle CBP \\ &= \left(90 - \frac{\angle B}{2}\right) - \frac{\angle A}{2} \\ &= \frac{\angle C}{2}\end{aligned}$$

so $\angle PBI_a = \angle PI_aB$ and hence $PI_a = PB$. A similar angle chase yields that $PB = PI$ so P is the midpoint of II_a. Now, looking at the triangle $I_aI_bI_c$ and noting that triangle ABC is the cevian triangle of I with respect to triangle $I_aI_bI_c$, amazingly we see that we have a special case of the configuration in **Delta 11.3**! This completes the proof. □

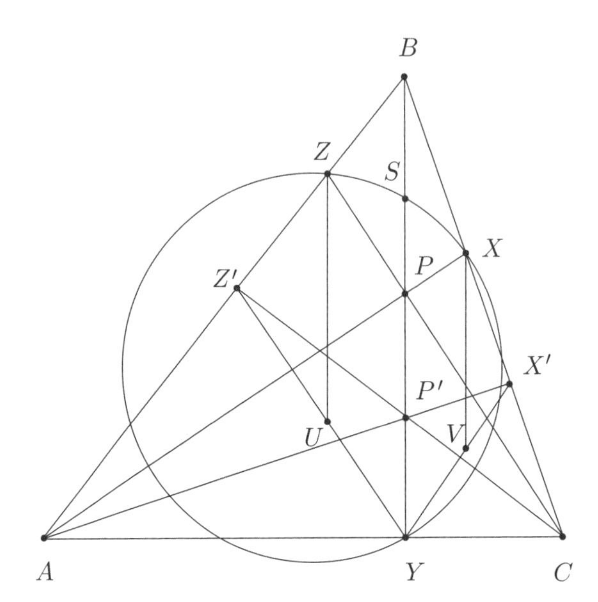

Delta 11.5. Let Y be the foot of the B-altitude of triangle ABC. Let P and P' be two points on line BY. Let $Z = CP \cap AB$, $X = AP \cap BC$, $Z' = CP' \cap AB$, and $X' = AP' \cap BC$. Let the perpendicular to the line CA through the point Z intersect the line YZ' at a point U. Let the perpendicular to the line CA through the point X intersect the line YX' at a point V. Then, the lines XU, ZV and BY concur.

Proof. We make use of Jacobi's Theorem! Assume without loss of generality we have that P is between B and P'. From Blanchet's Theorem applied twice we have that YB bisects angles $\angle XYZ$ and $\angle X'YZ'$ which implies that $\angle VYX = \angle UYZ$. Now, let the line BY intersect the circumcircle of triangle XYZ again at S. Then since $BY \parallel VX$ and BS bisects angle $\angle XYZ$ we have $\angle SXZ = \angle SYZ = \angle SYX = \angle VXY$. Similarly we find that $\angle SZX = \angle UZY$. Hence, by Jacobi's Theorem on triangle XYZ with the points U, V, S, we have that lines XU, ZV, YS concur. Since line YS coincides with line BY, this completes the proof. □

We propose a projective generalization of **Delta 11.5**:

Delta 11.6. Let ABC be a triangle, and let y be a line through the point B. Let P, P', Y, Y' be four points on the line y. Let $Z = CP \cap AB$, $X = AP \cap BC$, $Z' = CP' \cap AB$, and $X' = AP' \cap BC$. Also, let $U = ZY' \cap YZ'$ and $V = XY' \cap YX'$. Then, the lines XU, ZV, y concur.

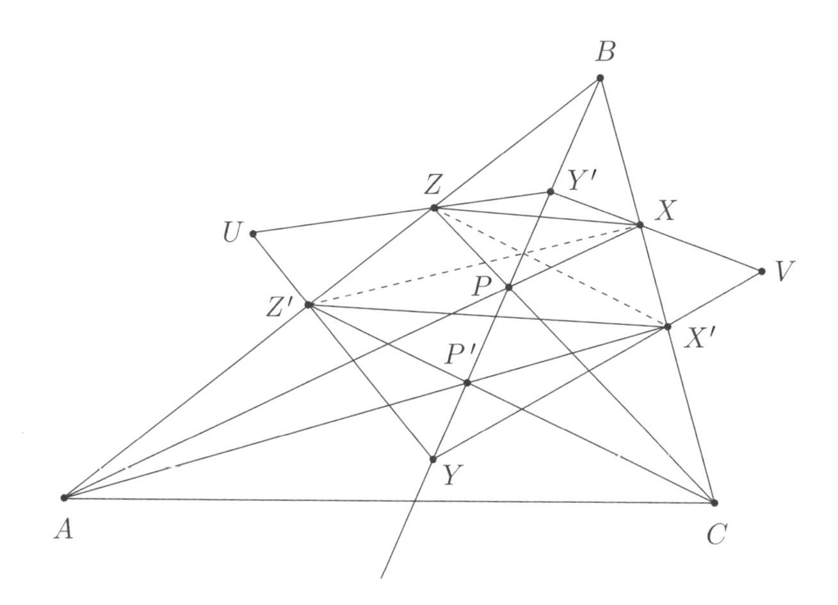

Proof. The lines YY', $Z'Z$, $X'X$ concur at B so by Desargues' Theorem (applied to triangles $YZ'X'$ and $Y'ZX$), the points $Z'X' \cap ZX, U, V$ are collinear. Also, by Pappus's Theorem on collinear points A, Z, Z' and C, X, X' we have that the point $ZX' \cap Z'X$ lies on y. Equivalently, the lines BY, ZX', XZ' concur. Then by Desargues' Theorem again (applied to triangles BZX and $YX'Z'$), points $ZX \cap X'Z'$, $XB \cap Z'Y$ and $BZ \cap YX'$ are collinear. But since we know that $ZX \cap X'Z'$ lies on line UV, we can rewrite these intersections as $ZX \cap X'Z' = ZX \cap VU$, $XB \cap Z'Y = XB \cap UY$ and $BZ \cap YX' = BZ \cap YV$. Thus we have shown that the points $ZX \cap VU$, $XB \cap UY$ and $BZ \cap YV$ are collinear. Now Desargues' Theorem one more time (applied to triangles BZX and YVU), yields that lines BY, ZV, XU concur. Since the line BY is the same as the line y, this completes the proof. □

Now, from **Delta 11.6** we can readily deduce **Delta 11.5**.

Consider the configuration of **Delta 11.5**. Let y be the line BY and let Y' be the point at infinity on line y. From $ZU \perp CA$ and $BY \perp CA$, we conclude that $ZU \parallel y$. Hence, the points Z, U, Y' are collinear. Similarly the points X, V, Y' are collinear so we have that $U = YZ' \cap Y'Z$ and $V = XY' \cap X'Y$. Essentially, we now have the following configuration; the line y passes through B. The points P, P', Y, Y' lie on the line y. We have $Z = CP \cap AB$, $X = AP \cap BC$, $Z' = CP' \cap AB$, and $X' = AP' \cap BC$. Also, $U = ZY' \cap YZ'$ and $V = XY' \cap YX'$. Hence, we can apply **Delta 11.6** to our configuration to see that lines XU, ZV and y concur. Since the line y was defined as the line BY, we conclude that the lines XU, ZV and BY concur. Thus, **Delta 11.5** is once again proven.

We conclude the section with an easy corollary of **Delta 11.5**.

Delta 11.7. Let Y be the foot of the B-altitude of triangle ABC. Let P be a point on the line BY. Let $Z = CP \cap AB$ and $X = AP \cap BC$. Let U and V be the projections of the points Z and X on line CA. Then, lines XU, ZV and BY concur.

Proof. Let $P' = Y$ in the configuration of **Delta 11.5**. Applying the result from **Delta 11.5** then completes the proof. □

Chapter 12

Poles and Polars

Definition. Let Γ be a circle and let P be a point in the plane of Γ. Furthermore, let XY be an arbitrary chord or secant passing through P with X and Y on Γ. The **polar** of the point P with respect to the circle Γ is the defined as the locus of the points Q in plane so that $(P, Q; X, Y)$ is harmonic (the chord XY passing though P is variable here). Surprisingly, this is a line (this requires proof of course) - and this line has a series of amazing properties that we shall soon see. But first, let's prove that the polar really is a line:

There are two cases - if P lies inside or outside of Γ. We handle each case separately.

If P lies outside of Γ then let PA, PB be the tangents to Γ from P. We claim that line AB is the polar of P with respect to Γ. Let XY be an arbitrary secant of Γ passing through P with X and Y on Γ and let XY intersect AB at Q - it suffices to show that $(P, Q; X, Y)$ is harmonic. We know that $AXBY$ is a harmonic quadrilateral and since $(P, Q; X, Y) \stackrel{A}{=} (A, B; X, Y) = -1$ we have the desired result.

If P lies inside Γ, let XY be an arbitrary chord of γ containing P. Let O be the center of Γ and let R be the second intersection of OP with the circumcircle of triangle XOY. Let AB be the diameter of Γ passing through P and assume without loss of generality that P lies between A and O. Let ℓ be the line through R perpendicular to OP. We claim that ℓ is fixed regardless of our choice of chord XY and that it is the polar of P with respect to Γ. Let S be the intersection if line XY and ℓ. Note that $\angle YRP = \angle YXO = \angle XYO = \angle XRP$ so line RP bisects angle $\angle XRY$. Moreover we have that $RS \perp RP$ and so $(S, P; X, Y)$ is harmonic. Now by Power of a Point we also have $PA \cdot PB = PX \cdot PY = PO \cdot PR$ and that $PA \cdot PB = OA^2 - OP^2$ so $OR \cdot OP = OP^2 + PO \cdot PR = OA^2$ and since O is the midpoint of AB this means that $(R, P; A, B)$ is harmonic. Therefore R is fixed regardless of our choice of XY and hence ℓ is fixed as well. But since $(S, P; X, Y)$ is harmonic, this implies that ℓ is indeed the polar of P with respect to Γ as desired.

Note that both constructions above show that if O is the center of Γ, then OP is perpendicular to the polar of P with respect to Γ.

//If ℓ is the polar of a point P with respect to a circle Γ, then the point P is called the **pole** of ℓ with respect to Γ.

We proceed with an incredible property of poles and polars that is at the heart of why this tool is so powerful.

Theorem 12.1. (La Hire's Theorem) Let P, Q be two points in the plane of the circle Γ. Then, P lies on the polar of Q with respect to Γ if and only if Q lies on the polar of P with respect to Γ.

Proof. It's clear that points P and Q can't both be inside Γ. If one point is outside of Γ and the other point is inside of Γ, let PQ intersect Γ at X and Y. Then since $(P, Q; X, Y)$ is harmonic if and only if $(Q, P; X, Y)$ is harmonic, we have the desired result by the definition of what a polar is. Otherwise, assume both P and Q are outside of Γ and that Q is on the polar of P. Let the tangents from P to Γ be PA and PB and let the tangents from Q to Γ be QC and QD. We know that the polar of P is AB and so Q lies on AB. Therefore quadrilateral $ACBD$ is harmonic and so P must lie on BD, which is the polar of Q. This completes the proof. $\square$

Let's see some easy applications.

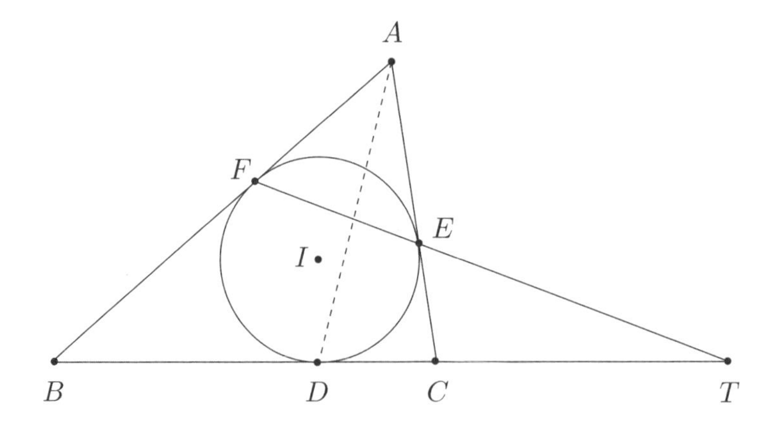

Delta 12.1. Let ABC be a triangle with incenter I and let X, Y, Z be the tangency points of the incircle with the sides BC, CA, AB respectively. Let T be the intersection of EF and BC. Prove that $TI \perp AD$.

Proof. Let ω be the incircle of triangle ABC - all poles and polars will be taken with respect to ω. It suffices to show that AD is the polar of T. Since TD is tangent to ω at D we know that D lies on the polar of T. Also, EF is the polar of A and since T lies on EF, by La Hire's Theorem we have that A lies on the polar of T. Therefore the polar of T is line AD as desired. $\square$

Corollary 12.1. (ELMO Shortlist 2012) Let ABC be a triangle with incenter I and let X, Y, Z be the tangency points of the incircle with the sides BC, CA, AB respectively. Let T be the intersection of EF and BC and let IT intersect AD at X. Then line XD bisects angle $\angle BXC$.

Proof. We know from **Delta 12.1** that $XT \perp XD$ and since lines AD, BE, CF concur at the Gergonne point of triangle ABC we have that $(B, C; D, T)$ is harmonic. Therefore XD bisects angle $\angle BXC$ as desired. $\square$

Delta 12.2. Let ABC be a triangle with incenter I and let D, E, F be the tangency points of the incircle with BC, CA, AB respectively. Prove that the lines ID and EF intersect on the A-median of triangle ABC.

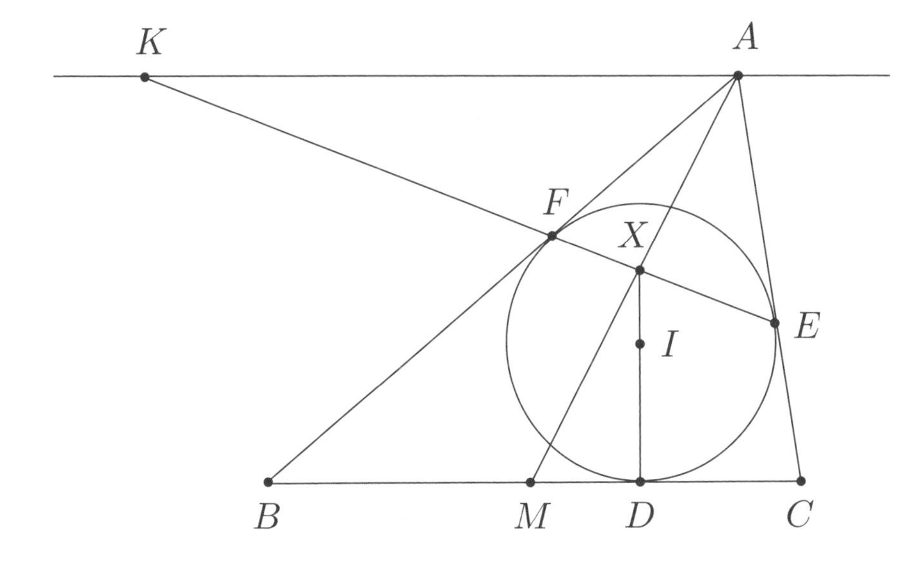

Proof. All poles and polars will be taken with respect to the incircle of triangle ABC. Let ℓ be the line passing through A parallel to BC and line EF intersect ℓ at K. Also let $X = ID \cap EF$. Line EF is the polar of A

and since X lies on EF we have by La Hire's Theorem that A lies on the polar of X. And since A lies on ℓ and $\ell \perp IX$, we must have that ℓ is the polar of X. Since K lies on ℓ by La Hire's Theorem again we must have that X lies on the polar of K. Therefore $(K, X; E, F)$ is harmonic. Now let $M = AX \cap BC$ and let P_∞ be the point at infinity on line BC. We have that $(P_\infty, M; C, B) \stackrel{A}{=} (K, X; E, F)$ so $(P_\infty, M; C, B)$ is harmonic and so M must be the midpoint of BC. This completes the proof, since X lies on AM. □

We proceed with some interesting lemmas that often come up in contests.

Theorem 12.2. (Brokard's Theorem) Let $ABCD$ be a cyclic quadrilateral whose circumcircle has center O and let $E = AC \cap BD$, $F = AB \cap CD$, and $G = AD \cap BC$. Prove that O is the orthocenter of triangle EFG.

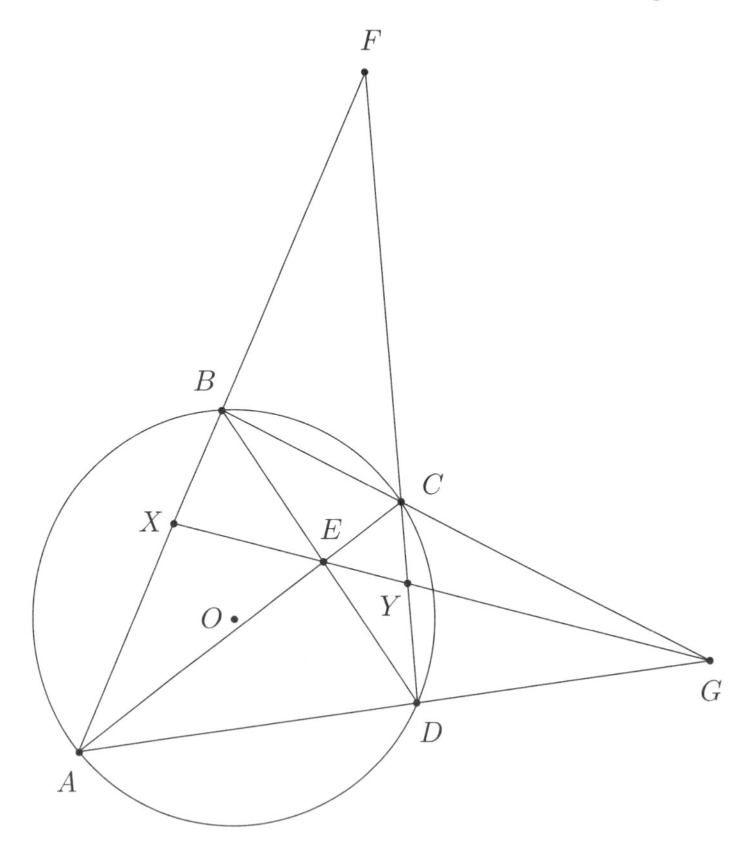

Proof. All poles and polars will be taken with respect to the circumcircle of $ABCD$. Let $X = GE \cap AB$ and $Y = GE \cap CD$. Since lines AC, BD, GX concur at E we have that $(A, B; X, F)$ is harmonic. Then since $(D, C; Y, F) \stackrel{G}{=} (A, B; X, F)$ we have that $(D, C; Y, F)$ is harmonic as well. Therefore both X and Y lie on the polar of F so EG is the polar of F. Similarly EF is the polar of G and so $FO \perp EG$ and $GO \perp EF$ so O is the orthocenter of triangle EFG as desired. □

Theorem 12.3. (Newton's Theorem) Let $ABCD$ be a quadrilateral which has an inscribed circle ω. Let M, N, P, Q be the tangency point of ω with AB, CD, DA, BC, respectively. Prove that

a) MP, NQ, BD are concurrent.

b) MN, PQ, AC, BD are concurrent.

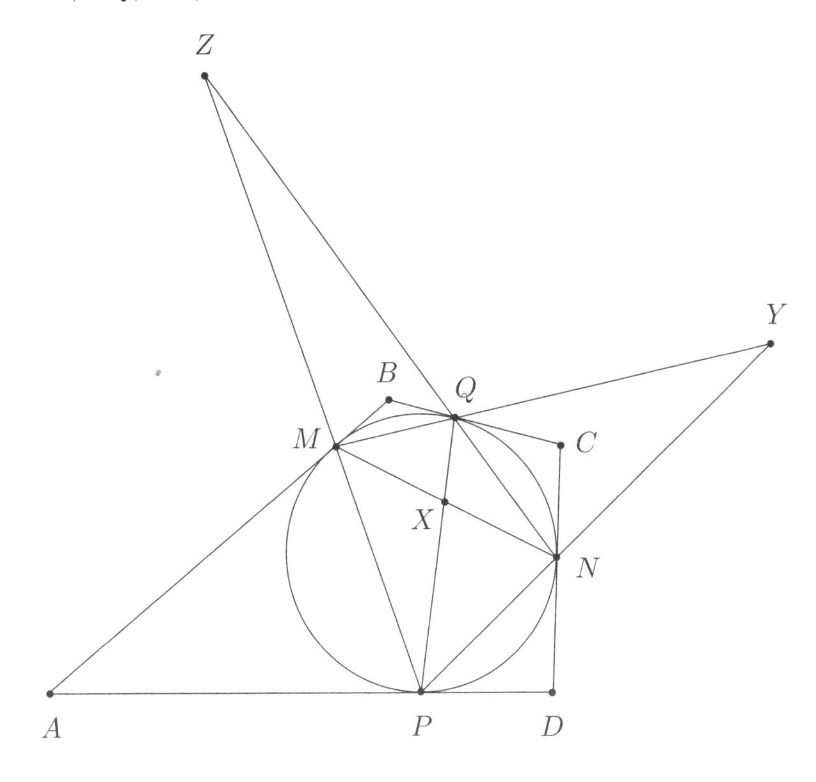

Proof. All poles and polars will be taken with respect to ω. Let $X = MN \cap PQ$, $Y = MQ \cap NP$, and $Z = MP \cap NQ$. We know from Brokard's Theorem that XZ is the polar of Y. Since MQ is the polar of B and Y lies on MQ, by La Hire's Theorem we have that B lies on XZ. Similarly D lies on XZ so lines MP, NQ, BD concur at Z. We also have that X lies on BD, and similarly X lies on AC so lines MN, PQ, AC, BD concur at X. This completes the proof. □

Delta 12.3. Let $ABCD$ be quadrilateral with an inscribed circle ω with center I and a circumscribed circle Ω with center O. Let $E = AC \cap BD$. Prove that points O, I, E are collinear.

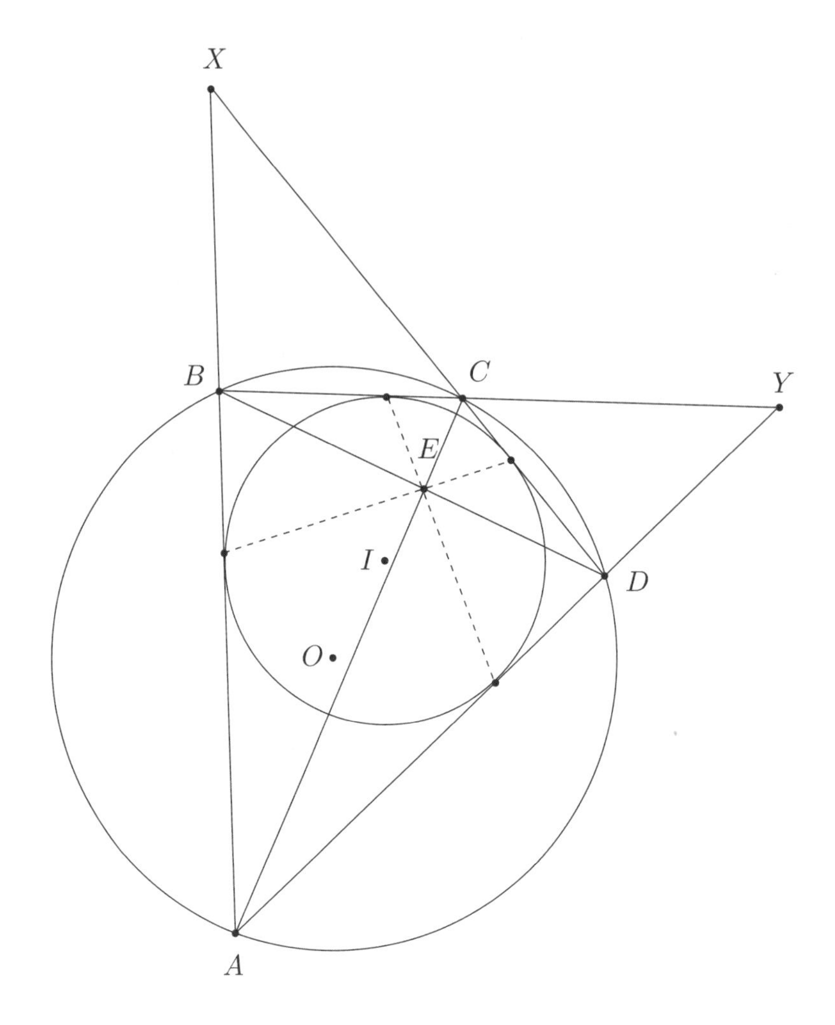

Proof. Let $X = AB \cap CD$ and $Y = DA \cap BC$. By Brokard's Theorem we have that $OE \perp XY$. Now note that the polars of X and Y with respect to ω both pass through E by Newton's Theorem, so XY is the polar of E with respect to ω. Therefore $IE \perp XY$ and so points O, I, E lie on a line perpendicular to XY. This completes the proof. □

Delta 12.4. Let $ABCD$ be a quadrilateral, which has an inscribed circle ω with center O. Let H be the projection of O onto line BD. Prove that $\angle AHB = \angle CHB$.

Proof. All poles and polars will be taken with respect to ω. Let ω be tangent to segments AB, BC, CD, DA at A', B', C', D' respectively. Let $X = A'C' \cap B'D'$, $Y = A'B' \cap C'D$, and $Z = D'A' \cap B'C'$. By Newton's Theorem applied twice we have that points A, C, X, Y are collinear and points

B, D, X, Z are collinear. Let $P = BD \cap A'B'$. Since by Brokard's Theorem XZ is the polar of Y we have that $(Y, P; A', B')$ is harmonic. And since $(Y, X; A, C) \stackrel{B}{=} (Y, P; A', B')$ we have that $(Y, X; A, C)$ is harmonic as well.

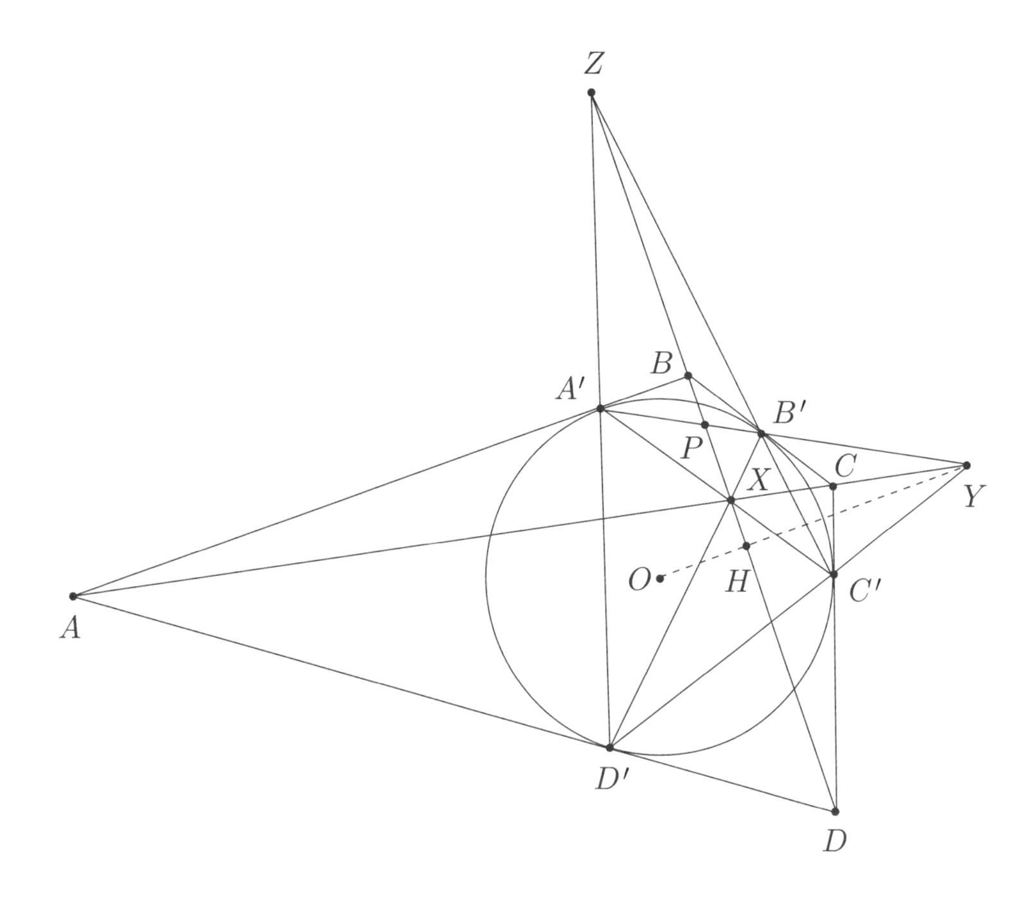

Furthermore, $YO \perp BD$ by Brokard's Theorem and $HO \perp BD$ as well so $HY \perp HX$. Since $(Y, X; A, C)$ is harmonic this implies HX bisects angle $\angle AHC$. Since line HX coincides with line HB, the proof is complete. □

Theorem 12.4. (Brianchon's Theorem) Let $ABCDEF$ be a hexagon with an inscribed circle ω. Then lines AD, BE, CF concur.

Proof. All poles and polars will be taken with respect to ω. Let ω be tangent to segments AB, BC, CD, DE, EF, FA at A', B', C', D', E', F' respectively. Let $X = A'B' \cap D'E'$, $Y = B'C' \cap E'F'$, and $Z = C'D' \cap F'A'$.

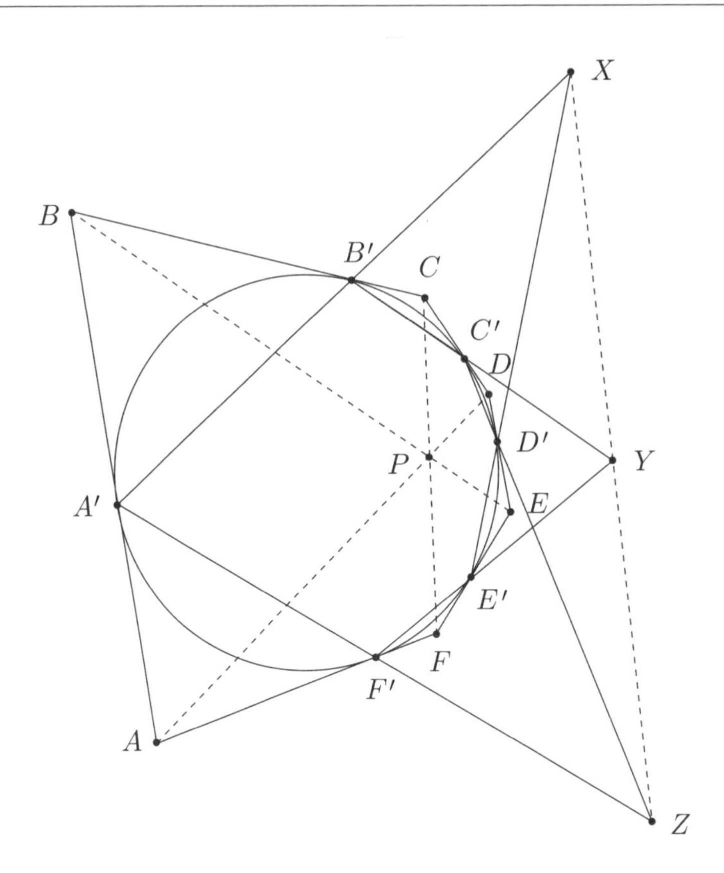

By Pascal's Theorem we know that points X, Y, Z are collinear. Let P be the pole of the line determined by points X, Y, Z. Since $A'B'$ is the polar of B and $D'E'$ is the polar of E by La Hire's Theorem we have that points B and E lie on the polar of X, so line BE is the polar of X. By La Hire's Theorem again, this implies that P lies on BE. Similarly P lies on AD and CF so we are done. □

//Brianchon's Theorem is what's known as the **projective dual** of Pascal's Theorem. Consider the two statements: every two distinct lines determine a point (possibly at infinity) and every two distinct points determine a line - these are projective duals, and switching the two statements in a configuration one obtains its projective dual. Another way of obtaining the projective dual of a configuration is by switching every pole with its polar and vice-versa.

Corollary 12.2. Let the incircle of triangle ABC touch sides BC, CA, AB at D, E, F respectively. Prove that lines AD, BE, CF concur.

Proof. Apply Brianchon's Theorem to degenerate hexagon $AFBDCE$. □

Theorem 12.5. (Salmon's Theorem) Let ω be a circle with center O and let P and Q be points in the plane of ω. Let ℓ_P and ℓ_Q be the polars of P and Q with respect to ω. Then

$$\frac{\delta(P, \ell_Q)}{\delta(Q, \ell_P)} = \frac{OP}{OQ}$$

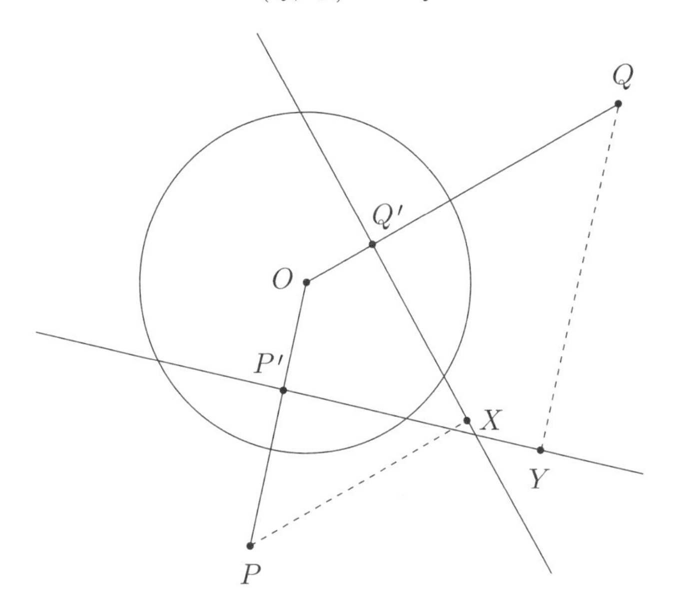

Proof. Let $P' = OP \cap \ell_P$ and $Q' = OQ \cap \ell_Q$. Also let X, Y be the projections from P, Q to ℓ_Q, ℓ_P respectively. If R is the radius of ω it's easy to derive that $OP \cdot OP' = OQ \cdot OQ' = R^2$ and so

$$\frac{OP}{OQ} = \frac{OQ'}{OP'}$$

This implies that quadrilaterals $OPXQ'$ and $OQYP'$ are similar, which implies the desired result. □

Delta 12.5. (Hartcourt's Theorem) Let ABC be a triangle with incircle ω. Let ℓ be a line tangent to ω at P, and let X, Y, Z be the projections of A, B, C respectively onto ℓ. Assume without loss of generality that B and C are on the same side of ℓ. Then if $a = BC, b = CA, c = AB, x = AX, y = BY, z = CZ$ we have that $by + cz - ax = 2[ABC]$

Proof. All poles and polars will be taken with respect to ω. Let I and r be the center and radius respectively of ω. Let R be the circumradius of triangle ABC. Let ω touch BC, CA, AB at D, E, F respectively. Since ℓ is the polar of P and since EF, FD, DE are the polars of A, B, C respectively, three applications of Salmon's Theorem on P and points A, B, C yield

$$\frac{IP}{IA} = \frac{\delta(P, EF)}{AX} \Longrightarrow ax = \frac{a \cdot IA \cdot \delta(P, EF)}{r}$$
$$\frac{IP}{IB} = \frac{\delta(P, FD)}{BY} \Longrightarrow by = \frac{b \cdot IB \cdot \delta(P, FD)}{r}$$
$$\frac{IP}{IC} = \frac{\delta(P, DE)}{CZ} \Longrightarrow cz = \frac{c \cdot IC \cdot \delta(P, DE)}{r}$$

So we can calculate:

$$\begin{aligned} by + cz - ax &= \frac{b \cdot IB \cdot \delta(P, FD)}{r} + \frac{c \cdot IC \cdot \delta(P, DE)}{r} - \frac{a \cdot IA \cdot \delta(P, EF)}{r} \\ &= \frac{2R \cdot IB \sin B \cdot \delta(P, FD)}{r} + \frac{2R \cdot IC \sin C \cdot \delta(P, DE)}{r} \\ &\quad - \frac{2R \cdot IA \sin A \cdot \delta(P, EF)}{r} \\ &= \frac{2R \cdot FD \cdot \delta(P, FD)}{r} + \frac{2R \cdot DE \cdot \delta(P, DE)}{r} \\ &\quad - \frac{2R \cdot EF \cdot \delta(P, EF)}{r} \\ &= \frac{4R}{r}([PFD] + [PDE] - [PEF]) \\ &= \frac{4R}{r}[DEF] \end{aligned}$$

So it suffices to show that $\frac{[DEF]}{[ABC]} = \frac{r}{2R}$ but this follows from Euler's Pedal Triangle Theorem and the fact that $R^2 - OI^2 = 2Rr$ where O is the circumcenter of triangle ABC. Hence, the proof is complete. □

Delta 12.6. Let $ABCD$ be a quadrilateral with an inscribed circle ω. Let O be the center of ω and let ℓ be a line tangent to ω. Let A', B', C', D' be the projections of A, B, C, D respectively onto ℓ. Then

$$\frac{AO \cdot CO}{BO \cdot DO} = \frac{AA' \cdot CC'}{BB' \cdot DD'}.$$

Proof. All poles and polars will be taken with respect to ω. Let ℓ be tangent to ω at K and let ω touch DA, AB, BC, CD at M, N, P, Q respectively. Let X, Y, Z, U be the projections of K onto lines MN, NP, PQ, QM respectively. Since ℓ is the polar of K and MN, NP, PQ, QM are the polars of A, B, C, D respectively, four applications of Salmon's Theorem on K and points A, B, C, D yield

$$\frac{AA'}{AO} = \frac{KX}{r}$$
$$\frac{BB'}{BO} = \frac{KY}{r}$$
$$\frac{CC'}{CO} = \frac{KZ}{r}$$
$$\frac{DD'}{DO} = \frac{KU}{r}$$

So it suffices to show that $KX \cdot KZ = KY \cdot KU$. But it's easy to see that $KX = KN \sin KNM$ and $KY = KN \sin KNP$ and $KZ = KQ \sin KQP$ and $KU = KQ \sin KQM$ and since $\angle KNM = \angle KQM$ and $\angle KNP = \angle KQP$ by multiplying we have that $KX \cdot KZ = KY \cdot KU$ as desired. This completes the proof. □

We proceed by destroying some Olympiad problems with these powerful tools!

Delta 12.7. (Iran TST 2002) Let ABC be a triangle. Its incircle touches the side BC at A' and the line AA' meets the incircle again at a point P. Let the lines CP and BP meet the incircle of triangle ABC again at N and M, respectively. Prove that the lines AA', BN and CM are concurrent.

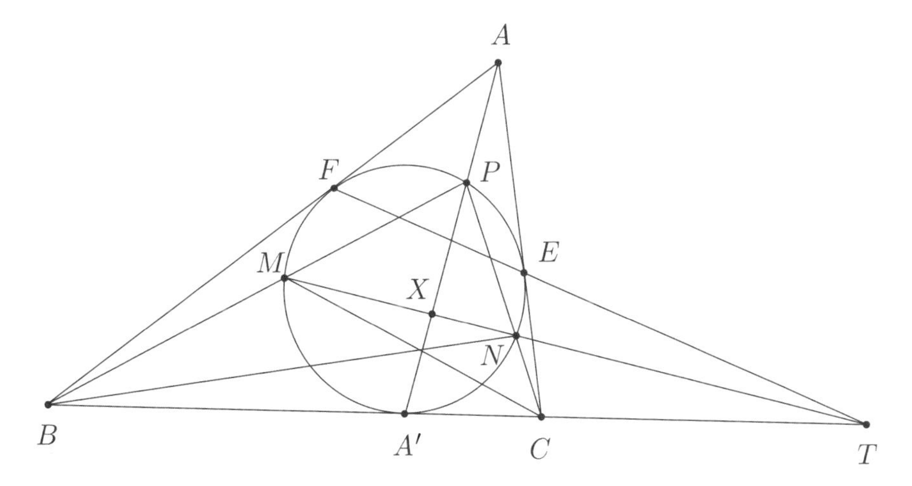

Proof. All poles and polars will be taken with respect to the incircle of triangle ABC. Let N' be the point on CP such that lines AA', BN', CM concur. Let $X = AA' \cap MN'$. Let the incircle of triangle ABC touch sides

CA, AB at E, F respectively and let $T = EF \cap BC$ and $T' = MN' \cap BC$. Since lines AA', BE, CF concur at the Gergonne point of triangle ABC we have that $(B, C; A', T)$ is harmonic. Moreover, since lines PA', BN', CM concur we have that $(B, C; A', T')$ is harmonic so $T = T'$. Since $(M, N'; X, T) \stackrel{P}{=} (B, C; A', T)$ this means that $(M, N'; X, T)$ is harmonic as well. Now since TA' is tangent to the incircle of ABC at A' we have that A' lies on the polar of T. Also EF is the polar of A and since T lies on EF we have that A lies on the polar of T so line AA' is the polar of T. But since $(M, N'; X, T)$ is harmonic this means that N' must lie on the incircle of triangle ABC. Therefore $N = N'$, which completes the proof. □

Delta 12.8. Let P be a point in the interior of triangle ABC and let the line through P perpendicular to PA intersect BC at point A_1. Define points B_1 and C_1 similarly. Prove that points A_1, B_1, C_1 are collinear.

Proof. Consider an arbitrary circle ω centered at P. All poles and polars will be taken with respect to ω. Let lines a, b, c, a_1, b_1, c_1 be the polars of points A, B, C, A_1, B_1, C_1 respectively. Since A_1 lies on line BC we have that the intersection $b \cap c$ lies on a_1. Moreover, Since $AP \perp a$ and $A_1P \perp a_1$ and $AP \perp A_1P$ we have that $a_1 \perp a$. This means that a_1 is an altitude of the triangle formed by lines a, b, c and similarly b_1 and c_1 are also altitudes of this triangle. Therefore lines a_1, b_1, c_1 concur at the orthocenter of the triangle formed by lines a, b, c and so their poles must be collinear. Hence, points A_1, B_1, C_1 are collinear as desired. □

We can say much more about this configuration. The following unexpected result was found by Luis Gonzalez and is left as an exercise for the die-hards:

Delta 12.9. Using the notation of **Delta 12.8**, let A', B', C' be the intersections of lines AP, BP, CP with BC, CA, AB respectively. Also let X, Y, Z be the projections of P onto lines $B'C', C'A', A'B'$ respectively. Prove that the line determined by points A_1, B_1, C_1 is the polar of P with respect to the circumcircle of triangle XYZ.

We end the section with a beautiful result given as the final problem in the 2012 Romanian Masters in Mathematics competition.

Delta 12.10. (RMM 2012) Let ABC be a triangle and let I and O denote its incenter and circumcenter respectively. Let ω_A be the circle through B and

C which is tangent to the incircle of the triangle ABC; the circles ω_B and ω_C are defined similarly. The circles ω_B and ω_C meet at a point A' distinct from A; the points B' and C' are defined similarly. Prove that the lines AA', BB' and CC' are concurrent at a point on the line IO.

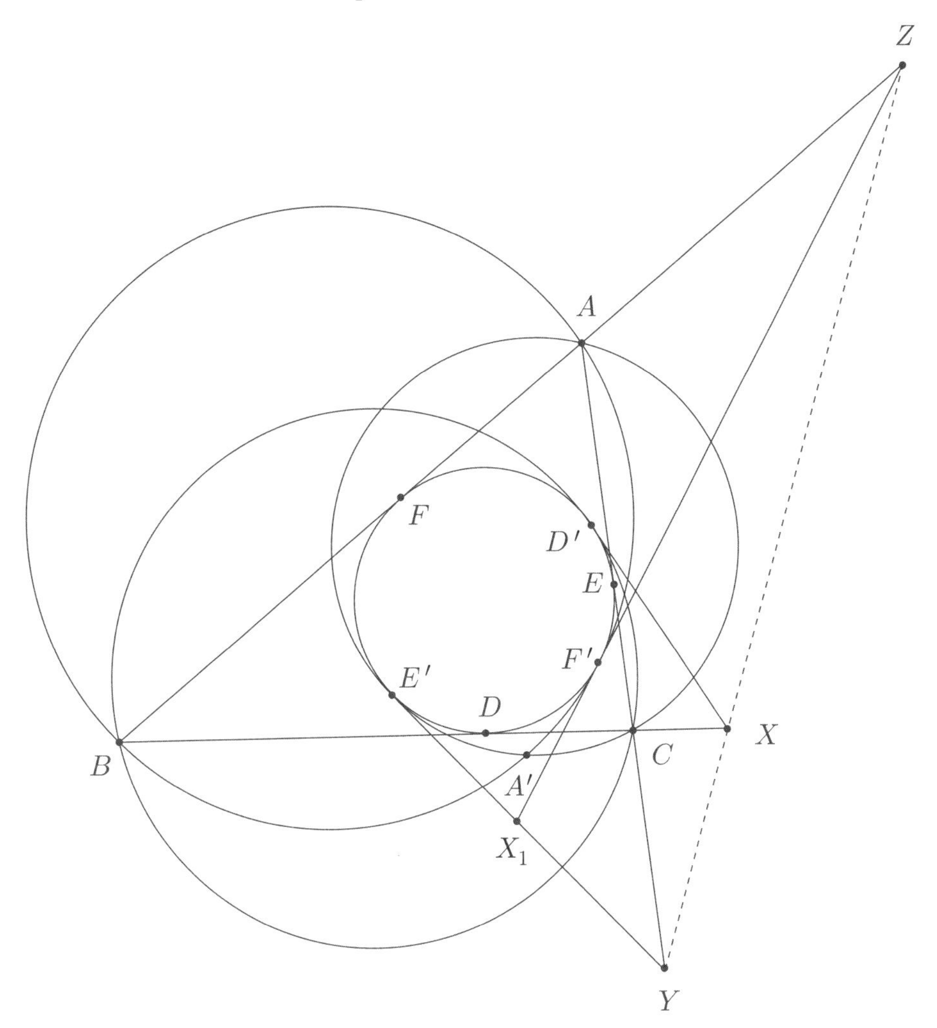

Proof. All poles and polars will be taken with respect to ω, the incircle of triangle ABC. Also, the line tangent to ω at a point P will be denoted by "line PP". Let ω touch BC, CA, AB at D, E, F respectively. Let ω touch $\omega_a, \omega_b, \omega_c$ at D', E', F' respectively. Now let lines $D'D', E'E', F'F'$ meet BC, CA, AB at X, Y, Z respectively. It's clear that X is the radical center of ω, ω_a, and Ω where Ω is the circumcircle of triangle ABC. Therefore X lies on the radical axis of ω and Ω and similarly Y and Z lie on this radical axis. Thus, some simple applications of Brokard's Theorem and La Hire's Theorem yield that the perspectrix of triangles DEF and $D'E'F'$ is the line determined by points

X, Y, Z - namely, the radical axis of ω and Ω. Now, the lines $AA', E'E', F'F'$ are concurrent at X_1, the radical center of $\omega, \omega_b, \omega_c$. Let a, b, c be the polars of A', B', C' respectively. Since A, X_1, A' are collinear, their polars, namely lines $EF, E'F', a$ are concurrent. Similarly lines $FD, F'D', b$ are concurrent and lines $DE, D'E', c$ are concurrent. Thus, triangle DEF and the triangle formed by lines a, b, c are perspective and have the same perspectrix as that of triangle DEF and triangle $D'E'F'$, which we know to be the radical axis of ω and Ω. This can be rewritten as follows: The triangle formed by the polars of A', B', C' and the triangle formed by the polars of A, B, C are perspective and have the radical axis of ω and Ω as their perspectrix. Therefore, taking the projective dual of the configuration, lines AA', BB', CC' concur at the pole of the radical axis of ω and Ω which clearly lies on line OI, as desired. This completes the proof. $\square$

Assigned Problems

Epsilon 12.1. Let ω be a semicircle with diameter CD and center O. Let E, F be two arbitrary points on ω and let the tangent lines at E, F to ω intersect each other at Q. Also let lines ED and FC intersect at P. Prove that $PQ \perp CD$.

Epsilon 12.2. (China 2006) Let AB be the diameter of a circle Γ with center O. C is a point on AB such that B is between A and C and a line through C intersects Γ at points D and E. Let F be a point such that OF is a diameter of the circumcircle of triangle BOD and let CF intersect the circumcircle of triangle BOD again at G. Prove that points O, E, A, G are concyclic.

Epsilon 12.3. Let ABC be a triangle and let A_1, B_1, C_1 be the feet of the altitudes from A, B, C, respectively. Let H be the orthocenter of ABC and let M be the midpoint of the side BC. Furthermore, let MH meet the line B_1C_1 at T and let the tangents at B and C with respect to the circumcircle of ABC meet at P. Prove that the points P, A_1, T are collinear.

Epsilon 12.4. Let ω be incircle of ABC. P and Q are on AB and AC, such that PQ is parallel to BC and is tangent to ω. AB, AC touch ω at F, E. Prove that if M is midpoint of PQ, and T is intersection point of EF and BC, then TM is tangent to ω.

Epsilon 12.5. Let ABC be a triangle and let ω be a circle which intersects side BC at points A_1 and A_2, side CA at B_1 and B_2, and side AB at C_1 and C_2. The tangents to ω at A_1 and A_2 intersect at X, and Y and Z are defined similarly. Prove that lines AX, BY, CZ concur.

Epsilon 12.6. Let triangle ABC have incircle ω with center I. Let M, N be the midpoints of segments CA, AB respectively. Prove that line MN is the polar of the orthocenter of triangle BIC with respect to ω.

Epsilon 12.7. Let A_1, B_1, C_1 be the feet of the A, B, C-altitudes respectively in acute-angled triangle ABC. A circle passes through B_1 and C_1 and touches minor arc BC of the circumcircle of triangle ABC at a point A_2. Points B_2 and C_2 are defined similarly. Prove that lines A_1A_2, B_1B_2, C_1C_2 concur on the Euler line of triangle ABC.

Chapter 13

Appendix B: An Incircle Related Perpendicularity

The following problem appears as a lemma in numerous other contests problems, so we decided that it deserves a small section of its own. You should think of this as an Appendix to Chapter 12.

Theorem 13.1. Let triangle ABC have incenter I, and let the incircle of triangle ABC touch sides BC, CA, AB at points D, E, F respectively. Let $P = BI \cap EF$. Then, lines PB and PC are perpendicular.

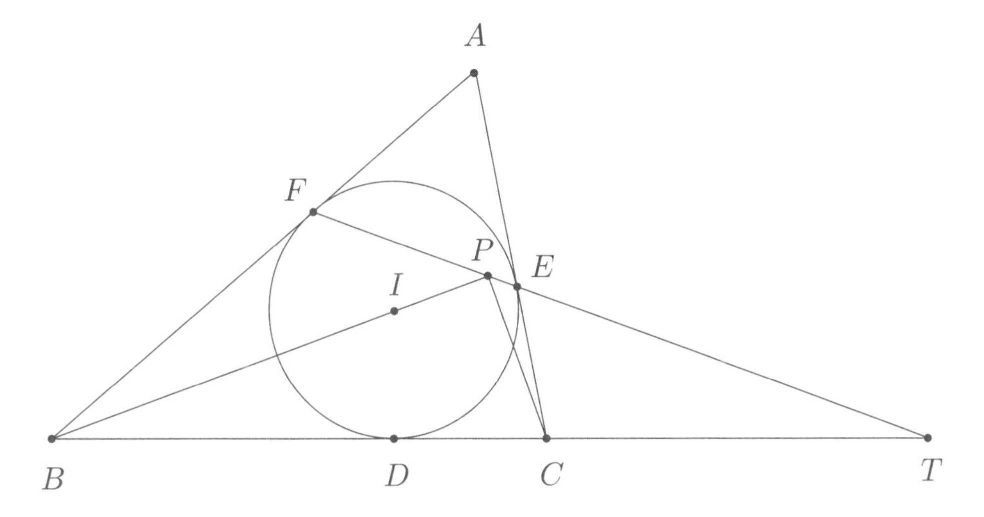

Proof. Let $T = EF \cap BC$. Since lines AD, BE, CF concur at the Gergonne point of triangle ABC we have that $(B, C; D, T)$ is harmonic. Now since quadrilateral $BFPD$ is a kite we have that $\angle PDB = \angle PFB = 180^\circ - \angle AFE = 180^\circ - \angle AEF = \angle CEP$ so quadrilateral $PECD$ is cyclic.

But since $CD = CE$ this means that line PC bisects angle $\angle DPT$ so since $(B, C; D, T)$ is harmonic we have $PB \perp PC$ as desired. □

We can use this result to derive another common configuration:

Theorem 13.2. Let triangle ABC have incenter I, and let the incircle of triangle ABC touch sides BC, CA, AB at points D, E, F respectively, Let M, N be the midpoints of sides BC, CA respectively. Prove that lines EF, MN, BI concur.

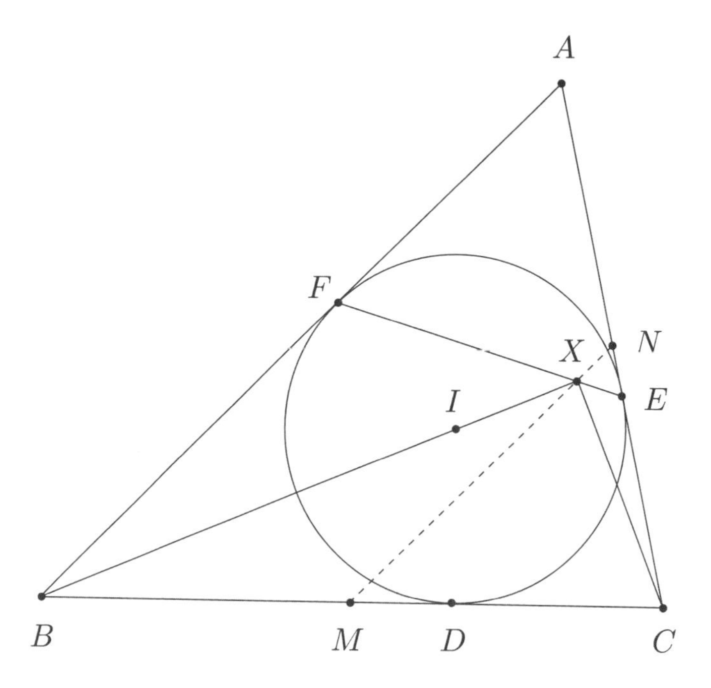

Proof. Let $X = BI \cap EF$. We know from **Theorem 13.1** that $\angle BXC = 90^\circ$ so M is the circumcenter of triangle BXC. Therefore $\angle XMC = 2\angle XBC = \angle ABC$ so $MX \parallel BC$ so X lies on the C-midline of triangle ABC as desired. □

The following two results are also incircle-related perpendicularities that involve medians and midlines so we include them below. However, they are not nearly as useful as **Theorem 13.1** or **Theorem 13.2**.

Delta 13.1. Let triangle ABC have incenter I, and let the incircle of triangle ABC touch sides BC, CA, AB at points D, E, F respectively. Let M, N be the midpoints of sides CA, AB respectively and let $X = BI \cap MN$ and $Y = CI \cap MN$. Prove that points A, E, F, X, Y lie on the same circle.

Proof. It's clear that points A, E, F all lie on the circle with diameter AI. Since $MN \parallel BC$ and since BI bisects angle $\angle ABC$ we have that $\angle NXB = \angle XBC = \angle FBX$ so $BN = NX = AN$.

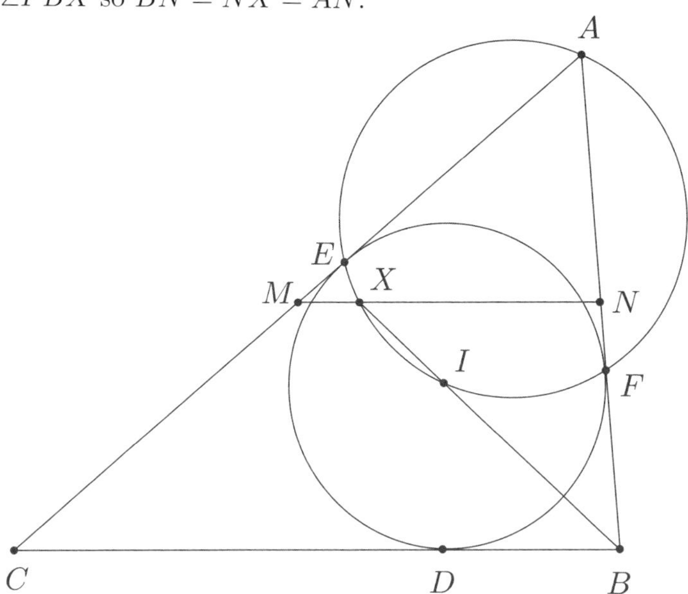

Consequently, $\angle AXI = 90°$ so X lies on the circle with diameter AI as well. Similarly, Y lies on this circle and hence the proof is complete. □

Delta 13.2. (A reminder of **Delta 12.2**) Let ABC be a triangle with incenter I and let D, E, F be the tangency points of the incircle with BC, CA, AB respectively. Prove that the lines ID and EF intersect on the A-median of triangle ABC.

Now, let's get to some Olympiad problems!

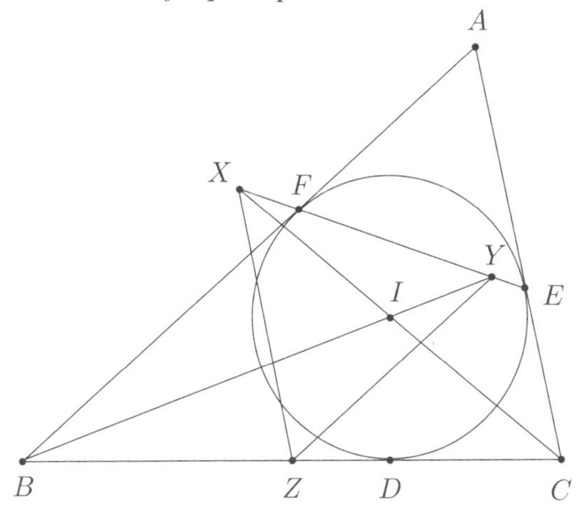

Delta 13.3. (BMO 2005) Let ABC be an acute-angled triangle whose inscribed circle touches AB and AC at D and E respectively. Let X and Y be the points of intersection of the bisectors of the angles $\angle ACB$ and $\angle ABC$ with the line DE and let Z be the midpoint of BC. Prove that the triangle XYZ is equilateral if and only if $\angle BAC = 60°$.

Proof. By **Theorem 13.1**, we know that $BY \perp CY$ and $BX \perp CX$; thus we have that $ZX = ZY$ since Z is the circumcenter of quadrilateral $BCYX$. Moreover, by **Theorem 13.2**, we have that lines YZ and XZ are the C and B-midlines respectively of triangle ABC; hence, $\angle YZX = \angle BAC$. Therefore we get that XYZ is equilateral if and only if $\angle YZX = \angle BAC = 60°$. □

Delta 13.4. (Peruvian TST 2007) Let P be an interior point of the semicircle whose diameter is AB. The incircle of triangle ABP touches AP and BP at M and N respectively. The line MN intersects the semicircle at X and Y. Prove that $\widehat{XY} = \angle APB$.

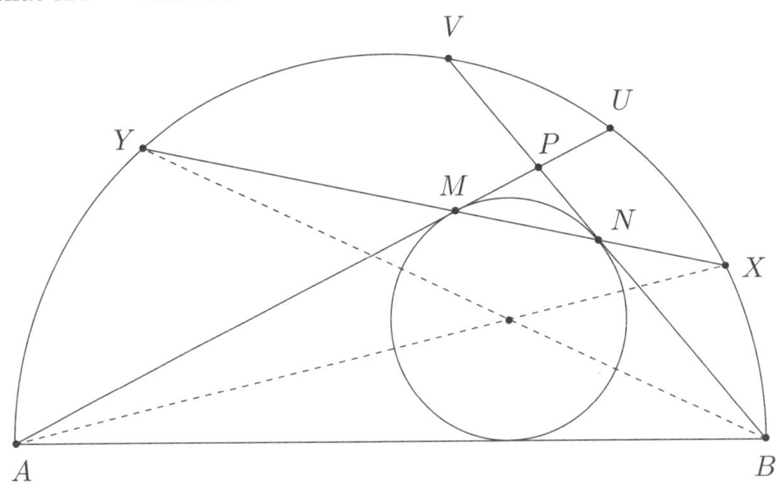

Proof. Since X, Y are the intersections of MN with the semicircle, we know that $AX \perp BX$ and $AY \perp BY$; thus, from **Theorem 13.1** it follows that AX and BY are the internal angle bisectors of angles $\angle PAB$ and $\angle PBA$ respectively. Now, let the lines AP and BP intersect semicircle again at U and V respectively. Since X, Y are the midpoints of the arcs BU and AV, it follows that

$$2\widehat{XY} = 2\widehat{UV} + \widehat{AY} + \widehat{YV} + \widehat{UX} + \widehat{XB} = 180° + \widehat{UV} = 2\angle APB$$

as desired. □

It is now time for some trickier applications. The following problem was posted by Virgil Nicula on the Art of Problem Solving Forum in 2008, where it didn't receive any solutions. We give a simple solution below.

Delta 13.5. The incircle of triangle ABC touches sides BC, CA, AB at points D, E, and F respectively. Let I be the incenter of triangle ABC. Let $X = CI \cap DE$ and let Y be the point on EF for which $IY \perp IC$. Prove that if Z is the intersection of XY with the line ID, then $CZ \perp BI$.

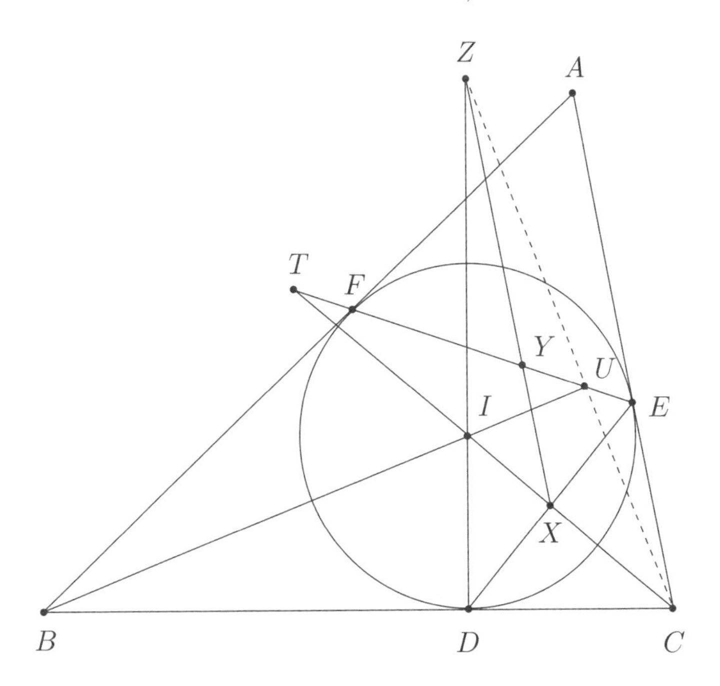

Proof. Let T and U be the intersections of EF with CI and BI respectively. By **Theorem 13.1**, we have that $TB \perp TC$ and $UB \perp UC$. Since lines DX, IY, BT are all perpendicular to CI, they concur at a point at infinity. Hence, triangles BID and TYX are perspective. Therefore, by Desargues' Theorem, the points U, Z, C are collinear and since $CU \perp BI$, it follows that $CZ \perp BI$ as desired. □

Delta 13.6. (IMO 2004 Shortlist) For a given triangle ABC, let X be a variable point on the line BC such that C lies between B and X and the incircles of the triangles ABX and ACX intersect at two distinct points P and Q. Prove that the line PQ passes through a point independent of X.

Proof. Let the incircles of triangles ABX and ACX touch BX at D and F, and AX at E and G, respectively. Clearly, $DE \| FG$. If the line PQ intersects BX at M and AX at N, then $MD^2 = MP \cdot MQ = MF^2$, i.e., $MD = MF$ and analogously $NE = NG$. It follows that line PQ lies directly in the middle of lines DE and FG.

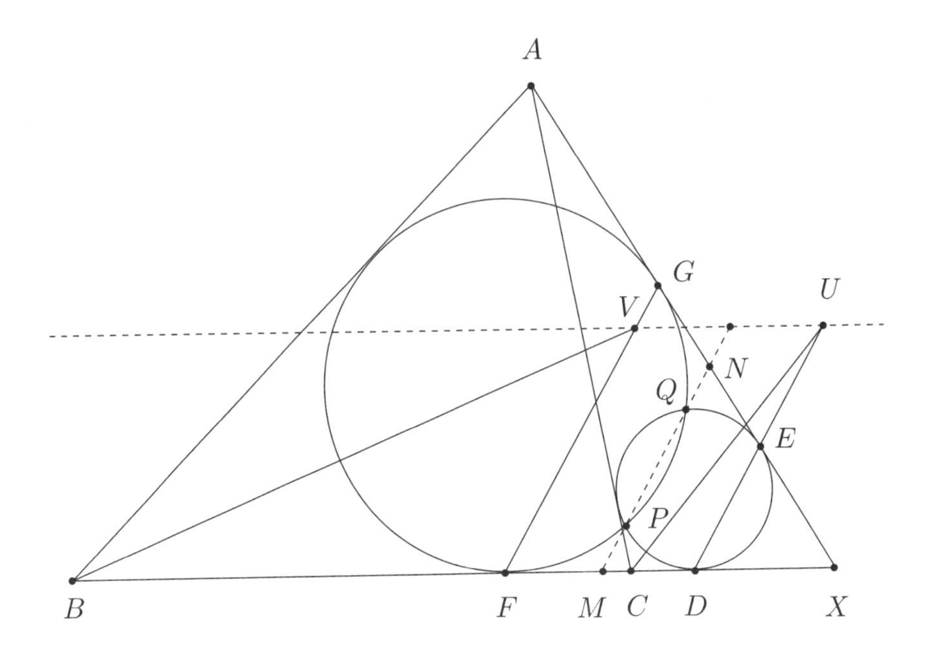

Now, let ℓ be the line passing through the midpoints of segments AB, AC, and AX - note that ℓ is fixed regardless of our choice of X. Let $U = DE \cap \ell$ and $V = FG \cap \ell$. By **Theorem 13.2**, U lies on the internal angle bisector of angle $\angle ABC$ and V lies on the external angle bisector of angle $\angle ACB$. Therefore points U and V are fixed regardless of our choice of X. Thus, the midpoint of segment UV is fixed regardless of our choice of X, and since line PQ lies directly in the middle of lines DE and FG, this midpoint lies on PQ. This completes the proof. □

Delta 13.7. (Romania TST 2007) Let ABC be a triangle and its incircle ω touch sides BC, CA, AB at D, E, F respectively. Let I be the center of ω and let M be the midpoint of BC. Let $N = AM \cap EF$ and let Γ be the circle with diameter BC. Let X, Y be the second intersections of lines BI, CI respectively with Γ. Prove that

$$\frac{NX}{NY} = \frac{AC}{AB}.$$

Proof. We have that $\angle BXC = \angle BYC = 90^\circ$ so by **Theorem 13.1** we can conclude that X and Y lie on line EF. Also, **Delta 13.2** guarantees that N lies on line DI. Now we also have that $\angle XIN = \angle BID = 90^\circ - \frac{\angle B}{2}$ and similarly $\angle YIN = 90^\circ - \frac{\angle C}{2}$. Also by Power of a Point we have that $IX \cdot IB = IY \cdot IC$.

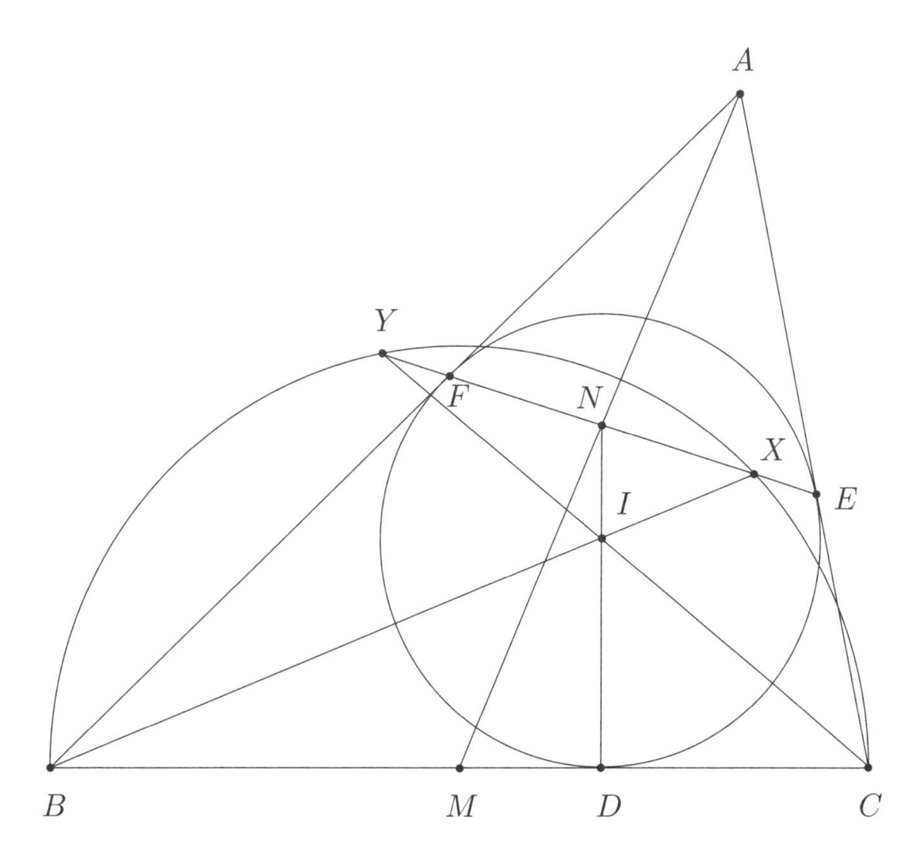

Putting everything together, we use the Ratio Lemma to write

$$\frac{NX}{NY} = \frac{IX}{IY} \cdot \frac{\sin XIN}{\sin YIN} = \frac{IC}{IB} \cdot \frac{\cos \frac{B}{2}}{\cos \frac{C}{2}} = \frac{\sin \frac{B}{2}}{\sin \frac{C}{2}} \cdot \frac{\cos \frac{B}{2}}{\cos \frac{C}{2}} = \frac{\sin B}{\sin C} = \frac{AC}{AB}$$

as desired. □

Delta 13.8. (USAJMO 2014) Let ABC be a triangle with incenter I, incircle γ and circumcircle Γ. Let M, N, P be the midpoints of sides BC, CA, AB respectively and let E, F be the tangency points of γ with CA and AB, respectively. Let U, V be the intersections of line EF with line MN and line MP, respectively, and let X be the midpoint of arc BAC of Γ.

(a) Prove that I lies on ray CV.

(b) Prove that line XI bisects segment UV.

Proof. Part (a) is an immediate consequence of **Theorem 13.2**. Now, for part (b), let Y be the midpoint of arc BC not containing A of Γ. We know that $YB = YI = YC$ so Y is the circumcenter of triangle BIC.

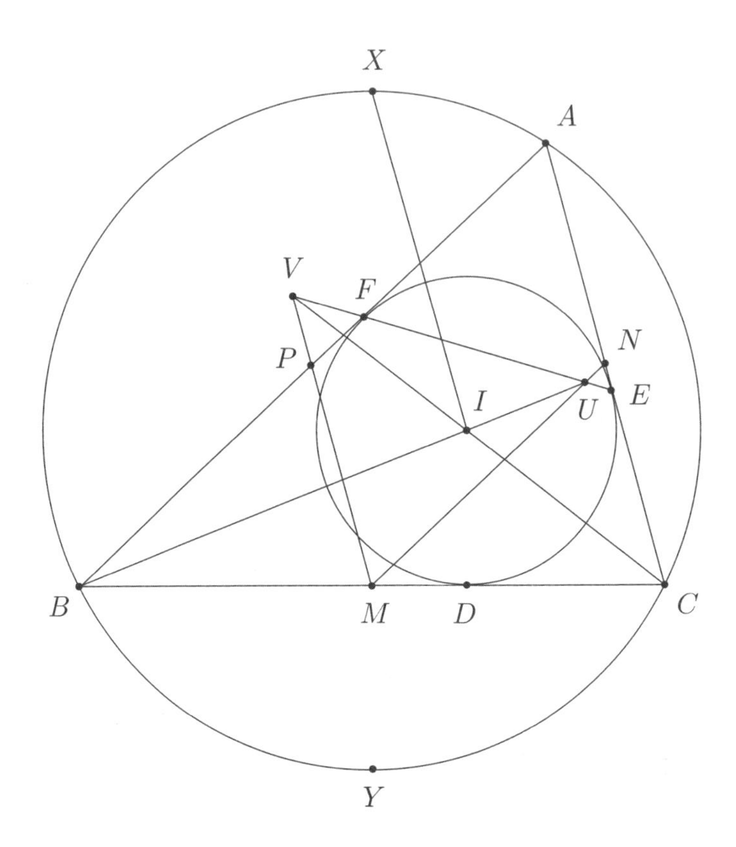

Moreover, since XY is a diameter of Γ we have that $\angle XBY = \angle XCY = 90^\circ$ so X is the intersection of the tangents to the circumcircle of triangle BIC at B and C. Therefore line IX is the I-symmedian of triangle BIC. Now from **Theorem 13.1** we know that points B, C, U, V all lie on the circle with diameter BC so UV is an anti-parallel to BC with respect to triangle BIC. Therefore IX bisects segment UV as desired. □

Delta 13.9. (USA TST 2015) Let ABC be a triangle with incenter I whose incircle is tangent to sides BC, CA, AB at D, E, F, respectively. Denote by M the midpoint of BC. Let Q be a point on the incircle such that $\angle AQD = 90^\circ$. Let P be the point inside the triangle on line AI for which $MD = MP$. Prove that if $AC > AB$ then $\angle PQE = 90^\circ$.

Proof. Let $P' = AI \cap DE$. We know from **Theorem 13.2** that $MP' \parallel CA$ so triangle DMP' is similar to triangle DCE. But since $CD = CE$, we have $MD = MP'$ and therefore $P' = P$. Hence, P lies on line DE. Now since $\angle AQD = 90^\circ$, line AQ passes through D', the point diametrically opposed to D on the incircle of triangle ABC. Let $X = AQ \cap BC$.

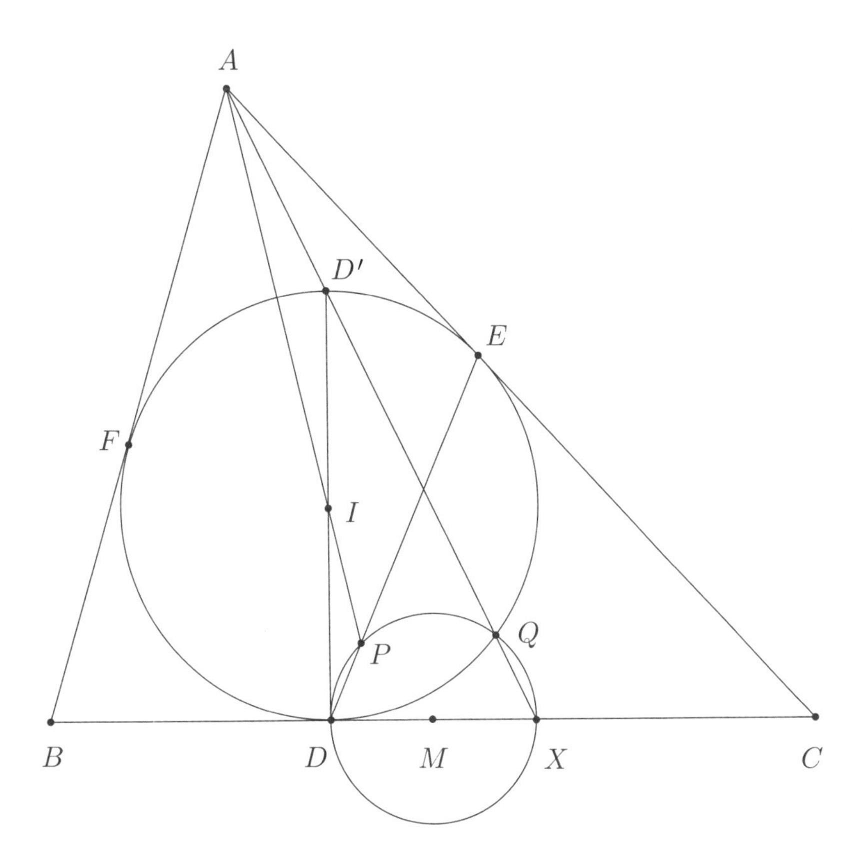

Consider the homothety centered at A that takes the incircle of triangle ABC to the A-excircle of triangle ABC. This homothety clearly takes D' to X and so X is the point where the A-excircle of triangle ABC touches side BC. Hence, since $BD = CX$, M is the midpoint of DX. Since $AQ \perp DQ$, Q lies on the circle with diameter DX. It follows that points D, P, Q, X are all on the circle with center M and radius MD. Since we have $\angle DQD' = 90^\circ$, it suffices to show $\angle DQP = \angle D'QE$. Since DD' is tangent to the circle with center M and radius MD, we have $\angle DQP = \angle D'DP$. We also have $\angle D'QE = \angle D'DE = \angle D'DP$ so $\angle DQP = \angle D'QE$ as desired and we are done. $\square$

Delta 13.10. (Eric Daneels, Forum Geometricorum) Let the incircle of triangle ABC have center I and touch sides BC, CA, AB at points D, E, F respectively. Let M, N, P be the midpoints of sides BC, CA, AB respectively. Let X, Y, Z be points on lines AI, BI, CI respectively. Prove that lines XD, YE, ZF concur if and only if lines XM, YN, ZP concur.

Proof. By **Theorem 13.2** lines AI, DE, MP concur, so we have

$$\frac{\delta(X, DE)}{\delta(X, MP)} = \frac{\delta(I, DE)}{\delta(I, MP)}$$

Considering the other five triples concurrent lines given by **Theorem 13.2** and multiplying the resulting similar expressions together we find that

$$\left(\frac{\delta(X,DE)}{\delta(X,DF)}\cdot\frac{\delta(Y,EF)}{\delta(Y,ED)}\cdot\frac{\delta(Z,FD)}{\delta(Z,FE)}\right)\cdot\left(\frac{\delta(X,MN)}{\delta(X,MP)}\cdot\frac{\delta(Y,NP)}{\delta(Y,NM)}\cdot\frac{\delta(Z,PM)}{\delta(Z,PN)}\right)=1$$

so

$$\frac{\delta(X,DE)}{\delta(X,DF)}\cdot\frac{\delta(Y,EF)}{\delta(Y,ED)}\cdot\frac{\delta(Z,FD)}{\delta(Z,FE)}=1$$

if and only if

$$\frac{\delta(X,MN)}{\delta(X,MP)}\cdot\frac{\delta(Y,NP)}{\delta(Y,NM)}\cdot\frac{\delta(Z,PM)}{\delta(Z,PN)}=1$$

which after applications of Ceva's Theorem and Trig Ceva is equivalent to what we wanted to prove. □

Chapter 14

Homothety

Definition. Consider a point P and a set of points $\mathcal{S}$, For each point $X \in \mathcal{S}$, let X' be the point on line PX such that $\frac{PX'}{PX} = k$ for some real number k (where we use directed lengths). Let $\mathcal{S}'$ be the set of points X'. Then we say that the **homothety** centered at P with ratio k takes $\mathcal{S}$ to $\mathcal{S}'$. By convention, a homothety centered at P takes P to itself. Homotheties are powerful because they preserve a lot of structure; namely, orientation and the similarity between figures. In fact, if any two figures are similar and oriented in the same way, then there exists a homothety that takes one to the other.

We begin with some interesting applications of this new tool:

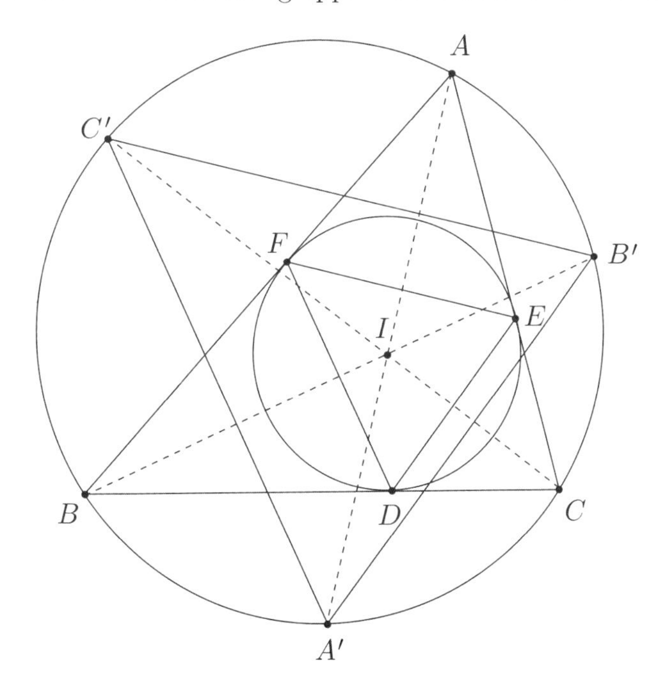

Delta 14.1. Let ABC be a triangle and let its incircle touch sides BC, CA, AB at points D, E, F respectively. Let I be the incenter of triangle ABC and let lines AI, BI, CI intersect the circumcircle of triangle ABC again at points A', B', C' respectively. Prove that lines $A'D, B'E, C'F$ concur.

Proof. A quick angle chase yields that $AA' \perp EF$ and $AA' \perp B'C'$ so we have that $EF \parallel B'C'$ and similarly the sides of triangle DEF are parallel to the corresponding sides of triangle $A'B'C'$. Therefore these two triangles are similar and oriented in the same way, and so there exists a homothety centered at some point P that takes triangle DEF to triangle $A'B'C'$. Hence, lines $A'D, B'E.C'F$ concur at P. This completes the proof. □

Delta 14.2. Let $ABCD$ be a trapezoid with $AB > CD$ and $AB \parallel CD$. Points K, L lie on segments AB, CD respectively such that $\frac{AK}{KB} = \frac{DL}{LC}$. Suppose there are points P, Q on line KL satisfying $\angle APB = \angle BCD$ and $\angle CQD = \angle ABC$. Prove that points P, Q, B, C are concyclic.

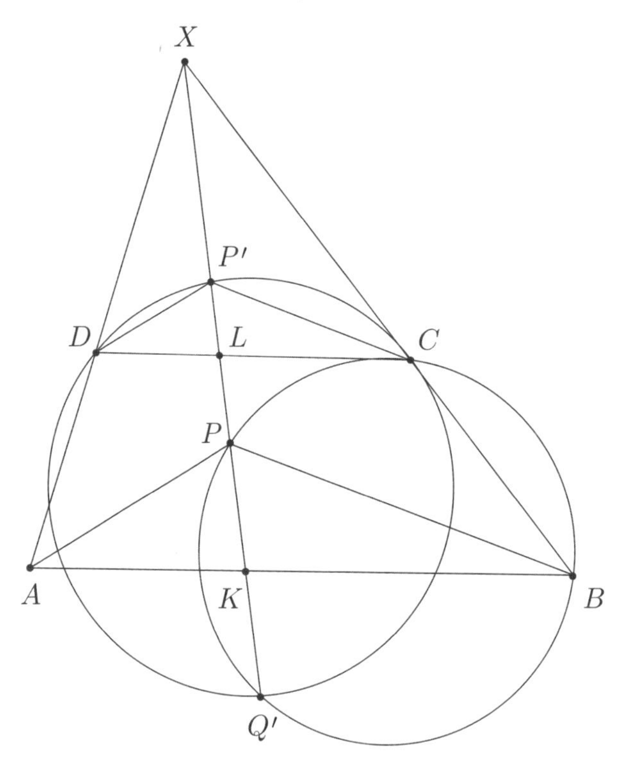

Proof. Let $X = AD \cap BC$. It is clear that there is a homothety centered at X taking segment DC to segment AB and since $\frac{AK}{KB} = \frac{DL}{LC}$ this homothety also takes L to K. Therefore X lies on line KL. Let Q' be the second intersection of line KL with the circumcircle of triangle PBC. Let the homothety centered at X that takes segment AB to segment DC take P to a point P'. Since quadrilateral $PQ'BC$ is cyclic we have that $\angle Q'CB = \angle Q'PB$ and since by definition $\angle APB = \angle BCD$ we have $\angle Q'CD = \angle Q'PA$. But $\angle Q'PA = \angle Q'P'D$ so quadrilateral $Q'CP'D$ is cyclic. Therefore $\angle P'Q'D = \angle P'CD = \angle PBA$ and since quadrilateral $PQ'BC$ is cyclic we also have $\angle PQ'C = \angle PBC$ so adding these two angle equalities we find that $\angle CQ'D = \angle ABC$. Therefore $Q' = Q$ and we are done. □

Now let's fry some bigger fish, in the form of one of the hardest problems ever given at the IMO.

Delta 14.3. (IMO 2011) Let ABC be an acute triangle with circumcircle Γ. Let ℓ be a tangent line to Γ, and let ℓ_a, ℓ_b and ℓ_c be the lines obtained by reflecting ℓ in the lines BC, CA and AB, respectively. Show that the circumcircle of the triangle determined by the lines ℓ_a, ℓ_b and ℓ_c is tangent to the circle Γ.

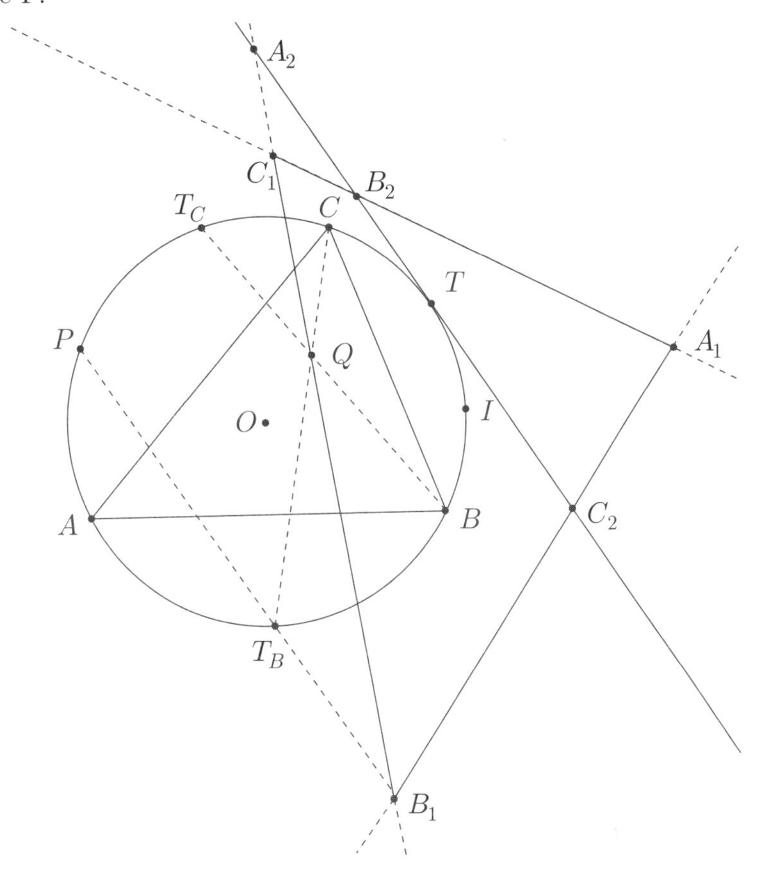

Proof. Let ℓ be tangent to Γ at T and let $A_1 = \ell_b \cap \ell_c$ and $B_1 = \ell_c \cap \ell_a$ and $C_1 = \ell_a \cap \ell_b$. Also, let $A_2 = \ell \cap \ell_a$ and $B_2 = \ell \cap \ell_b$ and $C_2 = \ell \cap \ell_c$. Without loss of generality assume that the order of points on line ℓ is C_2, T, B_2, A_2 (to avoid configuration issues). Let I be the incenter of triangle $A_1B_1C_1$. Consider triangle $A_2B_1C_2$. It is clear that line AB is the internal angle bisector of angle $\angle B_1C_2A_2$ and it is also clear that line BC is the internal angle bisector of angle $\angle B_1A_2C_2$. Therefore B is the incenter of this triangle; hence, line BB_1 bisects angle $\angle A_1B_1C_1$. Similarly line AA_1 bisects angle $\angle B_1A_1C_1$ so by symmetry we can conclude that lines AA_1, BB_1, CC_1 concur at I. Now, note that

$$\angle BIC = \angle B_1IC_1 = 90^\circ + \frac{\angle B_1A_1C_1}{2} = 180^\circ - \frac{\angle A_1C_2B_2}{2} - \frac{\angle A_1B_2C_2}{2}$$

$$= \angle AC_2B_2 + \angle AB_2C_2 = 180^\circ - \angle BAC$$

which implies that I lies on Γ.

Now, let O be the center of Γ and let T_B and T_C be the reflections of T about OB and OC respectively. Clearly T_B and T_C lie on Γ. Note that $\angle C_1A_2T = \angle BOT - \angle COT$ so $T_BT_C \parallel B_1C_1$. Letting T_A be the reflection of T about OA we have that $T_AT_B \parallel A_1B_1$ and $T_CT_A \parallel C_1A_1$ as well. Therefore there exists a homothety taking triangle $T_AT_BT_C$ to triangle $A_1B_1C_1$. It suffices to show that the center of this homothety lies on Γ, because then the homothety will take Γ to the circumcircle of triangle $A_1B_1C_1$ and so will be the tangency point between these two circles.

Now let Q be the reflection of T about BC. Clearly Q lies on line B_1C_1. Since $\angle TBQ = 2\angle TBC = \angle TBT_C$ we have that Q lies on line BT_C and similarly Q lies on line CT_B. Now let P be the second intersection of line B_1T_B with Γ and let $X = PT_C \cap IC_1$. It suffices to show that $X = C_1$. But by Pascal's Theorem on cyclic hexagon T_CBICT_BP we have that points Q, B_1, X are collinear so X lies on line B_1C_1 which implies that $X = C_1$ so P is the desired tangency point and we are done. □

We proceed with a famous result that appears in numerous Olympiad configurations.

Theorem 14.1. (Archimedes' Lemma) Let ω_2 be a circle internally tangent to a larger circle ω_1 at point A, let XY be a chord of ω_1 tangent to ω_2 at point B, and let C the midpoint of the arc XY not containing A of ω_1. Then:

(a) The points A, B, and C are collinear.

(b) $CA \cdot CB = CX^2$.

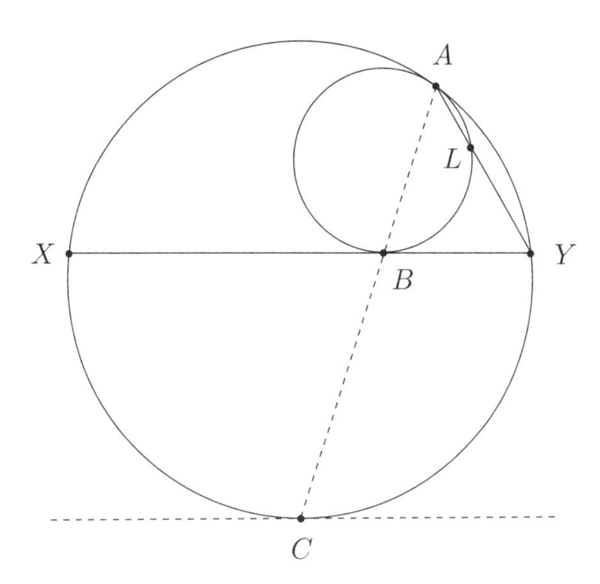

Proof. Consider the homothety centered at A which takes ω_2 to ω_1. This homothety clearly takes line XY to a line ℓ that is parallel to XY and tangent to ω_1. But it's clear that this is only possible if line ℓ touches ω_1 at point C, and since line XY touches ω_2 at B we can conclude that this homothety takes B to C and hence points A, B, C are collinear. This proves part (a). Now, this implies that line AC bisects angle $\angle XAY$. Therefore $\angle CAY = \angle CAX = \angle CYB$ so triangles CYB and CAY are similar, which immediately implies part (b). Hence, the proof is complete. □

Part (a) can have many other proofs. The most beautiful, in the authors' humble opinions, is through use of the Monge-D'Alembert circle theorem; we shall see it in a later section. As for now, let's see some applications!

Delta 14.4. (Russia NMO 2001) Using the same notation and diagram as in **Theorem 14.1**, prove that the circumradius of triangle CBY is independent of the position of B.

Proof. Let L be the second intersection of line AY with ω_2. The homothety centered at A which takes ω_2 to ω_1 also takes B to C by Archimedes' Lemma and L to Y, so we have that $BL \parallel CY$. Let R be the radius of ω_1, r the radius of ω_2, and r_1 the circumradius of triangle CBY. We have that $\frac{BY}{AY} = \frac{r_1}{R}$ since triangle CYB is similar to triangle CAY and $\frac{AL}{AY} = \frac{r}{R}$ because of the homothety. Since YB is tangent to ω_2 we have that $YB^2 = YL \cdot YA$. Consequently,

$$\left(\frac{r_1}{R}\right)^2 = \left(\frac{BY}{AY}\right)^2 = \frac{LY}{AY} = 1 - \frac{AL}{AY} = 1 - \frac{r}{R}$$

which yields that r_1 is fixed regardless of our choice of B, as desired. □

Delta 14.5. (Romania TST 2013) Circles Ω and ω are tangent at a point P (ω lies inside Ω). A chord AB of Ω is tangent to ω at C; the line PC meets Ω again at Q. Chords QR and QS of Ω are tangent to ω. Let I, X, and Y be the incenters of the triangles APB, ARB, and ASB, respectively. Prove that $\angle PXI + \angle PYI = 90°$.

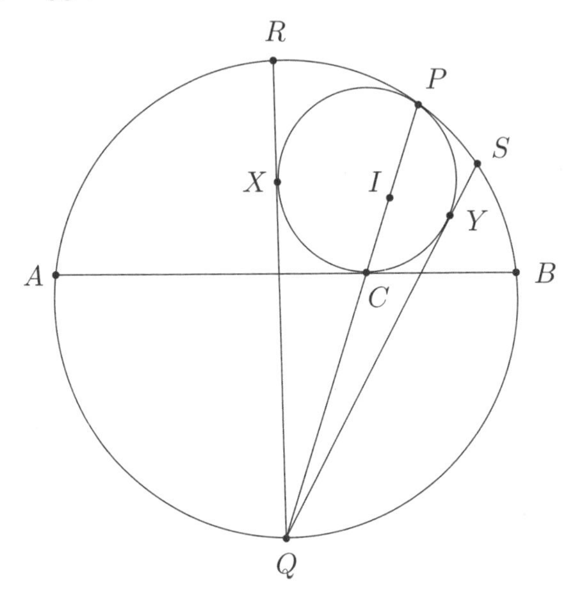

Proof. Let $T = \omega \cap QR$ and $U = \omega \cap QS$. Since Q is the midpoint of arc AB not containing P of Ω we know that $QA = QX$. Moreover, by part (b) of Archimedes' Lemma and Power of a Point we have that $QT^2 = QC \cdot QP = QA^2$ so $QT = QA = QX$. But line RQ bisects angle $\angle ARB$ so since X is the incenter of triangle ARB, X lies on this line. Therefore $X = T$ and similarly $Y = U$. Now we know that $QA = QI$ so Q is also the circumcenter of triangle XIY. Therefore $\angle XIY = 180° - \frac{\angle RQS}{2}$. Now by Archimedes' Lemma again we have that line PX bisects angle $\angle RPQ$ and similarly line PY bisects angle $\angle SPQ$ so $\angle XPY = \frac{\angle RPS}{2} = 90° - \frac{\angle RQS}{2}$. Therefore $\angle PXI + \angle PYI = \angle XIY - \angle XPY = \left(180° - \frac{\angle RQS}{2}\right) - \left(90° - \frac{\angle RQS}{2}\right) = 90°$ as desired. □

Now, we present one of the most common configurations in Olympiad geometry - you've already seen it in the proof of **Delta 13.9**!

Theorem 14.2. Let ABC be a triangle. Let ω and ω_a be its incircle and A-excircle respectively and let ω touch BC at D. Let D' be the point on ω diametrically opposite from D and let $X = AD' \cap BC$. Then X is the point where ω_a touches BC.

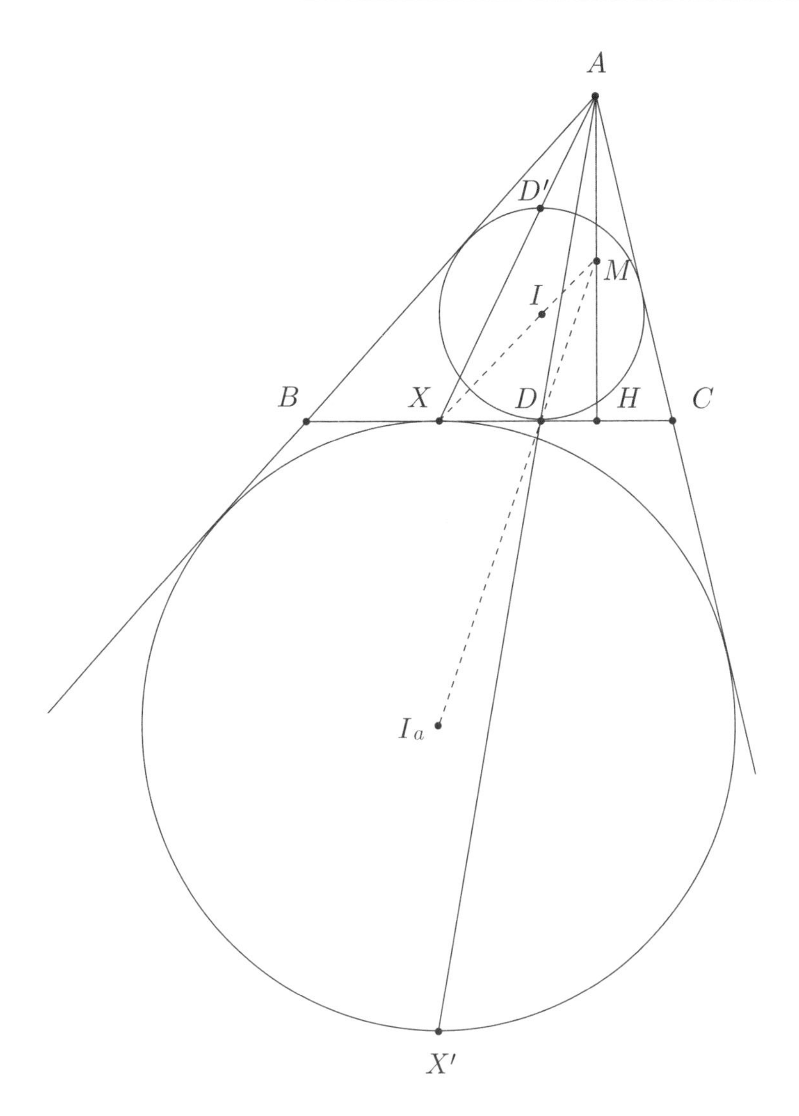

Proof. Consider the homothety centered at A that takes ω to ω_a. Since $DD' \perp BC$, it's clear that this homothety takes D' to the point where ω_a touches BC. But this homothety also takes D' to X, and hence we have the desired result. □

//Note that this implies $BD = CX$; in other words, the midpoint of BC is also the midpoint of DX. Do remember this!

Corollary 14.1. (Nagel's Lemma) Using the same notation and diagram as in **Theorem 14.2**, let I and I_a be the centers of ω and ω_a respectively. Let H be the foot of the A-altitude in triangle ABC and let M be the midpoint of AH. Then lines XI and DI_a concur at M.

Proof. Let X' be the point on ω_a diametrically opposite from X. We know from **Theorem 14.2** that the homothety centered at A that takes ω to ω_a also takes D' to X and by similar reasoning it takes D to X', so points A, D, X' are collinear and points A, D', X are collinear. Now since $DD' \parallel HA$ it's clear that there is a homothety centered at X that takes segment DD' to segment HA. Therefore it takes the midpoint of segment DD' to the midpoint of segment HA; hence, points X, I, M are collinear. Also, since $XX' \parallel HA$, there is a homothety centered at D that takes segment XX' to segment HA. Therefore it takes the midpoint of segment XX' to the midpoint of segment HA; hence, points D, I_a, M are collinear as well. This completes the proof. □

Delta 14.6. (the Nagel Line) Let ABC be a triangle. Let the A-excircle of triangle ABC touch BC at X. Similarly define Y on AC and Z on AB. Then AX, BY, CZ concur at a point N known as the **Nagel point** of triangle ABC. Let G be the centroid of triangle ABC and I the incenter of triangle ABC. Show that the points I, G, N lie in that order on a line, and moreover, that $GN = 2IG$.

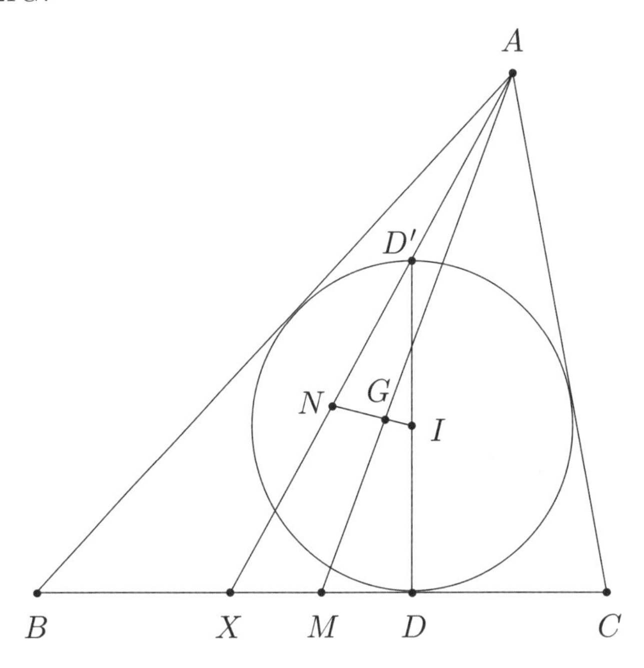

Proof. Let the incircle of triangle ABC touch BC at D, and let DD' be a diameter of this incircle. By **Theorem 14.2**, the points A, D', X are collinear. Let M be the midpoint of BC. Then, since M is the midpoint of DX, MI is a midline of triangle XDD', so $IM \parallel AX$. The homothety centered at G with ratio -2 takes M to A, and thus takes line IM to the line through A parallel to IM, namely the line AX. Hence the image of I under the homothety lies on the line AD. Analogously, it must also lie on BE and CF, and therefore the image of I is precisely N. This proves that I, G, N are collinear in that order with $GN = 2IG$. This completes the proof. □

Now, let's see some applications! We begin with a quick ruler and compass construction from Mathematical Reflections.

Delta 14.7. Starting with the incenter I of triangle ABC, the midpoint of the side BC and the foot of the altitude from A, reconstruct the triangle ABC using only straightedge and compass.

Proof. You are given the foot of the altitude from A and the midpoint M of BC, and the line determined by the two is precisely the line BC. Now, knowing the incenter we can draw the perpendicular from I to BC and obtain the incircle of ABC. Now, we are almost there! From **Theorem 14.2** we know that if D is the tangency point of the incircle with the side BC, and D' the antipode of D with respect to the incircle, then the points A, D', X are collinear, where X is the tangency point of the A-excircle with BC. In order to use this however, we need to find X! Fortunately, this is not a problem, since we know that $MD = MX$. So, we can get X. Now, just draw the lines XD' and the altitude from A (which we can draw since we have the foot of the altitude on BC and the line BC); they intersect at the vertex A. Then, just take the tangents from A to the incircle and intersect them with BC; this will give us the vertices B and C. Hence our construction is complete. □

Problems involving straightedge and compass construction are kind of old-fashioned and not many contest problems ask for such things. Nonetheless, we can't deny the beauty arising from the simplicity of the mechanism. In any case, let's finish with an easy IMO problem.

Delta 14.8. (IMO 1992). Let $\mathcal{C}$ be a circle, ℓ be a line tangent to the circle $\mathcal{C}$, and M a point on ℓ. Find the locus of all points P with the following property: there exists two points Q, R on ℓ such that M is the midpoint of QR and $\mathcal{C}$ is the inscribed circle of triangle PQR.

Proof. Let $\mathcal{C}$ touch ℓ at D, and let DE be a diameter of $\mathcal{C}$. For any such P, Q, R described in the problem, the line PE must intersect ℓ at a point F such that $MD = MF$ by **Theorem 14.2**. The point F depends only on M, ℓ, and $\mathcal{C}$. It follows that P must lie on the ray FE beyond E.

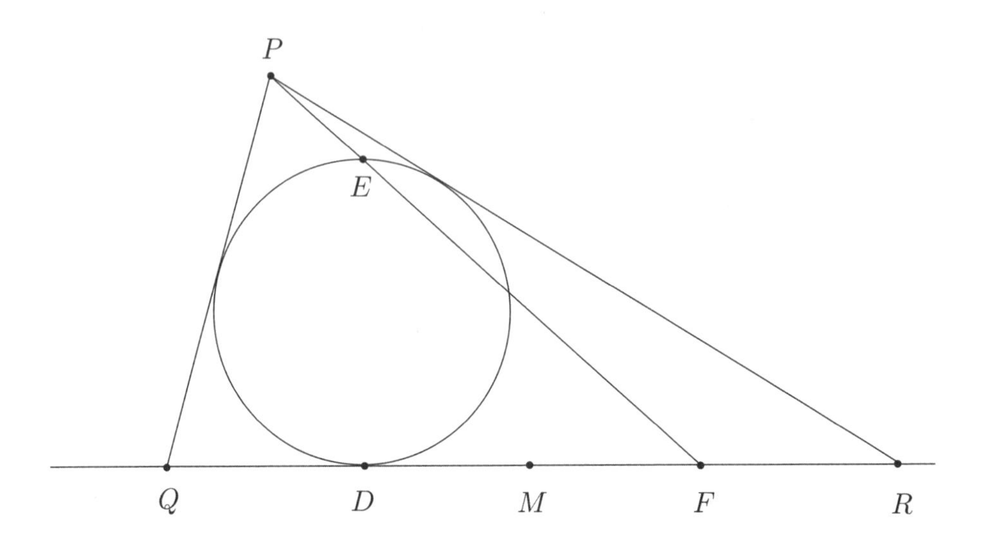

Conversely, given a point P lying on the ray FE beyond E, let the tangents from P to $\mathcal{C}$ meet ℓ at Q and R. We must have $QF = RD$, from which it follows that M is the midpoint of QR. Therefore, the locus is the ray FE beyond E. $\square$

Assigned Problems

Epsilon 14.1. Let ABC be a triangle. Circle k_1 is tangent to lines AB, AC, circle k_2 is tangent to lines BC, BA and circle k_3 is tangent to lines CA, CB. These three circles have equal radii and share a common point P. Prove that P lies on line IO where I and O are the incenter and circumcenter of triangle ABC respectively.

Epsilon 14.2. (Russia NMO 2000) Given two circles tangent internally at N, let AB and BC be two chords of the exterior circle that are tangent to the interior circle at K and M respectively. Let Q and P be the midpoints of the arcs AB and BC that contain the point N respectively. The circumcircles of triangles BQK and BPM intersect at B and B_1. Prove that the quadrilateral BPB_1Q is a parallelogram.

Epsilon 14.3. (IMO 1999) Let Γ_1 and Γ_2 be two intersecting circles that lie inside the circle Γ and which are tangent to Γ internally at points M and N, respectively ($M \neq N$). Suppose Γ_1 passes through the center of Γ_2 and let the line determined by the two common points of Γ_1 and Γ_2 intersect the circle Γ at points A and B. If C and D are the intersections of the lines MA and MB with Γ_1, prove that CD is tangent to Γ_2.

Epsilon 14.4. (Sharygin 2012) A circle ω with center I is inscribed into a segment of the disk, formed by an arc and a chord AB. Point M is the midpoint of this arc AB, and point N is the midpoint of the complementary arc. The tangents from N touch ω in points C and D. The opposite sidelines AC and BD of quadrilateral $ABCD$ meet at point X, and the diagonals of $ABCD$ meet at point Y. Prove that points X, Y, I and M are collinear.

Epsilon 14.5. (USAMO 1999) Let $ABCD$ be an isosceles trapezoid with $AB \| CD$. The inscribed circle ω of triangle BCD meets CD at E. Let F be a point on the (internal) angle bisector of $\angle DAC$ such that $EF \| CD$. Let the circumscribed circle of triangle ACF meet the line CD at C and G. Prove that triangle AFG is isosceles.

Epsilon 14.6. Let ABC be a triangle and let T_a be the tangency point of the incircle with BC, and M_a the midpoint of the A-altitude of ABC. Similarly, define T_b, T_c, and M_b, M_c, respectively. Prove that the lines T_aM_a, T_bM_b, and T_cM_c are concurrent.

Epsilon 14.7. (Tournament of Towns 2003) Again, let K be the tangency point of the incircle of triangle ABC with the side BC. Prove that if the line OI determined by the circumcenter and the incenter of triangle ABC is parallel to BC, then $AO \| HK$, where H denotes the orthocenter of triangle ABC.

Epsilon 14.8. (IMO 2005 Shortlist) In a triangle ABC satisfying $AB + BC = 3AC$ the incircle has center I and touches the sides AB and BC at D and E, respectively. Let K and L be the symmetric points of D and E with respect to I. Prove that the quadrilateral $ACKL$ is cyclic.

Epsilon 14.9. (USAMO 2001) Let ABC be a triangle and let ω be its incircle. Denote by D_1 and E_1 the points where ω is tangent to sides BC and AC, respectively. Denote by D_2 and E_2 the points on sides BC and AC, respectively, such that $CD_2 = BD_1$ and $CE_2 = AE_1$, and denote by P the point of intersection of segments AD_2 and BE_2. Circle ω intersects segment AD_2 at two points, the closer of which to the vertex A is denoted by Q. Prove that $AQ = D_2P$.

Epsilon 14.10. (IMO Shortlist 2006) Circles w_1 and w_2 with centers O_1 and O_2 are externally tangent at point D and internally tangent to a circle w at points E and F respectively. Line t is the common tangent of w_1 and w_2 at D. Let AB be the diameter of w perpendicular to t, so that A, E, O_1 are on the same side of t. Prove that lines AO_1, BO_2, EF and t are concurrent.

Chapter 15

Inversion

Definition. An **inversion** with respect to a circle with center O and radius r is a map that sends every point P to a point P' on ray OP such that $OP \cdot OP' = r^2$. By convention, O is taken to itself. Throughout the section, adding an apostrophe to a point will be used to denote its inverse with respect to the circle we are inverting about. Inversions are useful because they turn configurations with lots of nasty circles into configurations with lots of nice lines, and we will soon see why this is!

Consider a circle ω with center O and radius r, and let P and Q be points in its plane. Since $\angle POQ = \angle P'OQ'$ and since $OP \cdot OP' = OQ \cdot OQ' = r^2$ we have that triangle POQ is similar to triangle $Q'OP'$. This means that $\angle OPQ = \angle OQ'P'$ (remember this!). Moreover, this similarity shows that $P'Q' = \frac{r^2}{OP \cdot OQ} \cdot PQ$.

Now, what do lines and circles actually map to? Well, it's clear that a line passing through O inverts to itself. What about a line ℓ not passing through O? In this case, let P be the projection of O on ℓ and let Q be an arbitrary point on ℓ distinct from P. Since $\angle OQ'P' = \angle OPQ = 90$ we can conclude that line ℓ inverts to the circle with diameter OP'. Inverting back, this means that if ω is a circle passing through O with diameter OP then it inverts to the line through P' perpendicular to the line OP'. Finally, what if we have a circle Γ not passing through O? In this case, let A, B, C, D be any four points on Γ. Utilizing directed angles mod 180° to avoid configuration issues, we have that $\angle A'C'B' = \angle OC'B' - \angle OC'A' = \angle OBC - \angle OAC$. Analogously we have $\angle A'D'B' = \angle OBD - \angle OAD$ and so $\angle A'C'B' - \angle A'D'B' = \angle CBD - \angle CAD = 0$ so points A', B', C', D' are concyclic. Hence, Γ inverts to a circle.

To sum up:
(a) lines passing through O map to themselves

(b) lines not passing through O map to circles passing through O and vice-versa

(c) circles not passing through O map to circles not passing through O

Now let's see some applications! The next four problems can actually be done without the use of a diagram - they all follow from the basic properties of inversion!

Delta 15.1. Prove that if circles ω and γ are orthogonal, then the inversion about ω maps γ to itself.

Proof. Let circles ω and γ intersect at points A and B. Let O_1, O_2 be the centers of ω and γ respectively and let line O_1O_2 intersect γ at points C and D. Since γ inverts to a circle and since points A and B clearly invert to themselves, it suffices to show that point D inverts to point C. Now since $\angle O_1AO_2 = 90°$ because ω and γ are orthogonal, by Power of a Point and then the Pythagorean Theorem we have

$$O_1C \cdot O_1D = O_1O_2^2 - O_2A^2 = O_1A^2$$

which implies that D inverts to C as desired. □

Delta 15.2. (Ptolemy's Inequality) Let $ABCD$ be a quadrilateral. Prove that $AB \cdot CD + DA \cdot BC \geq AC \cdot BD$.

Proof. Invert about the circle centered at A with radius 1. Then $AB = \frac{1}{AB'}$ and $AC = \frac{1}{AC'}$ and $AD = \frac{1}{AD'}$ and $BC = B'C' \cdot AB' \cdot AC'$ and $BD = B'D' \cdot AB' \cdot AD'$ and $CD = C'D' \cdot AC' \cdot AD'$ so the inequality reduces to $B'C' + C'D' \geq B'D'$ which is true by the triangle inequality. Moreover, equality holds if and only if points B', C', D' are collinear - in other words, if quadrilateral $ABCD$ is cyclic. □

Delta 15.3. (IMO Shortlist 2003) Let Γ_1, Γ_2, Γ_3, Γ_4 be distinct circles such that Γ_1, Γ_3 are externally tangent at P, and Γ_2, Γ_4 are externally tangent at the same point P. Suppose that Γ_1 and Γ_2; Γ_2 and Γ_3; Γ_3 and Γ_4; Γ_4 and Γ_1 meet at A, B, C, D, respectively, and that all these points are different from P. Prove that

$$\frac{AB \cdot BC}{AD \cdot DC} = \frac{PB^2}{PD^2}.$$

Proof. Invert about the circle centered at P with radius 1. Circles Γ_1, Γ_2, Γ_3, Γ_4 go to lines $\ell_1, \ell_2, \ell_3, \ell_4$ and since inversion obviously preserves intersections, we also have that $A' = \ell_1 \cap \ell_2$ and $B' = \ell_2 \cap \ell_3$ and $C' = \ell_3 \cap \ell_4$ and $D' = \ell_4 \cap \ell_1$. Now since Γ_1 and Γ_3 are tangent at P we see that $\ell_1 \parallel \ell_3$ and similarly $\ell_2 \parallel \ell_4$ so quadrilateral $A'B'C'D'$ is a parallelogram. Hence, $A'B' = C'D'$, and since $A'B' = \frac{AB}{PA \cdot PB}$ and $C'D' = \frac{CD}{PC \cdot PD}$ by dividing we find that

$$\frac{AB}{CD} = \frac{PA \cdot PB}{PC \cdot PD}$$

Similarly we have

$$\frac{BC}{DA} = \frac{PB \cdot PC}{PD \cdot PA}$$

and upon multiplying we obtain the desired result. □

Delta 15.4. (IMO 1996) Let P be a point inside a triangle ABC such that $\angle APB - \angle ACB = \angle APC - \angle ABC$. Let D, E be the incenters of triangles APB, APC, respectively. Show that the lines AP, BD, CE meet at a point.

Proof. Let $X = BD \cap AP$ and let $Y = CE \cap AP$. By the Angle Bisector Theorem we have $\frac{AX}{PX} = \frac{AB}{PB}$ and $\frac{AY}{PY} = \frac{AC}{PC}$ so $X = Y$ if and only if $\frac{AB}{PB} = \frac{AC}{PC}$. Invert about the circle centered at A with radius 1. The angle condition becomes $\angle AB'P' - \angle AB'C' = \angle AC'P' - \angle AC'B'$ which is equivalent to $\angle C'B'P' = \angle B'C'P'$. Therefore triangle $B'C'P'$ is isosceles and so $B'P' = C'P'$. But $B'P' = \frac{BP}{AB \cdot AP}$ and $C'P' = \frac{CP}{AC \cdot AP}$ so we have that $\frac{AB}{PB} = \frac{AC}{PC}$ as desired. This completes the proof. □

Delta 15.5. Let the incircle of triangle ABC touch sides BC, CA, AB at points D, E, F respectively. Let I and O be the incenter and circumcenter of triangle ABC respectively. Prove that the orthocenter of triangle DEF lies on line IO.

Proof. Let M, N, P be the midpoints of segments EF, FD, DE respectively and let r be the inradius of triangle ABC. Invert about the incircle of triangle ABC. It's clear that points I, M, A are collinear and moreover we can calculate that $IA = \frac{r}{\sin \frac{A}{2}}$ and $IM = r \sin \frac{A}{2}$. Hence, since $IA \cdot IM = r^2$, we have that this inversion takes A to M and similarly takes B and C to N and P respectively.

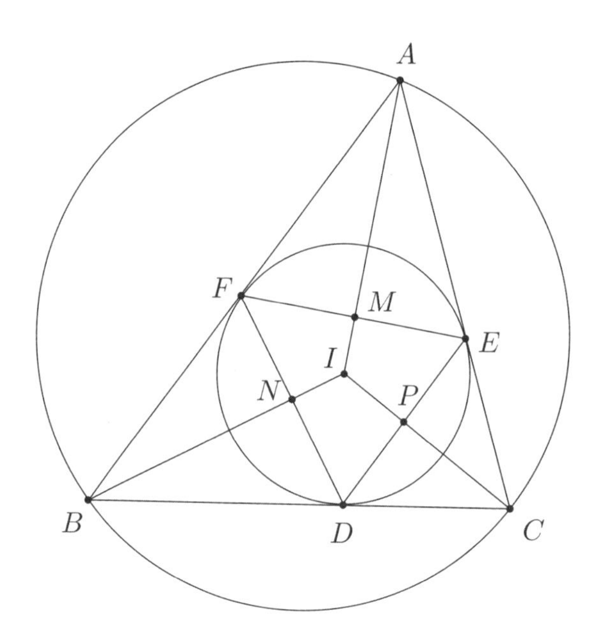

It follows that the inversion considered takes the circumcircle of triangle ABC to the circumcircle of triangle MNP - in other words, the nine-point circle of triangle DEF. Therefore the nine-point center of triangle DEF, I, and O are collinear. But the orthocenter of triangle DEF is clearly the reflection of I over the nine-point center of triangle DEF so we have the desired collinearity, as all three points lie on the Euler line of triangle DEF. □

Delta 15.6. Let ABC and XYZ be two triangles such that the circumcircles of triangles BCX, CAY, ABZ are concurrent at a point P. Prove that the circumcircles of triangles YZA, ZXB, XYC are concurrent as well.

Proof. Invert about a circle centered at P with arbitrary radius. the circumcircles of triangles BCX, CAY, ABZ map to the triangle $A'B'C'$ where points X', Y', Z' lie on sides $B'C', C'A', A'B'$ respectively. The circumcircles of triangles YZA, ZXB, XYC map to the circumcircles of triangles $Y'Z'A'$, $Z'X'B'$, $X'Y'C'$ so it suffices to show that these circles concur. But this is just Miquel's Pivot Theorem, introduced in the proof of The Droz-Farny Line Theorem (**Delta 8.8**)! Hence, the proof is complete. □

Delta 15.7. (China TST 2006) Let ω be the circumcircle of triangle ABC and P be an interior point of triangle ABC. A_1, B_1, C_1 are the second intersections of lines AP, BP, CP respectively with ω and A_2, B_2, C_2 are the reflections of points of A_1, B_1, C_1 over sides BC, CA, AB respectively. Show that the circumcircle of triangle $A_2B_2C_2$ passes through the orthocenter H of triangle ABC.

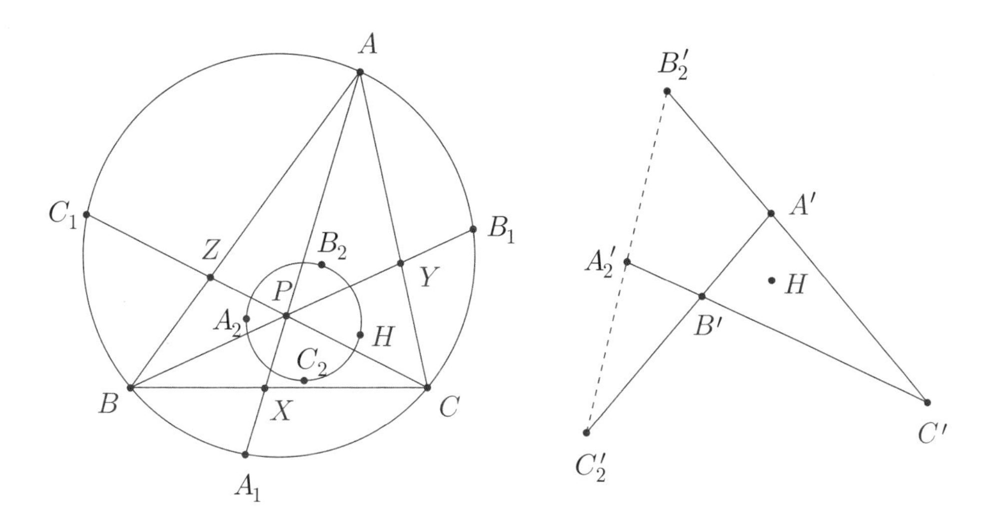

Proof. Let $X = AA_1 \cap BC$ and $Y = BB_1 \cap CA$ and $Z = CC_1 \cap AB$. First note that by multiple applications of the Ratio Lemma and finally Ceva's Theorem we have

$$\begin{aligned}\frac{A_2B}{A_2C}\cdot\frac{B_2C}{B_2A}\cdot\frac{C_2A}{C_2B} &= \frac{A_1B}{A_1C}\cdot\frac{B_1C}{B_1A}\cdot\frac{C_1A}{C_1B} \\ &= \left(\frac{XB}{XC}\cdot\frac{\sin CA_1A}{\sin BA_1A}\right)\cdot\left(\frac{YC}{YA}\cdot\frac{\sin AB_1B}{\sin CB_1B}\right)\cdot\left(\frac{ZA}{ZB}\cdot\frac{\sin BC_1C}{\sin AC_1C}\right) \\ &= \left(\frac{XB}{XC}\cdot\frac{YC}{YA}\cdot\frac{ZA}{ZB}\right)\cdot\left(\frac{\sin B}{\sin C}\cdot\frac{\sin C}{\sin A}\cdot\frac{\sin A}{\sin B}\right) \\ &= 1\end{aligned}$$

Now, invert about the circle centered at H with radius 1. Note that points B, C, H, A_2 all lie on the circle that, when reflected across line BC, becomes ω. Therefore points B', C', A_2' are collinear and similarly points C', A', B_2' and points A', B', C_2' are collinear. We also have

$$\frac{A_2'B'}{A_2'C'}\cdot\frac{B_2'C'}{B_2'A'}\cdot\frac{C_2'A'}{C_2'B'} = \left(\frac{A_2B}{A_2C}\cdot\frac{HC}{HB}\right)\cdot\left(\frac{B_2C}{B_2A}\cdot\frac{HA}{HC}\right)\cdot\left(\frac{C_2A}{C_2B}\cdot\frac{HB}{HA}\right) = 1$$

so by Menelaus' Theorem on triangle $A'B'C'$ with points A_2', B_2', C_2' we have that points A_2', B_2', C_2' are collinear. This implies that points H, A_2, B_2, C_2 are concyclic as desired. □

//The circumcircle of triangle $A_2B_2C_2$ is known as the P-Hagge Circle of triangle ABC.

Delta 15.8. (China TST 2015) Triangle ABC is isosceles with $AB = AC > BC$. Let D be a point in its interior such that $DA = DB + DC$. Suppose that the perpendicular bisector of segment AB meets the external angle bisector of angle $\angle ADB$ at P, and let Q be the intersection of the perpendicular bisector of segment AC and the external angle bisector of angle $\angle ADC$. Prove that points B, C, P, Q are concyclic.

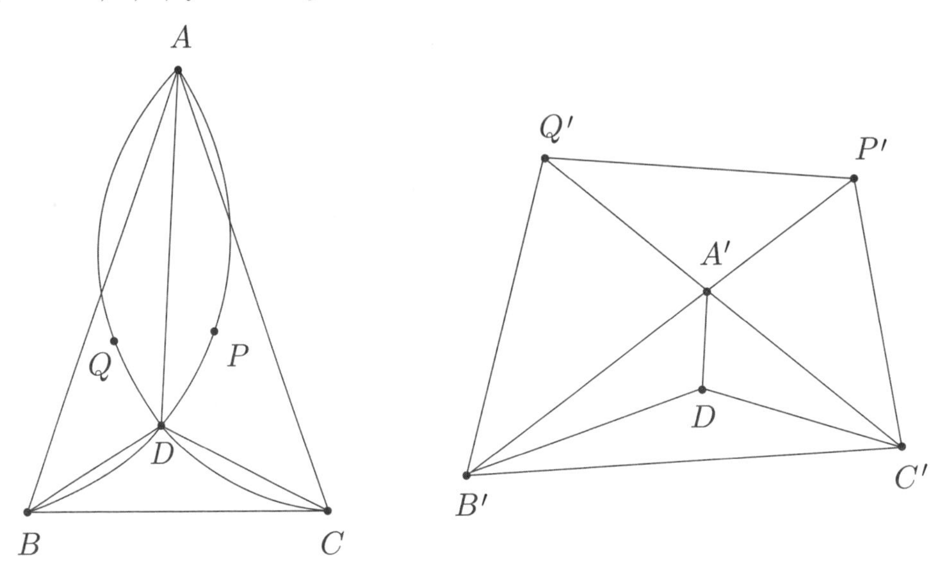

Proof. It is clear that P is the midpoint of arc BDA of the circumcircle of triangle BDA and similarly Q is the midpoint of arc CDA of the circumcircle of triangle CDA. Consider the inversion about the circle centered at D with radius 1. Because of the first observation, we have that points C', A', Q' collinear and points B', A', P' are collinear (in those orders). Now we proceed with four metric observations:

(1) : $\frac{1}{A'D} = \frac{1}{B'D} + \frac{1}{C'D}$ - this follows from the given condition

(2) : $\frac{A'B'}{A'C'} = \frac{B'D}{C'D}$ - this follows from the fact that triangle ABC is isosceles.

(3) : $\frac{A'P'}{B'P'} = \frac{A'D}{B'D}$ - this follows from the fact that triangle APB is isosceles.

(4) : $\frac{A'Q'}{C'Q'} = \frac{A'D}{C'D}$ - this follows from the fact that triangle AQC is isosceles.

It clearly suffices to show that quadrilateral $B'C'P'Q'$ is cyclic, which is equivalent to showing that $A'C' \cdot A'Q' = A'B' \cdot A'P'$. But by dividing (3) and (4) and then using (2) we have that $\frac{A'Q'}{A'P'} = \frac{A'B' \cdot C'Q'}{A'C' \cdot B'P'}$ so it suffices to show that $C'Q' = B'P'$. But by (3) and (4) again and the facts that $A'C' + A'Q' = C'Q'$ and $A'B' + A'P' = B'P'$ we find that $C'Q' = \frac{A'B'}{1 - \frac{A'D}{C'D}}$ and $B'P' = \frac{A'C'}{1 - \frac{A'D}{B'D}}$ and now by using (1) and (2) we can simplify to obtain the desired result. $\square$

Delta 15.9. (ELMO Shortlist 2014) Let $ABCD$ be a cyclic quadrilateral whose circumcircle ω has center O. Suppose the circumcircles of triangles AOB and COD meet again at G, while the circumcircles of triangles AOD and BOC meet again at H. Let ω_1 denote the circle passing through G as well as the feet of the perpendiculars from G to AB and CD. Define ω_2 analogously as the circle passing through H and the feet of the perpendiculars from H to BC and DA. Show that the midpoint of GH lies on the radical axis of ω_1 and ω_2.

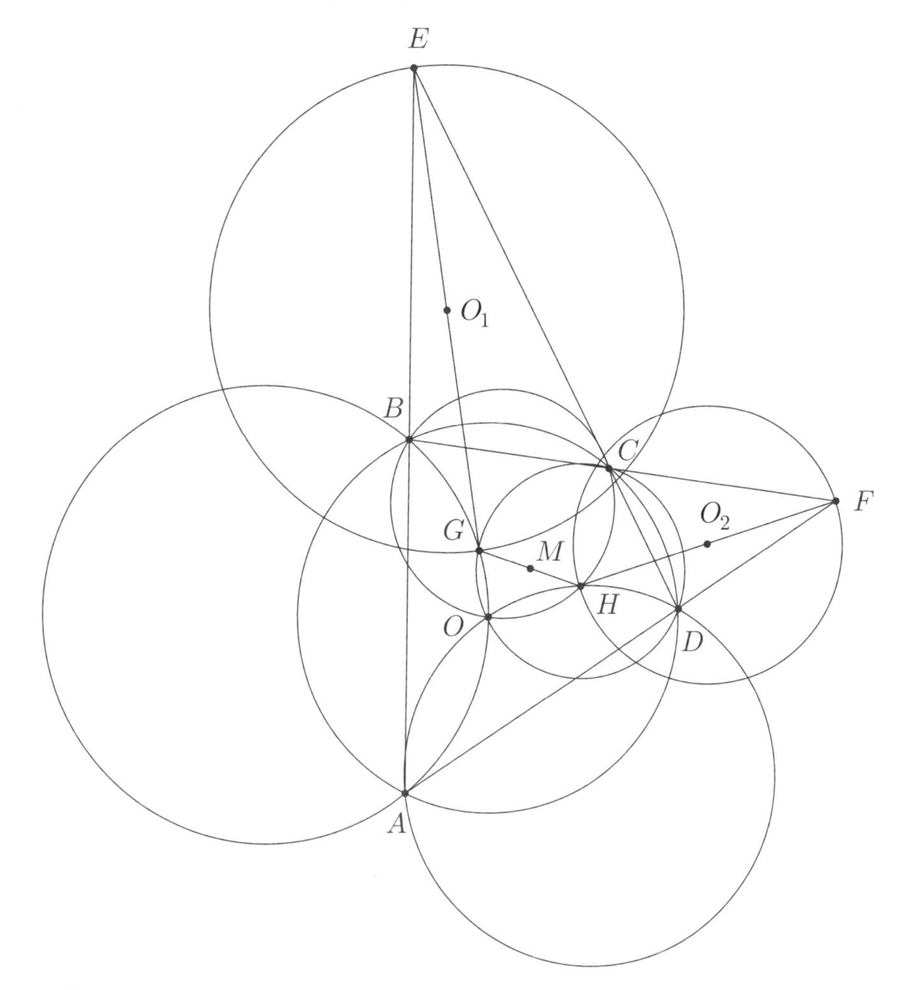

Proof. First, let $E = AB \cap CD$ and $F = BC \cap DA$. Let O_1, O_2 be the centers of ω_1, ω_2 respectively. Let M be the midpoint of segment GH. Consider the inversion about ω. It is clear that line AB inverts to the circumcircle of triangle AOB and that line CD inverts to the circumcircle of triangle COD so E inverts to G. Similarly F inverts to H. Moreover, note that ω_1 and ω_2 are the circles with diameters EG and FH respectively. Now since M is the midpoint of GH and since O_1 is the midpoint of GE we have that $O_1M \parallel HE$

and similarly $O_2M \parallel GF$.

Now since by Brokard's Theorem GF is the polar of E with respect to ω, we have that $GF \perp OE$ so $O_2M \perp OE$ which implies that $O_2M \perp OO_1$. Similarly $O_1M \perp OO_2$ and so M is the orthocenter of triangle OO_1O_2. This means that $OM \perp O_1O_2$ so to show that M is on the radical axis of ω_1 and ω_2 it suffices to show that O is on this radical axis. But recall that when we inverted about ω, G inverted to E and H inverted to F so we have that $OG \cdot OE = OH \cdot OF = R^2$ where R is the radius of ω and hence the powers of O with respect to ω_1 and ω_2 are equal. Therefore, O is on the radical axis of ω_1 and ω_2 as desired. This completes the proof. □

//In fact, the problem statement remains true even if O is replaced by any point on line OQ where $Q = AC \cap BD$!

We end the section with an amazing result first found by Karl Feuerbach in 1822.

Theorem 15.1. (Feuerbach's Theorem) Prove that the nine-point circle of triangle ABC is tangent to the incircle of triangle ABC

Proof. Let A' be the midpoint of segment BC. Let X, X' be the points where the incircle and A-excircle of triangle ABC respectively touch side BC. let I, J be the centers of the incircle and A-excircle of triangle ABC respectively. Let D be the foot of the A-altitude in triangle ABC and let $L = AI \cap BC$. Recall that $A'X = A'X'$. Now, consider the inversion about the circle with diameter XX'. Since $\angle IXA' = \angle JX'A' = 90°$ we have that the circle with diameter XX' is orthogonal to both the incircle and the A-excircle of triangle ABC and so by **Delta 15.1** we have that the incircle and the A-excircle of triangle ABC both invert to themselves. Also, since points A, I, L, J are collinear and since $AD \parallel IX \parallel JX'$ we easily find that

$$\frac{DX}{DX'} = \frac{LX}{LX'}$$

so $(D, L; X, X')$ is harmonic. But since A' is the midpoint of XX' this means that $AL \cdot AD = AX^2$ so D inverts to L. Since A' and D both lie on the nine-point circle of triangle ABC, we can conclude that this nine-point circle inverts to a line passing through L.

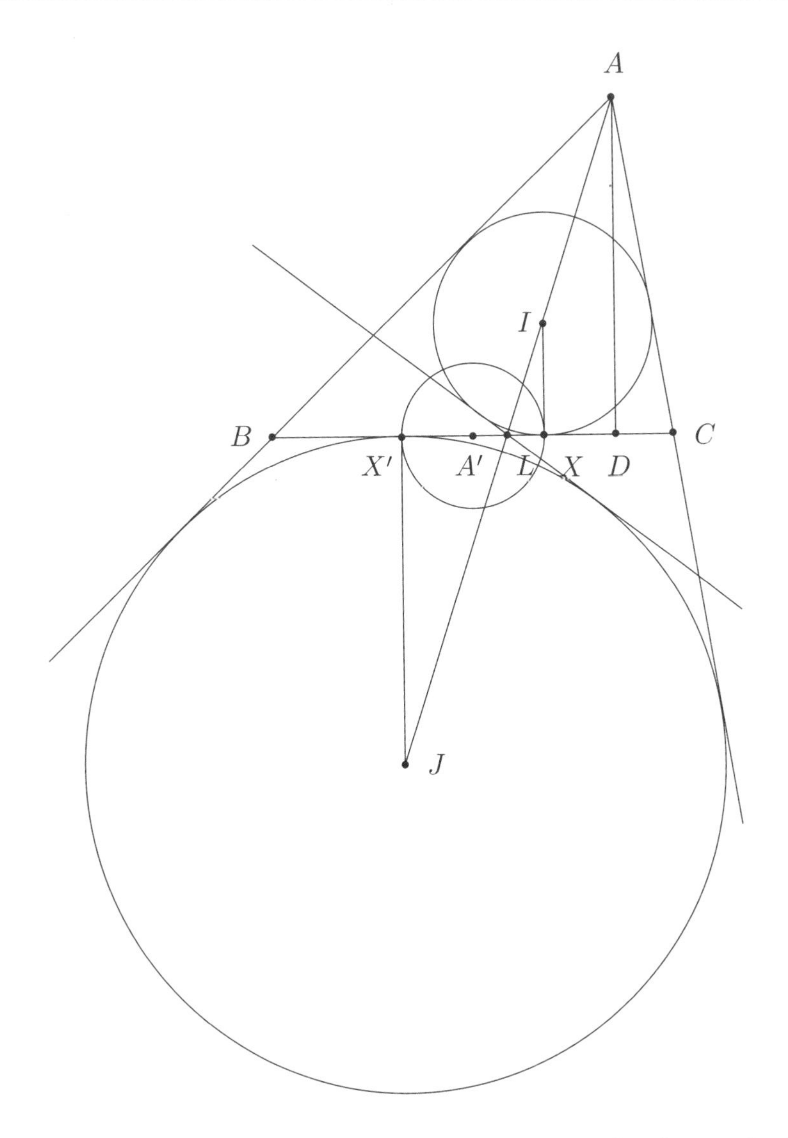

Now consider the line ℓ tangent to the nine-point circle of triangle ABC at A'. It's easy to see that $\angle(\ell, BC) = \angle(\ell, B'C') = |\angle B - \angle C|$ which, since the nine-point circle of triangle ABC must invert to a line parallel to ℓ, implies that the nine-point circle of triangle ABC inverts to a line that passes through L and makes an angle of $|\angle B - \angle C|$ with line BC. But an easy angle chase shows that this is precisely the line symmetric to line BC with respect to line IJ! Therefore the nine-point circle of triangle ABC inverts to an internal tangent of the incircle and A-excircle of triangle ABC and since inversion preserves tangency, we are done. $\square$

//In fact, we've proven more: the nine-point circle of triangle ABC is tangent to the excircles of triangle ABC as well.

Assigned Problems

Epsilon 15.1. Consider four circles $\omega_1, \omega_2, \omega_3, \omega_4$ such that ω_2 and ω_4 are each tangent to both ω_1 and ω_4 at A, B and C, D respectively. Prove that the points A, B, C, D are concyclic.

Epsilon 15.2. Let points A, B, C lie on a line in this order. Semicircles $\omega, \omega_1, \omega_2$ are drawn on segments AC, AB, BC respectively on the same side of line AB. A sequence of circles (k_n) is constructed as follows: k_0 is the circle determined by ω_2, and k_n is the circle externally tangent to ω, ω_1, and k_{n-1} for all integers $n \geq 1$. Prove that the distance from the center of k_n to line AB is $2n$ times the radius of k_n for all nonnegative integers n.

Epsilon 15.3. Let s be the semi-perimeter of a triangle ABC and let E and F be the points on line AB such that $CE = CF = s$. Prove that the circumcircle of triangle CEF is tangent to the C-excircle of triangle ABC.

Epsilon 15.4. Let ω be the incircle of triangle ABC let ω touch sides BC, CA, AB at points D, E, F respectively. Also let I be the incenter of triangle ABC. Prove that the circumcircles of triangles AID, BIE, CIF concur again on the Euler line of triangle DEF.

Epsilon 15.5. Let ω be a semicircle with diameter PQ. A circle k is internally tangent to ω and to segment PQ at C. Let AB be the tangent to k with $AB \perp PQ$ and A on k and B on segment CQ. Show that line AC bisects angle $\angle PAB$.

Epsilon 15.6. (China TST 2012) Given two circles ω_1, ω_2, let $\mathcal{S}$ denotes all triangles ABC such that ω_1 is the circumcircle of triangle ABC and ω_2 is the A- excircle of triangle ABC. For some such triangle ABC, let ω_2 touch sides BC, CA, AB at points D, E, F respectively. If $\mathcal{S}$ is not empty, prove that the centroid of triangle DEF is fixed regardless of the choice of triangle ABC.

Epsilon 15.7. (ELMO Shortlist 2013) In triangle ABC, a point D lies on line BC. The circumcircle of triangle ABD meets AC at F (other than A), and the circumcircle of triangle ADC meets AB at E (other than A). Prove that as D varies, the circumcircle of triangle AEF always passes through a fixed point other than A, and that this point lies on the median from A to BC.

Epsilon 15.8. (IMO 2015) Let ABC be an acute triangle with $AB > AC$. Let Γ be its circumcircle, H its orthocenter, and F the foot of the altitude from A. Let M be the midpoint of BC. Let Q be the point on Γ such that $\angle HQA = 90^\circ$ and let K be the point on Γ such that $\angle HKQ = 90^\circ$. Assume that the points A, B, C, K and Q are all different and lie on Γ in this order. Prove that the circumcircles of triangles KQH and FKM are tangent to each other.

Epsilon 15.9. (APMO 1994) Is there an infinite set of concyclic points such that the distance between any two points is rational? (Comment: ok ok, this is a number theory problem. But what happens if you invert the line $x = 1$ about the unit circle? Specifically, what happens to the points of the form $\left(1, \frac{2s}{s^2-1}\right)$ where $s \in \mathbb{Q}$?)

Chapter 16

The Monge-D'Alembert Circle Theorem

Definition. Consider two circles ω_1 and ω_2 with centers O_1 and O_2 and radii r_1 and r_2 respectively. Then the **exsimilicenter** of ω_1 and ω_2 is the point E on line O_1O_2 that satisfies

$$\frac{EO_1}{EO_2} = \frac{r_1}{r_2}$$

and the **insimilicenter** of ω_1 and ω_2 is the point I lying on line O_1O_2 that satisfies

$$\frac{IO_1}{IO_2} = -\frac{r_1}{r_2}$$

where we are using directed lengths.

Delta 16.1. If the common external tangents to circles ω_1 and ω_2 intersect at E, and the common internal tangents to circles ω_1 and ω_2 intersect at I, prove that E and I are the exsimilicenter and insimilicenter of ω_1 and ω_2 respectively.

Among numerous beautiful theorems in geometry, some stand out for their simplicity and broad applicability in various problems where it is often hard to obtain the same result with equal elegance using other techniques. One such result is the **Monge-D'Alembert Circle Theorem** (which we will refer to as Monge's Theorem for short), named after the renowned French geometers Gaspard Monge and Jean-le-Rond D'Alembert.

Theorem 16.1. (The Monge-D'Alembert Circle Theorem) The pairwise exsimilicenters of three distinct circles, all lying in the same plane, are collinear. In particular, for three non-intersecting circles, the pairwise intersections of their common external tangents are collinear.

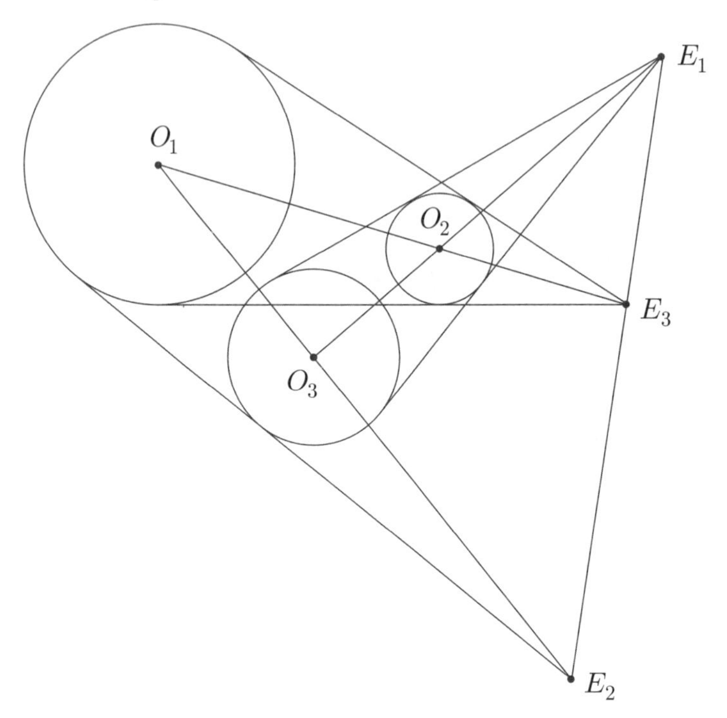

First Proof. Denote our three circles by $\omega_1, \omega_2, \omega_3$ and let them have centers O_1, O_2, O_3 and radii r_1, r_2, r_3 respectively. Denote the exsimilicenter of ω_2 and ω_3 by E_1 and define E_2 and E_3 similarly. Then points E_1, E_2, E_3 lie on lines O_2O_3, O_3O_1, O_1O_2 respectively such that

$$\frac{E_1O_2}{E_1O_3} = \frac{r_2}{r_3}, \quad \frac{E_2O_3}{E_2O_1} = \frac{r_3}{r_1}, \quad \frac{E_3O_1}{E_3O_2} = \frac{r_1}{r_2}.$$

Hence, by Menelaus' theorem it follows that the points E_1, E_2, E_3 are collinear as desired. □

Second Proof. Assume that none of circles lie inside one another. Using the same notation as in the first proof, consider the spheres determined $\Omega_1, \Omega_2, \Omega_3$ by circles $\omega_1, \omega_2, \omega_3$ respectively. Let the plane of circles $\omega_1, \omega_2, \omega_3$ be denoted by P_1 and let the other plane tangent to $\Omega_1, \Omega_2, \Omega_3$ with all three spheres on the same side of the plane be denoted by P_2. These planes intersect at a line ℓ. E_1, E_2, E_3 are also the pairwise exsimilicenters of the spheres (where we extend the definition of exsimilicenters to spheres in the obvious way) and it's clear that these points all must lie on line ℓ. This completes the proof. □

Note that the first proof can easily be adapted to show the following variation, which we will also refer to (in short) as Monge's Theorem:

Theorem 16.2. Two of the insimilicenters determined by three distinct circles, all lying in the same plane, are collinear with the exsimilicenter of the last pair of circles.

Let us go into some applications of this remarkable theorem! We begin with Paul Yiu's main result from [37].

Delta 16.2. Let Ω be the circumcircle of triangle ABC and let ω_a be the circle tangent to segment CA, segment AB, and Ω. Define ω_b and ω_c similarly. Let $\omega_a, \omega_b, \omega_c$ touch Ω at points A', B', C' respectively. Then lines AA', BB', CC' concur on line OI where O and I are the circumcenter and incenter of triangle ABC respectively.

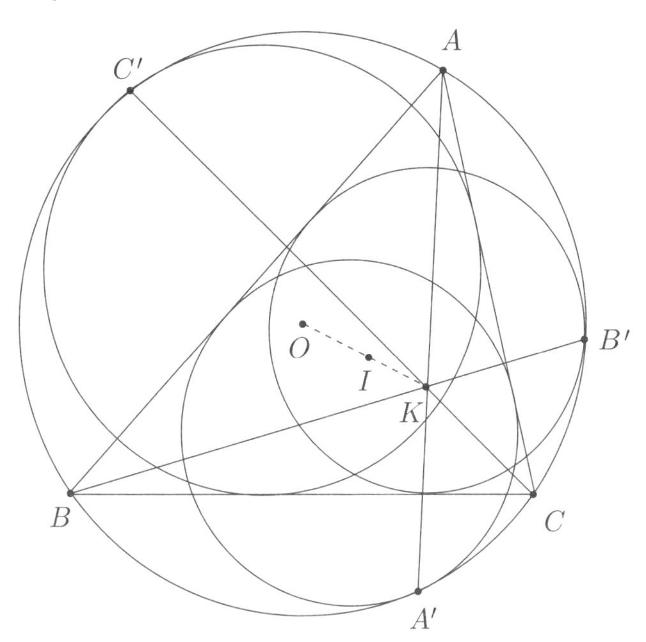

Proof. Let ω be the incircle of triangle ABC and let K be the exsimilicenter of ω and Ω. Note that lines CA and AB are the common external tangents to circles ω and ω_a so by **Delta 16.1**, A is the exsimilicenter of these two circles. Also, since ω_a and Ω are tangent at A', it's clear that A' is the exsimilicenter of these two circles. Therefore by Monge's Theorem on Ω, ω, and ω_a we have that K lies on line AA'. Similarly K lies on lines BB' and CC' and since by definition K lies on line IO we have the desired result. □

We will return to this configuration later in the material where we will make use of inversion to show a few more interesting and difficult properties. We now pass to another quick application of the Monge-D'Alembert Theorem: Archimedes' Lemma! We are now ready to see the second proof of that result.

Delta 16.3. (Archimedes' Lemma) Let ω_2 be a circle internally tangent to a larger circle ω_1 at point A, let XY be a chord of ω_1 tangent to ω_2 at point B, and let C the midpoint of the arc XY not containing A of ω_1. Then points A, B, and C are collinear.

Proof. Consider the three circles ω_1, ω_2, and the degenerate circle line XY. The exsimilicenter of ω_1 and ω_2 is A, the exsimilicenter of ω_2 and line XY is B, and last but not least, the exsimilicenter of line XY and ω_1 is C (this is harder to visualize; try to convince yourselves it makes sense). Thus, the conclusion follows by Monge's Theorem. $\square$

Now, something slightly harder: a problem from the 2007 Romanian Team Selection Test.

Delta 16.4. (Romania TST 2007) Given a triangle ABC, let Γ_A be a circle tangent to sides AB and AC, let Γ_B be a circle tangent to sides BC and BA, and let Γ_C be a circle tangent to sides CA and CB. Suppose $\Gamma_A, \Gamma_B, \Gamma_C$ are all tangent to one another. Let D be the tangency point between Γ_B and Γ_C, E the tangency point between Γ_C and Γ_A, and F the tangency point between Γ_A and Γ_B. Prove that the lines AD, BE, CF are concurrent.

Proof. Let X be the exsimilicenter of Γ_B and Γ_C. Since E is the insimilicenter of Γ_C and Γ_A and F is the insimilicenter of Γ_C and Γ_A by Monge's Theorem on $\Gamma_A, \Gamma_B, \Gamma_C$ we have that $X = EF \cap BC$. Similarly if $Y = FD \cap CA$ and $Z = DE \cap AB$ we have that Y is the exsimilicenter of Γ_C and Γ_A and Z is the exsimilicenter of Γ_A and Γ_B. Therefore by Monge's Theorem again on $\Gamma_A, \Gamma_B, \Gamma_C$ we have that points X, Y, Z are collinear. Hence, by Desargues' Theorem on triangles ABC and DEF we obtain the desired concurrency. $\square$

//These three circles are known as the **Malfetti Circles** of triangle ABC. Though given in a Romanian IMO Team Selection Test from 2007 as indicated, this result is in fact rather known in literature. The common point of AD, BE and CF appears as X_{179} in [23], with trilinear coordinates

$$\left(\sec^4 \frac{A}{4} : \sec^4 \frac{B}{4} : \sec^4 \frac{A}{4}\right),$$

and is called the first Ajima-Malfatti point.

The next exercise is a difficult proposal of Poland for the 48th edition of the IMO, hosted by Vietnam in 2007. This problem was sent by Waldemar Pompe and appeared as problem G8 on the IMO Shortlist. Again, a very elegant solution is possible, using the Monge-D'Alembert Circle Theorem. With that said, let's again scale the G8 summit!

Delta 16.5. (IMO Shortlist 2007) Point P lies on side AB of a convex quadrilateral $ABCD$. Let ω be the incircle of triangle CPD, and let I be its incenter. Suppose that ω is tangent to the incircles of triangles APD and BPC at points K and L, respectively. Let lines AC and BD meet at E, and let lines AK and BL meet at F. Then, the points E, I, and F are collinear.

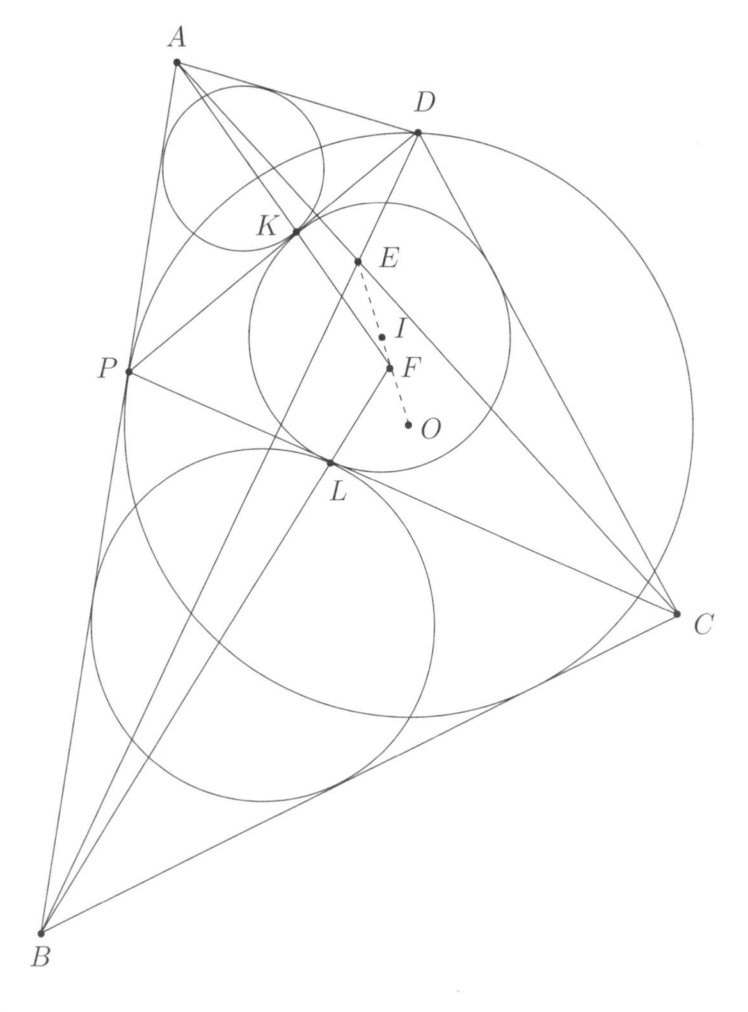

Proof. We begin with an important claim.

Claim. (Pithot's Theorem) A convex quadrilateral $ABCD$ has an inscribed circle if and only if $AB + CD = AD + BC$.

Proof. The direct implication is immediate. Assume that $ABCD$ has an inscribed circle and let X, Y, Z, T be the tangency points of the incircle with sides AB, BC, CD, DA respectively. By equal tangents we have that $AX = AT$, $BX = BY$, $CY = CZ$, $DZ = DT$; thus, we immediately get that $AB + CD = AD + BC$, as claimed.

Conversely, let assume for sake of contradiction that $ABCD$ does not have an inscribed circle, and let Γ be the circle tangent to sides DA, AB, and BC. According to our assumption, this circle is not tangent to CD, so if we take the tangent from D to Γ different from DA and intersect it with BC at a point C', then $C' \neq C$. But, then quadrilateral $ABC'D$ has an incircle, so by the direct implication we have that $AB + C'D = AD + BC'$; thus, by subtracting this identity from its analog about quadrilateral $ABCD$, it follows that $CD - C'D = CC'$, which means, via the triangle inequality, that the points C, D, and C' are collinear. However, by constructions, this shows that $C' = C$, which is a contradiction. Hence, Γ needed to be the inscribed circle of quadrilateral $ABCD$ as desired. This proves the converse.

Returning to the problem, consider the circle Γ with center O tangent to the sides AB, BC, AD of the quadrilateral $ABCD$ and denote by ω_1, ω_2, ω the incircles of triangles APD, BPC, and CPD respectively. Since by **Delta 16.1** A is the exsimilicenter of ω_1 and Γ and K is the insimilicenter of ω_1 and ω, from Monge's Theorem on Γ, ω, ω_1, we know that the line AK intersects OI at the insimilicenter of Γ and ω. Analogously, we get that the line BL intersects OI at the insimilicenter of Γ and ω. Hence F is the insimilicenter of Γ and ω and thus it remains to prove that E lies on the line OI.

Using the simple fact that the tangents from a point to a circle are equal in length, a quick calculation yields that $BC+PD = BP+CD$ and $AD+PC = AP+CD$ so by Pithot's Theorem, the quadrilaterals $APCD$ and $PBCD$ have inscribed circles. Denote by ω_a and ω_b their respective incircles. Now A is the exsimilicenter of ω_a and Γ and C is the exsimilicenter of ω_a and ω, and thus, by Monge's Theorem on Γ, ω, ω_a, the line AC intersects the line OI at the exsimilicenter of the circles ω and Γ. Similarly, the line BD intersects line OI at the exsimilicenter of ω and Γ. Therefore, we conclude that E and F are the insimilicenter and exsimilicenter of the circles Γ and ω respectively. This completes the proof, and moreover shows that the $(O, I; E, F)$ is harmonic. □

As a final application of the Monge-D'Alembert Circle Theorem, we chose one of the most difficult IMO problems ever: the last problem from the 2008

IMO. It was proposed by Vladimir Shmarov, and selected as problem 6 in the contest. The problem was solved by an exceptionally low number of students. Only 12 out of the 535 contestants managed to give perfect solutions!

Delta 16.6. (IMO 2008). Let $ABCD$ be a convex quadrilateral with $BA \neq BC$. Denote the incircles of triangles ABC and ADC by k_1 and k_2 respectively. Suppose that there exists a circle k tangent to ray BA beyond A and to ray BC beyond C, which is also tangent to the lines AD and CD. Prove that the common external tangents to k_1 and k_2 intersect on k.

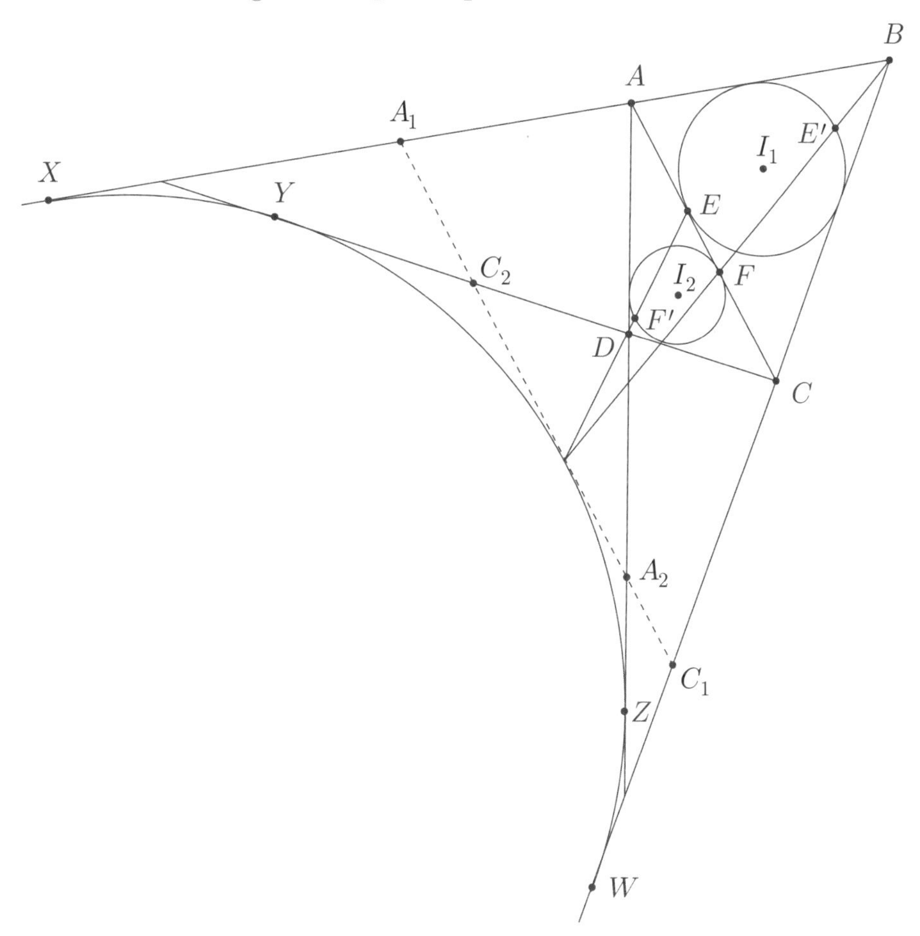

Proof. Let K be the intersection of these tangents. Let I_1, I_2 and I be the centers of k_1, k_2 and k, respectively. Let k_1 and k_2 touch AC at E and F, respectively. Let E' and F' be the points diametrically opposite to E and F on k_1 and k_2 respectively. Let BC, BA, DC and DA touch k at W, X, Y and Z respectively. Then

$$2CE = CA + CB - AB = CA + AX - CW = CA + AD - DC = 2AF$$

and so $CE = AF$. As a result, points C, E', F are collinear and points D, F', E are collinear and so since $I_1E' \parallel I_2F$ and $I_2F' \parallel I_1E$, we see that $BF \cap DE = K$. Now, draw a line ℓ parallel to AC through K and let it intersect lines BA, BC, DA, DC at A_1, C_1, A_2, C_2 respectively. Consider the homothety centered at B that takes segment AC to segment A_1C_1. We know it takes F to K and since F is the point where the B-excircle of triangle ABC touches AC, we can conclude that K is the point where the B-excircle of triangle BA_1C_1 touches ℓ. Similarly, K is the point where the D-excircle of triangle DA_2C_2 touches ℓ. Denote these excircles by ω_1 and ω_2, respectively. By Monge's Theorem on k, ω_1 and ω_2 we see that points B, D and K must be collinear, unless two of the circles share their center (in which case we cannot use Monge's Theorem). But it's easy to see that if points B, D, K are collinear then $BA = BC$, contradiction. Therefore we immediately have that $k = \omega_1 = \omega_2$ and hence the proof is complete. $\square$

Assigned Problems

Epsilon 16.1. Let k_1 and k_2 be two given circles. Consider all circles k externally tangent to both of them and denote the tangency points by T_1 and T_2 respectively. Prove that line T_1T_2 passes through a fixed point.

Epsilon 16.2. (ELMO Shortlist 2011) Let ABC be a triangle. Draw circles ω_A, ω_B, and ω_C such that ω_A is tangent to AB and AC, and ω_B and ω_C are defined similarly. Let P_A be the insimilicenter of ω_B and ω_C. Define P_B and P_C similarly. Prove that AP_A, BP_B, and CP_C are concurrent.

Epsilon 16.3. (China 2013) Let non-intersecting circles $\omega_1, \omega_2, \omega_3$ all be internally tangent to a circle Ω at points A, B, C respectively. Let ℓ_1 and ℓ_2 and ℓ_3 be common external tangents to circles ω_2, ω_3 and ω_3, ω_1 and ω_1, ω_2 and let $X = \ell_2 \cap \ell_3$, $Y = \ell_3 \cap \ell_1$, and $Z = \ell_1 \cap \ell_2$. Prove that lines AX, BY, CZ concur on line IO where I is the incenter of triangle XYZ and O is the center of Ω.

Epsilon 16.4. Given a triangle ABC, let Γ_A be a circle tangent to sides AB and AC, let Γ_B be a circle tangent to sides BC and BA, and let Γ_C be a circle tangent to sides CA and CB. Suppose $\Gamma_A, \Gamma_B, \Gamma_C$ are all tangent to one another. Let E be the tangency point between Γ_C and Γ_A and let F be the tangency point between Γ_A and Γ_B. Prove that the lines BF and CE concur on the A-internal angle bisector of triangle ABC.

Epsilon 16.5. Given circle ω. Let ω_1 and ω_2 be two circles that internally touch ω at A and B respectively; d_1, d_2 be two tangents from A to ω_2, d_3, d_4 be two tangents from B to ω_1. Prove that lines d_1, d_2, d_3, d_4 determine a quadrilateral with an inscribed circle.

Epsilon 16.6. Let k_1, k_2 be two circles and let ω be a circle externally tangent to both k_1 and k_2 at A, B respectively. Let Ω be a circle orthogonal to both k_1 and k_2 and let C be one of the intersections of Ω and k_1 and let D be one of the intersections of Ω and k_2. Then the exsimilicenter X of k_1 and k_2 is on the radical axis of ω and Ω.

Epsilon 16.7. Circles k_1 and k_2 are tangent to one of their common external tangents at T_1 and T_2 respectively. A circle k is externally tangent to k_1 ad k_2 at points L_1, L_2 respectively. Prove that lines L_1T_1 and L_2T_2 concur on k.

Epsilon 16.8. (ELMO Shortlist 2011) Let $\omega, \omega_1, \omega_2$ be three mutually tangent circles such that ω_1, ω_2 are externally tangent at P, ω_1, ω are internally tangent at A, and ω, ω_2 are internally tangent at B. Let O, O_1, O_2 be the centers of $\omega, \omega_1, \omega_2$, respectively. Given that X is the foot of the perpendicular from P to AB, prove that $\angle O_1XP = \angle O_2XP$.

Chapter 17

Mixtilinear and Curvilinear Incircles

We begin with the so-called **mixtilinear incircles** of a triangle. Geometric constructions, properties and relations between them (together with their nice history) can be found for example in [28] and [37]. Beware - this section is one of the most difficult in the book. Nonetheless, we promise that spending time on it will turn out to be extremely rewarding.

Definition. The A-mixtilinear incircle of triangle ABC is the circle internally tangent to the circumcircle of triangle ABC and tangent to segments AB and AC. Note that we've already encountered mixtilinear incircles in problems like **Delta 16.2**!

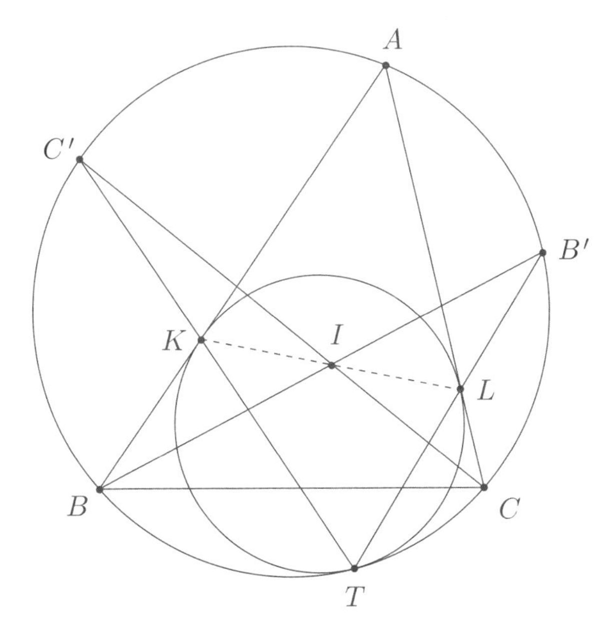

Delta 17.1. Let Ω be the circumcircle of a triangle ABC and let the A-mixtilinear of this triangle be tangent to Ω, AB, AC at points T, K, L respectively. If I is the incenter of triangle ABC, show that I is the midpoint of segment KL.

Proof. Let B' be the midpoint of arc AC not containing B of Ω and let C' be the midpoint of arc AB not containing C of Ω. It's clear that $I = BB' \cap CC'$ and by Archimedes' Lemma applied twice we also have that $T = B'L \cap C'K$. Therefore by Pascal's Theorem on cyclic hexagon $BB'TC'CA$ we have that points I, K, L are collinear. Also, since AK and AL are the tangents from A to the A-mixtilinear incircle of triangle ABC we have that $AK = AL$. But since AI is the A-internal bisector of triangle AKL, we may conclude that I is the midpoint of KL as desired. □

Corollary 17.1. Using the same notation as in the proof of **Delta 17.1**, we have that $\angle KTI = \angle LTA$.

Proof. Note that the tangents at K and L to the circumcircle of triangle KLT (the A-mixtilinear incircle of triangle ABC) intersect at A so line TA is the T-symmedian of triangle KLT. But we know from **Delta 17.1** that TI is the T-median in triangle KLT and so lines TA and TI are isogonal with respect to triangle KLT as desired. □

Delta 17.2. Using the same notation as in **Delta 17.1**, show that line TI bisects angle $\angle BTC$.

Proof. Since $AI \perp KL$ an easy angle chase shows that $KL \parallel B'C'$. Now, notice that

$$\angle TKI = \angle TKL = \angle TC'B' = \angle TBB' = \angle TBI$$

so quadrilateral $TBKI$ is cyclic. Similarly quadrilateral $TCLI$ is cyclic and so we have

$$\angle BTI = \angle AKL = \angle ALK = \angle CTI$$

which implies the desired result. □

The next solution will be a rare gem in Olympiad geometry. Don't forget it!

Delta 17.3. (EGMO 2013) Let Ω be the circumcircle of triangle ABC. The circle ω is tangent to the sides AB and AC, and it is internally tangent to the circle Ω at the point T. A line ℓ parallel to BC intersecting the interior of triangle ABC is tangent to ω at Q. Prove that $\angle BAT = \angle CAQ$.

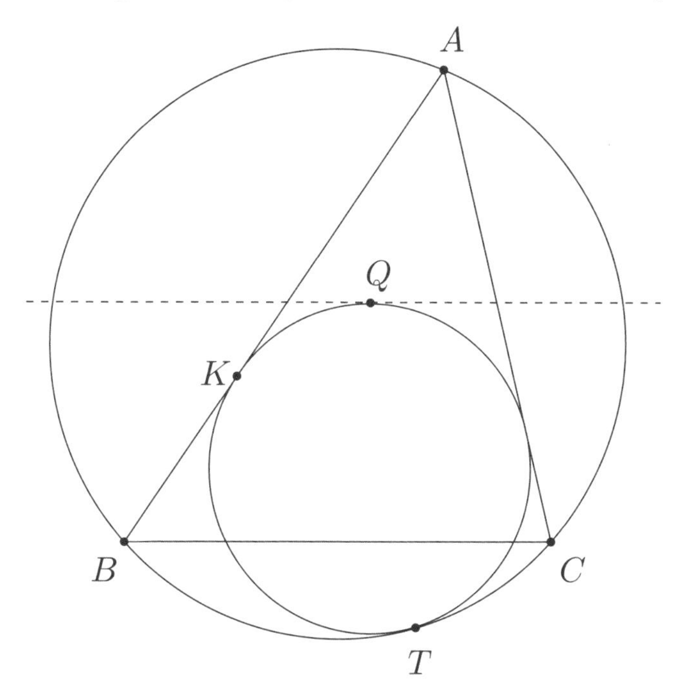

Proof. Let ω touch side AB at point K. Consider the composition of the inversion about the circle centered at A with radius AK and the reflection over the A-internal bisector of triangle ABC. It's clear that ω maps to itself. Moreover, Ω inverts to a line that is an anti-parallel to side BC with respect to triangle ABC and so after the reflection it maps to a line parallel to BC. But inversion and reflection both preserve tangency so Ω must map to line ℓ! Hence, point T maps to point Q which immediately implies the desired result. $\square$

//By considering the homothety centered at A that sends ω to the A-excircle of triangle ABC we find that line AQ passes through the point where the A-excircle of triangle ABC touches side BC. Therefore, utilizing the result in **Delta 16.2**, we can conclude that the Nagel point and the exsimilicenter of the incircle and the circumcircle are isogonal conjugates with respect to triangle ABC.

Delta 17.4. Let ω and I be the incircle and incenter respectively of triangle ABC and let ω touch BC at point D. Let A' be the midpoint of arc BC not containing A of the circumcircle of triangle ABC. If M is the midpoint of segment ID, show that $A'M$ is the radical axis of the B and C-mixtilinear incircles of triangle ABC.

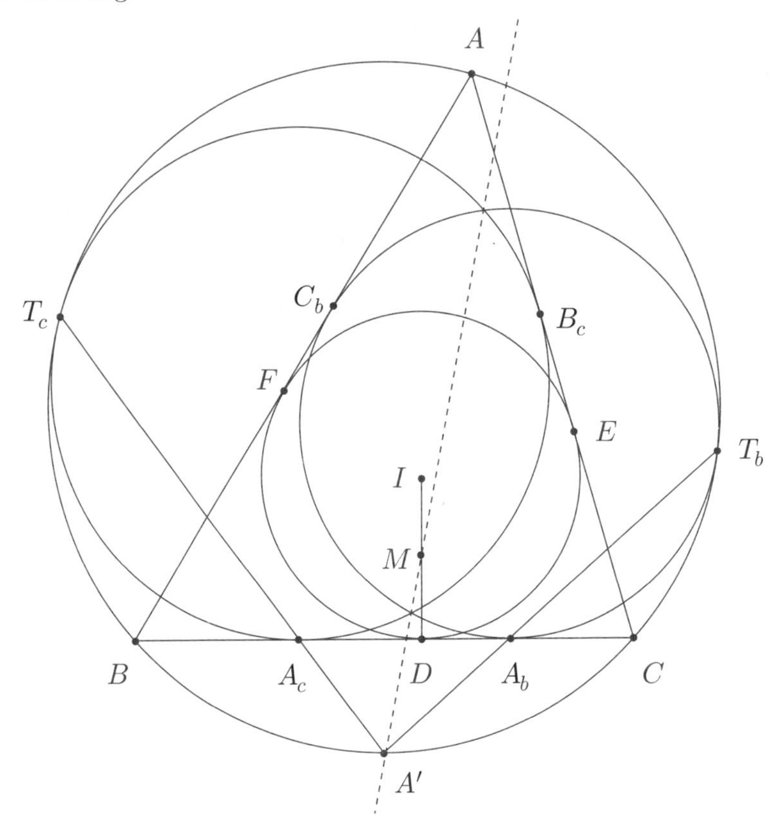

Proof. Let Ω be the circumcircle of triangle ABC and let ω_b, ω_c be the B and C-mixtilinear incircles of triangle ABC respectively. Let ω_b be tangent to Ω, BC, and BA at T_b, A_b, C_b respectively and let ω_c be tangent to Ω, CA, and CB at T_c, B_c, A_c respectively. We know from Archimedes' Lemma that $A' = A_bT_b \cap A_cT_c$ and moreover that $A'A_b \cdot A'T_b = A'B^2 = A'C^2 = A'A_c \cdot A'T_c$ so the powers of A' with respect to the ω_b and ω_c are equal. Now, let ω touch sides AC and AB at E and F respectively. Since the ω_b touches sides BC and BA at points A_b and C_b respectively and since ω touches sides BC and AB at points D and F respectively we have that $DF \parallel A_bC_b$ and hence by **Delta 17.1** that $DF \parallel IC_b$. It's clear that the radical axis of ω and ω_b is the line directly in the middle of lines DF and IC_b so it passes through the midpoint of segment DI; namely, through M. Similarly the radical axis of ω and ω_c passes through M so M is the radical center of $\omega, \omega_b, \omega_c$. Therefore M lies on the radical axis of ω_b and ω_c as well - hence, we have the desired result.

We proceed with one of the author's favorite problems - in fact, a monetary reward was given to one of the authors for the following solution!

Delta 17.5. (ELMO Shortlist 2015) Let ABC be a triangle with incenter I and incircle ω. Suppose E and F lie on segments AB and AC, respectively such that $EF \parallel BC$ and EF is tangent to ω at D'. Let ω' be the incircle of AEF, tangent to EF at G. Prove that if GI meets ω' again at T, there is a line passing through T tangent to ω' and the B and C-mixtilinear incircles of triangle ABC.

Proof. Let ω_B and ω_C be the B, C-mixtilinear incircles of triangle ABC respectively and let them touch side BC at Y, Z respectively. Let their centers be O_B and O_C respectively. Let K be the their exsimilicenter and let M be the midpoint of arc BC not containing A of the circumcircle of triangle ABC. Let ω touch side BC at D and let I' be the center of circle ω'. Consider the inversion about the circle centered at K orthogonal to the circumcircle of triangle ABC. This inversion sends ω_B to ω_C and B to C. Therefore the inversion takes the circle with diameter BY to the circle with diameter CZ and since $\angle BIY = \angle CIZ = 90°$ by **Delta 17.1**, this implies that I inverts to itself. Hence, line KI must be tangent to the circumcircle of triangle BIC.

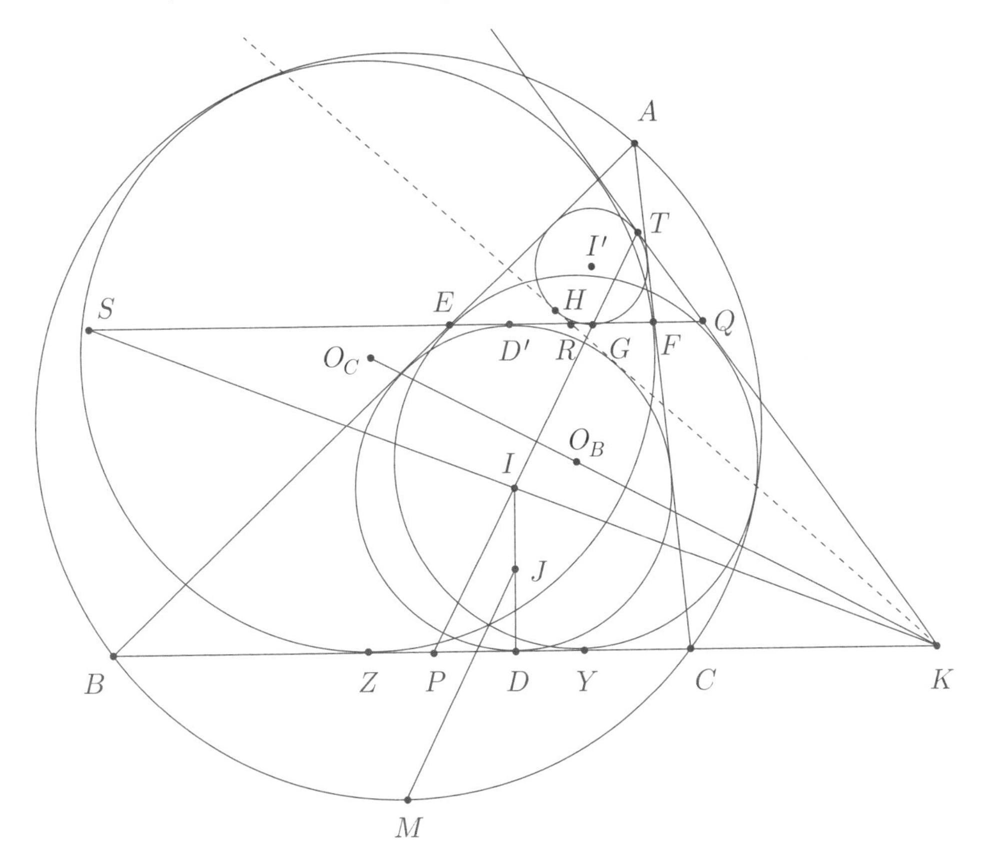

Since the center of this circle is M, we have that $KI \perp IM$ and thus $KI \perp II'$. Now let $R = II' \cap EF$ and $S = KI \cap EF$. Since $EF \parallel BC$ and $ID = ID'$, triangles KID and SID' are congruent and so $IS = IK$. But $RI \perp SK$ so triangle KRS is isosceles. Thus $\angle RKI = \angle RSI = \angle DKI$ and hence line KR is an internal common tangent to circles ω and ω'. Now let line KR be tangent to ω' at H. Since $\angle KII' = \angle KHI' = 90^\circ$ we have that quadrilateral $KIHI'$ is cyclic so

$$\angle I'KI = 180^\circ - \angle I'HI = 180^\circ - \angle I'GI = \angle TGI' = \angle I'TI'$$

where we used the fact that triangle $TI'G$ is isosceles for the last equality. Therefore quadrilateral $KITI'$ is cyclic as well and hence $\angle KTI' = \angle KII' = 90^\circ$ so line KT is tangent to ω'.

Now, let J be the midpoint of segment ID and let E_a be the A-excenter of triangle ABC. Note that the homothety centered at A that takes G to D also takes I to E_a so we have that $GI \parallel DE_a$. Since J is the midpoint of DI and since M is the midpoint of E_aI we have that $MJ \parallel DE_a$ and since by **Delta 17.4** the radical axis of ω_B and ω_C is line MJ, we can conclude that $GI \perp O_BO_C$.

Now, let $P = IG \cap BC$. To finish off this problem, it suffices to show that the reflection of line KP over line O_BO_C is line KT. Since $PT \perp O_BO_C$, it suffices to show that $\angle KPT = \angle KTP$. Letting Q be the intersection of line EF with the tangent to ω' at T, it suffices to show that $\angle QGT = \angle QTG$ which is immediate since lines QG and QT are both tangents to ω'. Hence, the proof is complete. Phew! □

Delta 17.6. (ELMO Shortlist 2014) In triangle ABC with incenter I and circumcenter O, let A', B', C' be the points of tangency of its circumcircle with its A, B, C-mixtilinear circles, respectively. Let ω_A be the circle through A' that is tangent to AI at I, and define ω_B, ω_C similarly. Prove that $\omega_A, \omega_B, \omega_C$ have a common point X other than I, and that $\angle AXO = \angle OXA'$.

Proof. Let A_1 be the midpoint of arc BAC of the circumcircle of triangle ABC and define B_1 and C_1 similarly. Note that by **Delta 17.2**, we have that $I = A'A_1 \cap B'B_1 \cap C'C_1$. Now, consider the composition of the inversion about the circle centered at I with radius

$$\sqrt{A'I \cdot A_1I} = \sqrt{B'I \cdot B_1I} = \sqrt{C'I \cdot C_1I}$$

and a reflection over I. We have that ω_A inverts to the line parallel to AI passing through the reflection of A_1 over I so ω_A maps to the line through A_1

parallel to AI. Denote this line by ℓ_A and define ℓ_B, ℓ_C similarly. Now, let E_A, E_B, E_c be the centers of the A, B, C-excircles of triangle ABC respectively. Since $AI \perp E_BE_C$ and $CI \perp E_CE_A$ we have that I is the orthocenter of triangle $E_AE_BE_C$ and that the circumcircle of triangle ABC is the nine-point circle of triangle $E_AE_BE_C$. But since $AA_1 \perp AI$ this means that A_1 is the midpoint of segment E_BE_C and so lines ℓ_A, ℓ_B, ℓ_C concur at the circumcenter of triangle $E_AE_BE_C$. This establishes the existence of point X and moreover proves that X lies on line IO, the Euler line of triangle $E_AE_BE_C$.

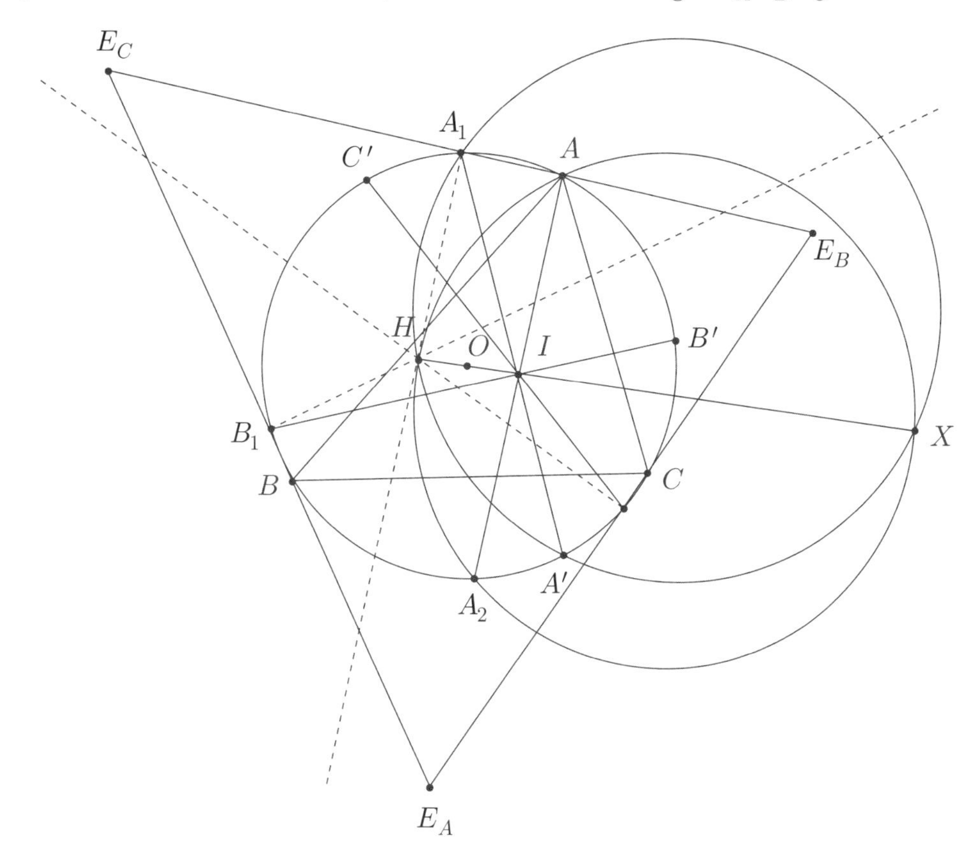

Now denote the circumcenter of triangle $E_AE_BE_C$ by H. It's clear that H is the reflection of I over O. Let A_2 be the midpoint of arc BC not containing A of the circumcircle of triangle ABC. Since

$$IA \cdot IA_2 = IA' \cdot IA_1 = IH \cdot IX$$

we have that points A, A_2, H, X are concyclic and points A', A_1, H, X are concyclic. Moreover, since lines IH and A_1A_2 bisects each other at point O we have that A_1HA_2X is a parallelogram. Therefore

$$\angle AXO = \angle AXH = \angle AA_2H = \angle A'A_1H = \angle A'XH = \angle OXA'$$

as desired. This completes the proof. □

The next problem was given as the last problem on one day of the Taiwanese Team Selection Test in 2014. It was fully solved by none of the contestants, which is especially surprising since that year the Taiwan team placed third at the IMO and had one of only three perfect scorers!

Delta 17.7. (Cosmin Pohoata, Taiwan TST 2014) Let M be any point on the circumcircle of triangle ABC. Suppose the tangents from M to the incircle of triangle ABC meet line BC at two points X_1 and X_2. Prove that the circumcircle of triangle MX_1X_2 intersects the circumcircle of triangle ABC again at the tangency point of the A-mixtilinear incircle of triangle ABC with the circumcircle of triangle ABC.

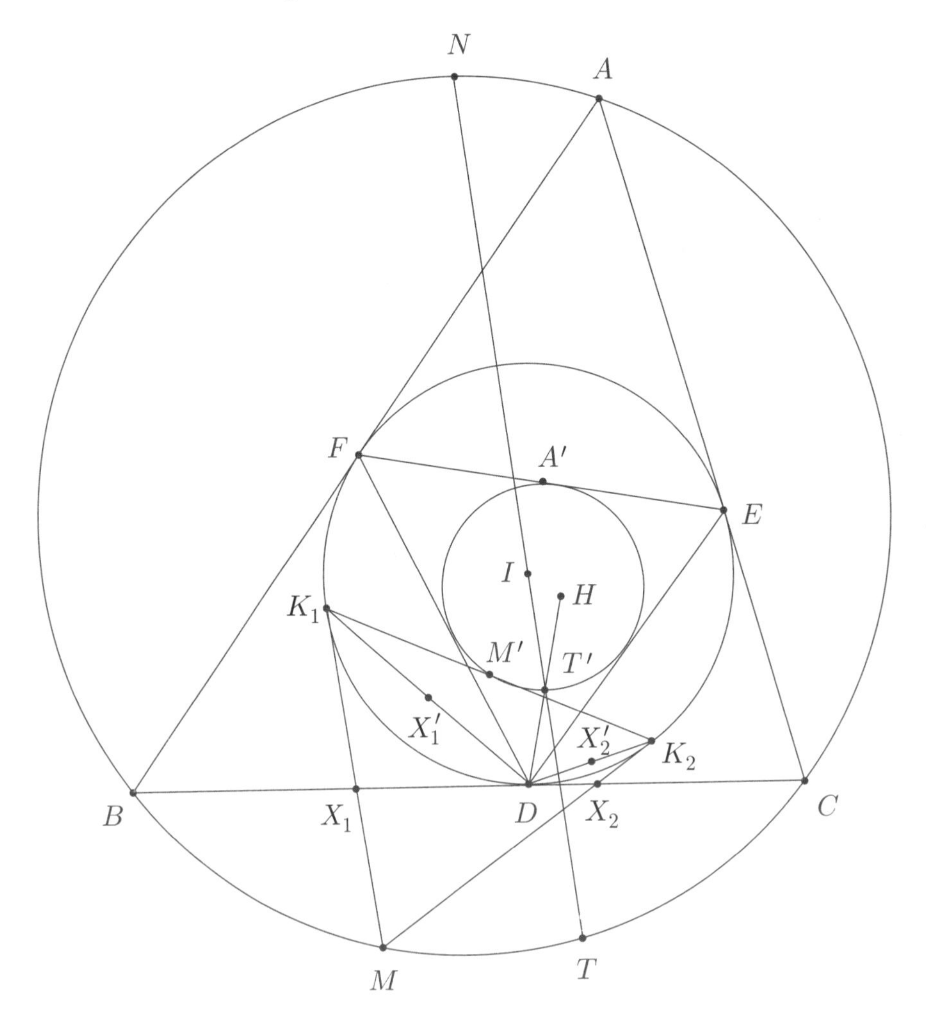

Proof. Let ω be the incircle of triangle ABC and let ω touch sides BC, CA, AB at points D, E, F respectively. Let the tangents to the incircle be MK_1 and MK_2. Let the A-mixtilinear incircle of triangle ABC touch the circumcircle of triangle ABC at T and let H be the orthocenter of triangle DEF. Invert about the incircle of triangle ABC - inverses of points will be denoted by the original point name with an apostrophe added. We know from **Delta 15.5** that A' is the midpoint of segment EF thus the circumcircle of triangle $A'B'C'$ is the nine-point circle of triangle DEF. Let N be the midpoint of arc BAC of the circumcircle of triangle ABC. Then we have that $\angle IN'A' = \angle IAN = 90°$. But since by **Delta 17.2** points N, I, T are collinear, this implies that T' is diametrically opposite from A' on the nine-point circle of triangle DEF. Hence, T' is the midpoint of segment DH. Moreover, it's easy to see that M' is the midpoint of segment K_1K_2, X_1' is the midpoint of segment DK_1, and X_2' is the midpoint of segment DK_2. Now, let D_1 and H_1 be the reflections of D and H respectively over M'. Points K_1, K_2, D, H_1 all lie on the incircle of triangle ABC and so points K_1, K_2, D_1, H all lie on the circle obtained by reflecting the incircle over line K_1K_2. Now, consider the homothety centered at D with ratio $\frac{1}{2}$ - this homothety sends points K_1, K_2, D_1, H to X_1', X_2', M', T' respectively. Hence points X_1', X_2', M', T' are concyclic, and so the proof is complete. □

We proceed by generalizing the concept of a mixtilinear incircle, as well as generalizing some of the results already given in this section.

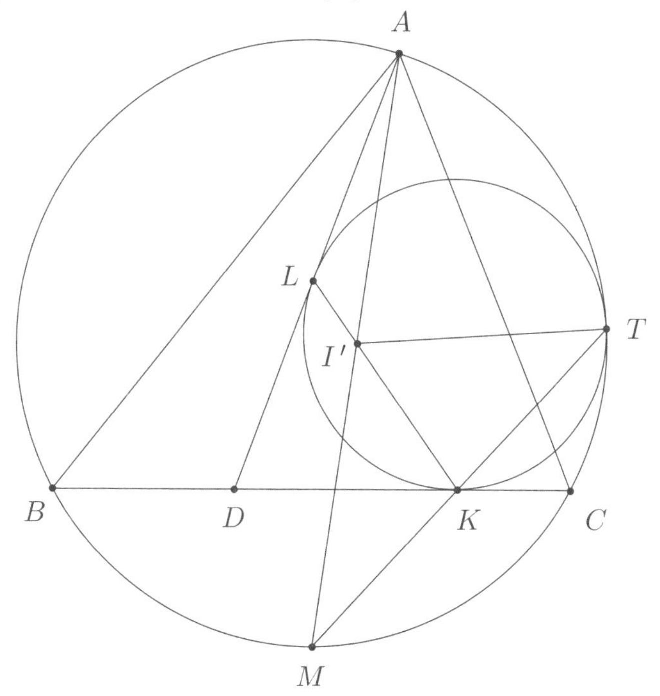

Definition. Let ABC be a triangle inscribed in a circle Ω and let D be a point on segment BC. The circle internally tangent to Ω and tangent to segments AD and CD is called a **curvilinear incircle**. These curvilinear incircles satisfy numerous interesting properties, as we shall soon see.

Theorem 17.1. (Sawayama's Theorem) Let ABC be a triangle with incenter I and circumcircle Ω. Let D be a point on segment BC. Consider circle internally tangent to Ω at T and tangent to segments CD and AD at K, L respectively. Then points K, L, I are collinear.

Proof. Let M be the midpoint of arc BC not containing A of Ω and let $I' = KL \cap AM$. By Archimedes' Lemma, we have that M lies on line TK. Notice that by homothety arc TK not containing L of the curvilinear incircle is equal in measure to arc TCM of Ω so $\angle TLI' = \angle TAI'$, hence quadrilateral $TALI'$ is cyclic. Therefore

$$\angle MKI' = 180^\circ - \angle TKL = 180^\circ - \angle TLA = 180^\circ - \angle TI'A = \angle MI'T$$

and so triangle $MI'K$ is similar to triangle MTI'. Thus by part (b) of Archimedes' Lemma we have $MI'^2 = MK \cdot MT = MC^2$ and so we must have $I' = I$. This completes the proof. □

Notice that **Delta 17.1** is just a degenerate case of Sawayama's Theorem! We proceed with a related result, discovered by the French mathematician Victor Thebault and first proved by none other than J. Sawayama himself in 1905. The proof unexpectedly utilizes Pappus's Theorem, and is again another gem in Olympiad geometry.

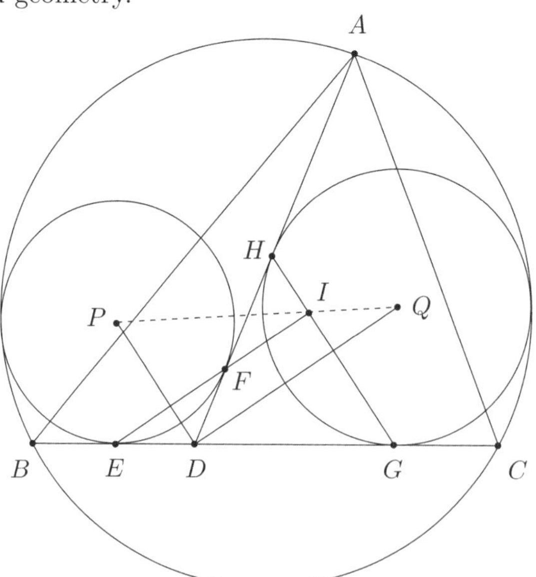

Theorem 17.2. (Thebault's Theorem) Let ABC be a triangle with incenter I and circumcircle Ω. Let D be a point on segment BC. Let k_1 be the circle internally tangent to Ω and tangent to segments DB at E and DA at F and let k_2 be the circle internally tangent to Ω and tangent to segments DC at G and DA at H. Let the centers of k_1 and k_2 be P and Q respectively. Then points P, Q, I are collinear.

Proof. Since $EP \perp BC$ and $GQ \perp BC$ we have that $EP \parallel GQ$. Also since line DP bisects angle $\angle BDA$ and line DQ bisects angle $\angle CDA$ we have that $DP \perp DQ$ which immediately yields that $EF \parallel DQ$ and $GH \parallel DP$. Hence, by Sawayama's Theorem, we have that $EI \parallel DQ$ and $GI \parallel DP$. Now, consider hexagon $PEIGQD$. All pairs of opposite sides are parallel and so their pairwise intersections are collinear on the line at infinity. Hence, by the converse of Pappus's Theorem, points P, Q, I must be collinear as desired. □

//The legitimacy of the converse of Pappus's Theorem comes from the fact that 5 points determine a conic (which, in this case, is two intersecting lines).

Assigned Problems

Epsilon 17.1. Let ABC be a scalene triangle. If the B-mixtilinear incircle of triangle ABC is tangent to side AB at M and the C-mixtilinear incircle of triangle ABC is tangent to side AC at N, prove that the circumcircle of triangle AMN is tangent to the A-mixtilinear incircle of triangle ABC.

Epsilon 17.2. Prove that the radical center of the three mixtilinear incircles of triangle ABC lies on the line IO where I, O are the incenter and circumcenter of triangle ABC respectively.

Epsilon 17.3. (Romania TST 1997) Let I denote the incenter of triangle ABC which has circumcircle Γ and let D be the intersection of the A-internal angle bisector with the side BC. Consider the circles $\mathcal{T}_1$ and $\mathcal{T}_2$ that are tangent to AD, DB, Γ, and AD, DC, Γ, respectively. Prove that $\mathcal{T}_1$ and $\mathcal{T}_2$ are tangent at I.

Epsilon 17.4. (Ehrmann-Pohoata) Let I denote the incenter of triangle ABC which has circumcircle Γ and let D be the point where the A-excircle of triangle ABC touches side BC. Consider the circles $\mathcal{T}_1$ and $\mathcal{T}_2$ that are tangent to AD, DB, Γ, and AD, DC, Γ, respectively. Prove that $\mathcal{T}_1$ and $\mathcal{T}_2$ are congruent.

Epsilon 17.5. (Romania TST 2006) Let ABC be an acute triangle with $AB \neq AC$. Let D be the foot of the altitude from A and ω the circumcircle of the triangle. Let ω_1 be the circle tangent to AD, BD and ω. Let ω_2 be the circle tangent to AD, CD and ω. Let ℓ be the interior common tangent to both ω_1 and ω_2, different from AD. Prove that ℓ passes through the midpoint of BC if and only if $2BC = AB + AC$.

Epsilon 17.6. D is an arbitrary point lying on side BC of triangle ABC. Circle ω_1 is tangent to segments AD , BD and the circumcircle of triangle ABC and circle ω_2 is tangent to segments AD , CD and the circumcircle of triangle ABC. Let X and Y be the touch points of ω_1 and ω_2 with BC respectively and let M be the midpoint of segment XY. Let T be the midpoint of the arc BC which does not contain A of the circumcircle of triangle ABC. If I is the incenter of triangle ABC, prove that TM passes through the midpoint of segment ID.

Epsilon 17.7. (Mathematical Reflections) Let ABC be a triangle with incircle ω and circumcircle Ω. Let ω touch sides BC, CA, AB at D, E, F respectively and let line EF intersect Ω at points X_1 and X_2. Prove that the circumcircle of triangle DX_1X_2 intersects Ω at the point where the A-mixtilinear incircle of triangle ABC touches Ω.

Epsilon 17.8. Let $ABCD$ be a cyclic quadrilateral. Prove that there exists a line mutually tangent to the D-mixtilinear incircles of triangles DAB, DAC, and DBC.

Chapter 18

Ptolemy and Casey

The classical theorem of Ptolemy (used for computations by Claudius Ptolemaeus of Alexandra, 2nd century AD, but probably known even before him [see, for example, ? and ??]) states that if A, B, C, D are, in this order, the vertices of a cyclic quadrilateral, then

$$AB \cdot CD + AD \cdot BC = AC \cdot BD.$$

This very simple identity not only generalizes the Pythagorean Theorem (when $ABCD$ is chosen to be a rectangle), but also provides numerous interesting identities within particular cyclic quadrilaterals. Moreover, this result also has a converse, which can be of valuable aid when trying to prove concyclities. We proved it (and the stronger Ptolemy's Inequality) in **Section 15**, but we also provide a simpler proof below:

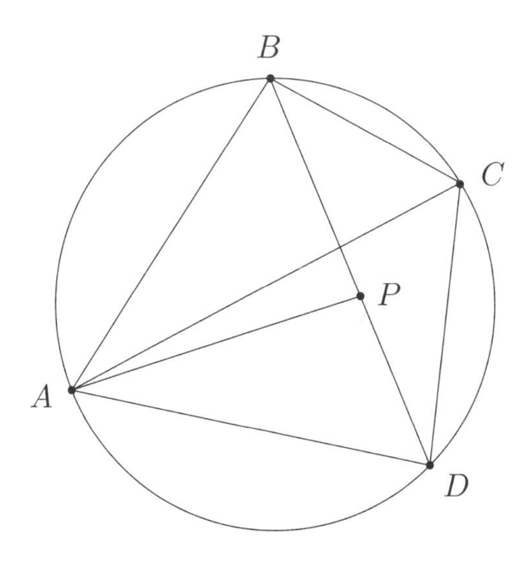

Theorem 18.1. (Ptolemy's Theorem) Let A, B, C, D be, in this order, four points in plane. Then, quadrilateral $ABCD$ is cyclic if and only if

$$AB \cdot CD + AD \cdot BC = AC \cdot BD.$$

Proof. Construct the point P in the interior of quadrilateral $ABCD$ such that triangles CAB and DAP are similar. We then have that

$$\frac{AB}{AP} = \frac{AC}{AD} = \frac{BC}{PD},$$

and so $AC \cdot PD = AD \cdot BC$.

However, $\angle BAC = \angle PAD$, so $\angle BAP = \angle CAD$. But we also know from the above that $\frac{AB}{AP} = \frac{AC}{AD}$; hence, triangles BAP and CAD are also similar. Thus, it follows that

$$\frac{AB}{AC} = \frac{BP}{CD}, \text{ or equivalently, } AC \cdot BP = AB \cdot CD.$$

Adding the two relations that we got, we deduce that

$$AC \cdot (BP + PD) = AD \cdot BC + AB \cdot CD.$$

Now, if we assume that $ABCD$ is cyclic, then $\angle PDA = \angle BCA = \angle BDA$, so P needs to lie on the diagonal BD. In this case $BP + PD = BD$, and we immediately get that

$$AB \cdot CD + AD \cdot BC = AC \cdot BD,$$

as intended. Conversely, if we assume that

$$AB \cdot CD + AD \cdot BC = AC \cdot BD$$

holds true, we need to have $BP+PD = BD$, in which case, the points B, P, D are collinear from the triangle inequality. Hence, $\angle BDA = \angle PDA = \angle BCA$, and so $ABCD$ is cyclic. This completes the proof. □

Now, let's see some applications!

Delta 18.1. In an acute-angled triangle ABC, let h_b, h_c denote the lengths of its $B-$ and $C-$altitudes. Prove that

$$\frac{h_b h_c}{a^2} = \cos A + \cos B \cos C.$$

Proof. Let BE and CF be the $B-$ and $C-$ altitudes of triangle ABC with E and F on the sides CA and AB respectively. We know that quadrilateral $BCEF$ is cyclic, so by Ptolemy's Theorem, we have that

$$h_b h_c = BC \cdot EF + BF \cdot CE.$$

But recall that $BF = a\cos B$, $CE = a\cos C$, and $EF = HA\sin A$; thus, keeping in mind also that $HA = 2R\cos A$, we get that

$$h_b h_c = 2aR\sin A\cos A + a^2\cos B\cos C.$$

Dividing by a^2 and using the Extended Law of Sines (i.e. $a = 2R\sin A$), we obtain the desired result. □

Delta 18.2. (Mathematical Reflections) In triangle ABC let B' and C' be the feet of the angle bisectors of $\angle B$ and $\angle C$ respectively. Prove that

$$B'C' \geq \frac{2bc}{(a+b)(a+c)}\left[(a+b+c)\sin\frac{A}{2} - \frac{a}{2}\right].$$

Proof. First, note that by Ptolemy's Inequality for the cyclic quadrilateral $BCB'C'$,

$$B'C' \geq \frac{BB' \cdot CC' - BC' \cdot CB'}{BC}.$$

Now, recall that $BB' = \frac{2ac}{a+c}\cos\frac{B}{2}$ and that $\cos\frac{B}{2}\cos\frac{C}{2} = \frac{s}{a}\sin\frac{A}{2}$. These imply that

$$BB' \cdot CC' = \frac{4a^2bc}{(a+b)(a+c)}\cos\frac{B}{2}\cos\frac{C}{2} = \frac{2abc(a+b+c)}{(a+b)(a+c)}\sin\frac{A}{2}.$$

In addition, $BC' = \frac{ac}{a+b}$ and $CB' = \frac{ab}{a+c}$, and multiplying yields

$$BC' \cdot CB' = \frac{a^2bc}{(a+b)(a+c)}.$$

Consequently, by combining the two identities above, it follows that

$$B'C' \geq \frac{2bc}{(a+b)(a+c)}\left[(a+b+c)\sin\frac{A}{2} - \frac{a}{2}\right],$$

as desired. This completes the proof. □

We proceed with a more Olympiad-esque problem from the 1997 IMO Shortlist.

Delta 18.3. (IMO 1997 Shortlist) The lengths of the sides of a convex hexagon $ABCDEF$ satisfy $AB = BC$, $CD = DE$, $EF = FA$. Prove that:

$$\frac{BC}{BE} + \frac{DE}{DA} + \frac{FA}{FC} \geq \frac{3}{2}.$$

Proof. Let's look at quadrilateral $ABCE$. By Ptolemy's inequality, we have that

$$CE \cdot AB + AE \cdot BC \geq AC \cdot BE,$$

thus, since $AB = BC$, we can write

$$\frac{BC}{BE} \geq \frac{AC}{CE + AE}.$$

Similarly, we get that

$$\frac{DE}{DA} \geq \frac{CE}{EA + CA} \text{ and } \frac{FA}{FC} \geq \frac{EA}{AC + EC}.$$

By the extremely famous Nesbitt's Inequality, this implies that

$$\frac{BC}{BE} + \frac{DE}{DA} + \frac{FA}{FC} \geq \frac{3}{2},$$

as claimed. Hence our inequality is proven. □

Delta 18.4. (Romania TST 2009) Prove that the quadrilateral $ABCD$ is cyclic if and only if

$$\delta(E, AB) \cdot \delta(E, CD) = \delta(E, AC) \cdot \delta(E, BD) = \delta(E, AD) \cdot \delta(E, BC),$$

for any point E in the plane, where $\delta(X, YZ)$ denotes the distance from point X to the line YZ.

Proof. Let

$$k = \delta(E, AB) \cdot \delta(E, CD) = \delta(E, AC) \cdot \delta(E, BD) = \delta(E, AD) \cdot \delta(E, BC).$$

The concyclicity of points A, B, C, D is equivalent to

$$AC \cdot BD = AB \cdot CD + BC \cdot DA.$$

Multiplying both sides by k, this equation can be rewritten as

$$[EAC] \cdot [EBD] = [EAB] \cdot [ECD] + [EBC] \cdot [EDA],$$

where $[\mathcal{P}]$ is the unsigned area of the convex polygon $\mathcal{P}$. Expressing these areas differently, we get the new equivalent relation:

$$\sin AEC \sin BED = \sin AEB \sin CED + \sin BEC \sin DEA$$

(we have canceled the term $EA \cdot EB \cdot EC \cdot ED$ from both sides). Now denote $\angle AEB = x$, $\angle BEC = y$, $\angle CED = z$. In this case, the last relation can be rewritten as

$$\sin(x+y)\sin(y+z) = \sin x \sin z + \sin y \sin(x+y+z),$$

which can be easily verified. Hence quadrilateral $ABCD$ is cyclic as desired. □

Now, we move to Casey's Theorem, which represents a very powerful generalization of Ptolemy.

Theorem 18.2. (Casey's Theorem) Circles α, β, γ, δ are internally tangent to a fifth circle at points A, B, C, D, respectively, and $ABCD$ is a convex quadrilateral. Let $t_{\alpha\beta}$ be the length of a common external tangent to α and β. Define $t_{\beta\gamma}$ etc. similarly. Then,

$$t_{\alpha\beta}t_{\gamma\delta} + t_{\beta\gamma}t_{\delta\alpha} = t_{\alpha\gamma}t_{\beta\delta}.$$

Moreover, the converse is also true! More precisely, given four circles α, β, γ, δ satisfying the identity

$$\pm t_{\alpha\beta}t_{\gamma\delta} \pm t_{\beta\gamma}t_{\gamma\alpha} \pm t_{\alpha\gamma}t_{\beta\delta} = 0,$$

there is a fifth circle tangent to all of α, β, γ, and δ.

This result was stated for the first time by John Casey in 1881. We won't give a complete proof here, as the converse is pretty tedious and defeats the purpose of the material (nonetheless, the reader is advised to consult R. A. Johnson, *Advanced Euclidean Geometry*, Dover, 2007, pp. 121-127 for the full discussion). In what follows, however, we include the proof of the direct statement, since it involves a very nice and useful lemma.

Proof. We begin with claim:

Claim. Let ω_1 and ω_2 be two circles internally tangent to a circle ω at points A and B respectively. Let R, r_1, r_2 be the radii of circles $\omega, \omega_1, \omega_2$ respectively and assume without loss of generality that $r_1 \geq r_2$. Also let

O, O_1, O_2 be the centers of circles $\omega, \omega_1, \omega_2$ respectively. Let t be the length of a common external tangent to ω_1 and ω_2. Then

$$t = \frac{AB}{R}\sqrt{(R - r_1)(R - r_2)}$$

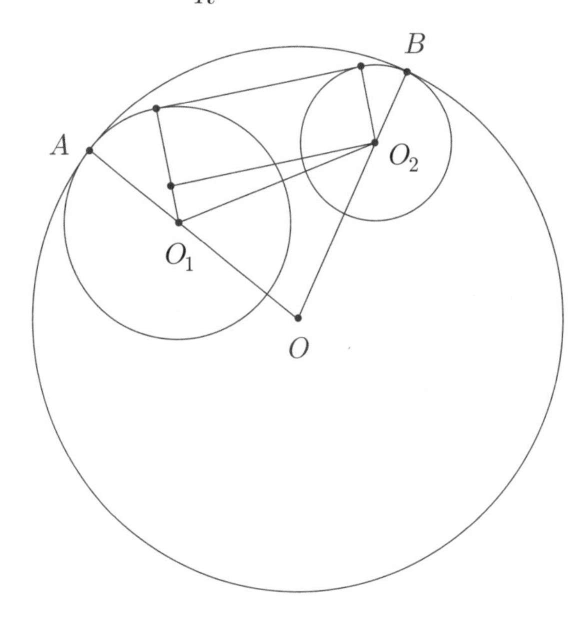

Proof. It's easy to see that

$$t^2 = (O_1O_2)^2 - (r_1 - r_2)^2.$$

Now by the Law of Cosines in triangle OO_1O_2 we have that

$$(O_1O_2)^2 = (R - r_1)^2 + (R - r_2)^2 - 2(R - r_1)(R - r_2)\cos O_1OO_2$$

Also, by the Law of Cosines in triangle OAB we have

$$(AB)^2 = 2R^2(1 - \cos O_1OO_2)$$

and combining these results and simplifying we obtain the desired identity.

Returning to the problem, let R be the radius of the large fifth circle circles $\alpha, \beta, \gamma, \delta$ are internally tangent to and let r_1, r_2, r_3, r_4 be the radii of circles $\alpha, \beta, \gamma, \delta$ respectively. Then, using the claim, multiplying both sides by R^2, and dividing both sides by $\sqrt{(R - r_1)(R - r_2)(R - r_3)(R - r_4)}$, we have that

$$t_{\alpha\beta}t_{\gamma\delta} + t_{\beta\gamma}t_{\delta\alpha} = t_{\alpha\gamma}t_{\beta\delta}$$

is equivalent to

$$AB \cdot CD + DA \cdot BC = AC \cdot BD$$

Which is just Ptolemy's Theorem! This completes the proof. □

//We can also use Casey's Theorem if some circles are externally tangent to a larger circle as well. In that case, if any two circles are both internally or externally tangent to the larger circle, we take their common external tangent length and if one is externally tangent to the larger circle and one is internally tangent to the larger circle than we take their internal common tangent length. With regards to the identity in out claim, if a circle with radius r is externally tangent to a larger circle with radius R, then if we were to use the claim, we would actually use $R+r$ rather than $R-r$ (convince yourself that this makes sense - think about the distance between the centers of the circles).

This theorem is of incredible usefulness in Olympiad geometry problems, so let's see some applications! Keep in mind the fact that we can use Casey on degenerate circles (which are just points), as this idea will show up often.

Delta 18.5. Triangle ABC is isosceles with $AB = AC = \ell$. A circle ω is tangent to BC and the arc BC not containing A of the circumcircle of triangle ABC. A tangent line from A to ω touches ω at P. Prove that the locus of P as the circle ω varies is an arc of a circle.

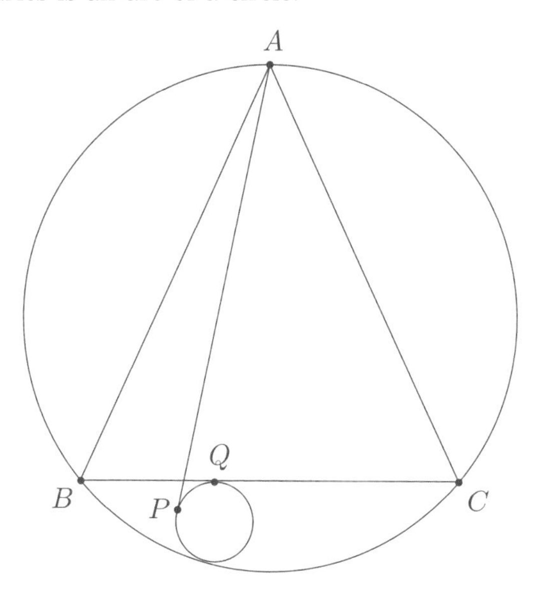

Proof. We use Casey's Theorem on ω and degenerate circles $(A), (B), (C)$, all internally tangent to the circumcircle of triangle ABC. Thus, if ω touches BC at Q we obtain:

$$BQ \cdot \ell + CQ \cdot \ell = AP \cdot BC \Longrightarrow AP = \ell$$

so the locus of points P is an arc of the circle centered at A with radius ℓ. This completes the proof. □

Delta 18.6. Let ω be a circle with diameter AB. Let P and Q be points on ω on opposite sides of line AB and let T be the projection of Q onto AB. Let ω_1, ω_2 be the circles with diameters TA and TB respectively and let PC and PD be tangents from P to ω_1 and ω_2 respectively. Show that $PC+PD = PQ$.

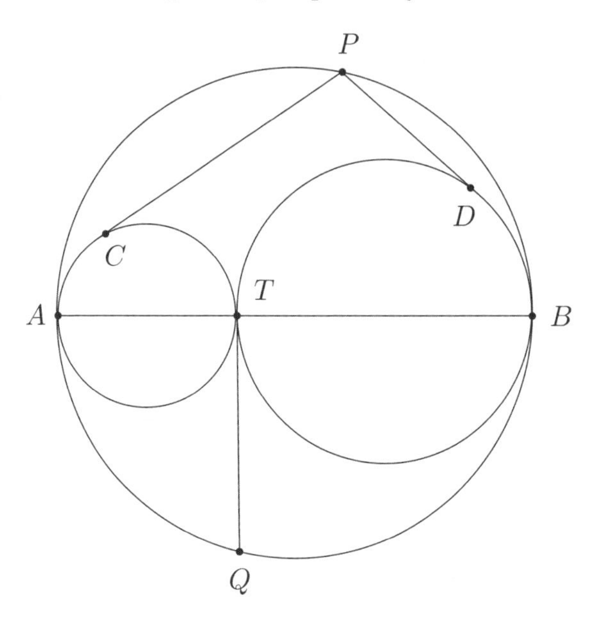

Proof. Let t be the length of the external common tangent of ω_1 and ω_2. We use Casey's Theorem on ω_1, ω_2, and degenerate circles $(P), (Q)$, all internally tangent to ω. This yields

$$PC \cdot QT + PD \cdot QT = PQ \cdot t \Longrightarrow PC + PD = \frac{t}{QT} \cdot PQ$$

so it suffices to show that $t = QT$. But it's easy to see that

$$QT = \sqrt{TA \cdot TB}$$

and from the claim in Casey's Theorem we can also verify that

$$t = \sqrt{TA \cdot TB},$$

hence, we have the desired result. □

Delta 18.7. In triangle ABC with circumcircle Ω, let ω_A be the circle internally tangent to Ω and tangent to BC at the midpoint of side BC. Define ω_B and ω_C similarly. Let t_{BC}, t_{CA}, t_{AB} denote the lengths of common external tangents of circles ω_B and ω_C, ω_C and ω_A, ω_A and ω_B respectively. Show that

$$t_{BC} = t_{CA} = t_{AB} = \frac{a+b+c}{4}$$

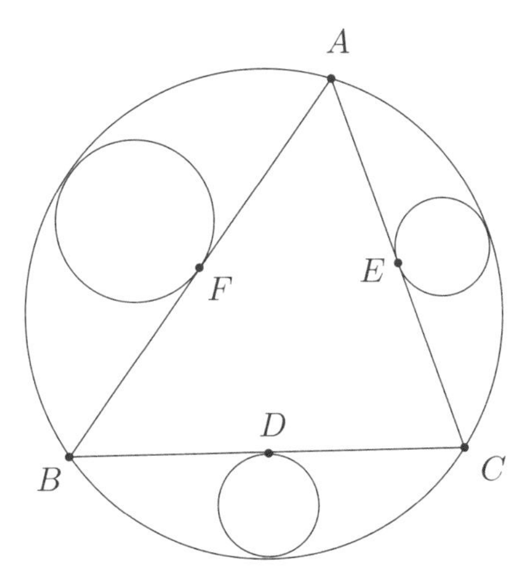

Proof. Let D, E, F be the midpoints of sides BC, CA, AB respectively. Also let t_A, t_B, t_C be the lengths of the common external tangents from points A, B, C to circles $\omega_A, \omega_B, \omega_C$ respectively. By Casey's Theorem on ω_A and degenerate circles $(A), (B), (C)$, all internally tangent to Ω, we obtain

$$a \cdot t_A = b \cdot BD + c \cdot CD \Longrightarrow t_A = \frac{b+c}{2}.$$

Similarly,

$$t_B = \frac{c+a}{2} \text{ and } t_C = \frac{a+b}{2}.$$

Now, by Casey's Theorem on ω_B and ω_C and degenerate circles $(B), (C)$, all internally tangent to Ω, we obtain

$$t_B t_C = a \cdot t_{BC} + BF \cdot CE \Longrightarrow t_{BC} = \frac{\left(\frac{a+c}{2}\right)\left(\frac{a+b}{2}\right) - \frac{bc}{4}}{a} = \frac{a+b+c}{4}$$

and since we can do the same for t_{CA} and t_{AB}, we are done. □

Delta 18.8. Let Ω be a circle passing through vertices B and C in triangle ABC and let ω be a circle tangent to segments AB and AC at points P and Q respectively and externally tangent to Ω at point T. Let M be the midpoint of arc BTC of Ω. Show that lines BC, PQ, MT concur.

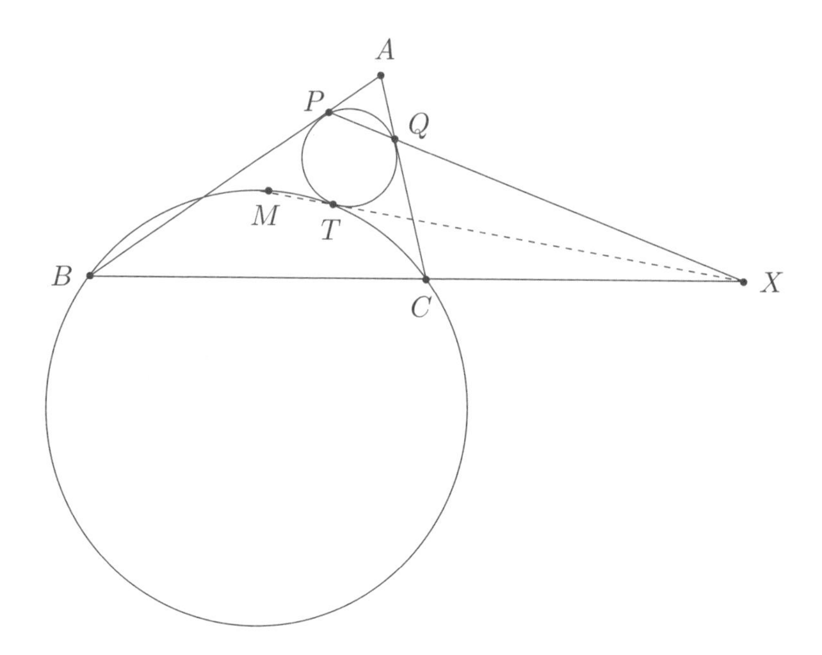

Proof. Let R, r be the radii of Ω, ω respectively. Using the claim in the proof of Casey's Theorem on ω and degenerate circle (B), both externally tangent to Ω, we obtain

$$BP = \frac{TB}{R}\sqrt{R(R+r)}$$

and similarly we have

$$CQ = \frac{TC}{R}\sqrt{R(R+r)}.$$

Thus, by dividing these two expressions, we have

$$\frac{TB}{TC} = \frac{BP}{CQ}.$$

Now let $X = BC \cap PQ$. By Menelaus' Theorem for triangle ABC at points P, Q, X we have that

$$\frac{PB}{PA} \cdot \frac{QA}{QC} \cdot \frac{XC}{XB} = 1$$

and since $AP = AQ$ as both segments are tangents from A to ω, this means that

$$\frac{XC}{XB} = \frac{QC}{PB} = \frac{TC}{TB}.$$

Hence, by the Angle Bisector Theorem in triangle BTC, X lies on the external angle bisector TM of angle $\angle BTC$ as desired. $\square$

Delta 18.9. Let ω be the incircle of triangle ABC. Let S and T be points on sides AB and AC respectively such that line ST is tangent to ω and parallel to BC. Let ω' be the incircle of triangle AST. Prove that the circle passing through points B and C tangent to ω is also tangent to ω'.

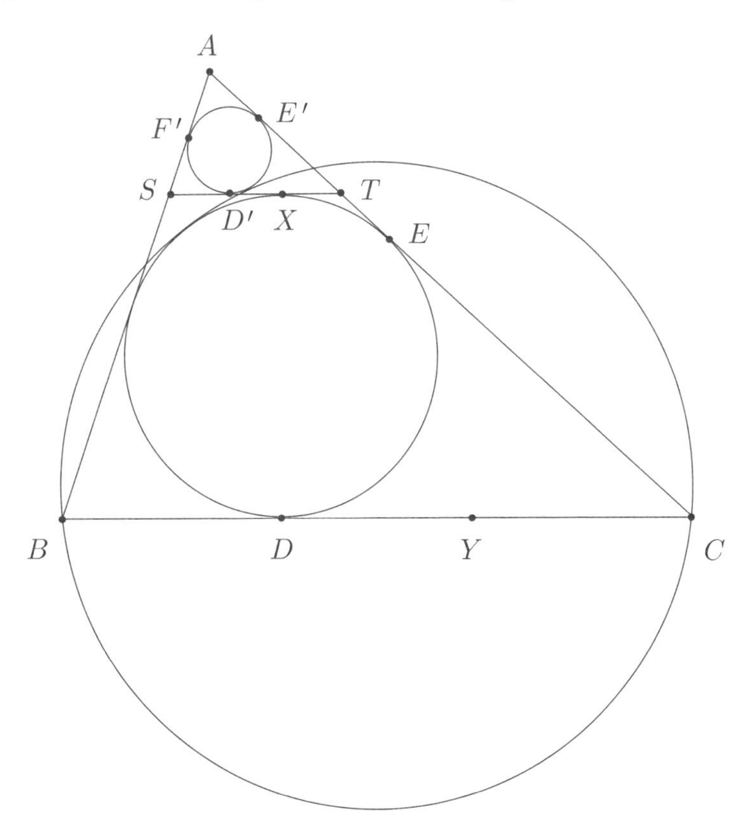

Proof. Assume without loss of generality that $CA \geq AB$. Let ω touch BC, CA, AB, ST at D, E, F, X respectively and let ω' touch ST, AT, AS at D', E', F' respectively. By the converse of Casey's Theorem on ω and ω' and degenerate circles $(B), (C)$, it suffices to show that

$$BF' \cdot CE = a \cdot D'X + BF \cdot CE'.$$

Noting that the homothety centered at A with ratio $\frac{s-a}{s}$ takes ω to ω', we see that

$$AE' = \frac{s-a}{s} \cdot AE = \frac{(s-a)^2}{s} \Longrightarrow CE' = b - \frac{(s-a)^2}{s}$$

and similarly

$$BF' = c - \frac{(s-a)^2}{s}.$$

Now, let Y be the point where the A-excircle of triangle ABC touches BC. We have that

$$D'X = \frac{s-a}{s} \cdot DY = \frac{(b-c)(s-a)}{s}$$

and since $BF = s - b$ and $CE = s - c$ we can easily verify the sufficient identity. $\square$

Assigned Problems

Epsilon 18.1. (Pompeiu's Theorem) Let ABC be a triangle with $AB = AC$ and let P be a point lying on the arc BC of the circumcircle of ABC, not containing the vertex A. Prove that

$$2PA \sin \frac{A}{2} = PB + PC.$$

Epsilon 18.2. Let $ABCD$ be a cyclic quadrilateral. Prove that

$$\frac{AC}{BD} = \frac{AB \cdot AD + CB \cdot CD}{BA \cdot BC + DA \cdot DC}.$$

Epsilon 18.3. (MOP 1997) Let Q be a quadrilateral whose side lengths are a, b, c, d in that order. Show that the area of Q does not exceed $(ac + bd)/2$.

Epsilon 18.4. (IMO Shortlist 2001) Let ABC be a triangle with centroid G. Determine, with proof, the position of the point P in the plane of ABC such that $AP \cdot AG + BP \cdot BG + CP \cdot CG$ is a minimum, and express this minimum value in terms of the side lengths of ABC.

Epsilon 18.5. (IMO 1997) It is known that $\angle BAC$ is the smallest angle in the triangle ABC. The points B and C divide the circumcircle of the triangle into two arcs. Let U be an interior point of the arc between B and C which does not contain A. The perpendicular bisectors of AB and AC meet the line AU at V and W, respectively. The lines BV and CW meet at T. Show that $AU = TB + TC$.

Epsilon 18.6. Prove Sawayama's Theorem (**Theorem 17.1**) using Casey's Theorem and Menelaus.

Epsilon 18.7. (Vladimir Zajic) Let ABC be a triangle with centroid G, incenter I, incircle ω, and nine-point circle Γ. Let the line IG meet BC at P and let the common tangent of ω and Γ meet BC at Q. Prove that the midpoint of BC is also the midpoint of PQ.

Epsilon 18.8. Prove Feuerbach's Theorem (**Theorem 15.1**) with Casey's Theorem. (Hint: use the converse of Casey's Theorem on the midpoints of the sides of the triangle [which are degenerate circles] and the incircle of the triangle.)

Epsilon 18.9. (Lev Emelyanov, Forum Geometricorum) Let D, E, F be points on sides BC, CA, AB of triangle ABC respectively such that lines AD, BE, CF concur. Let Ω be the circumcircle of triangle ABC and let ω_A be the circle internally tangent to Ω and tangent to BC at D. Define circles ω_B and ω_C similarly. Show that there exists a circle tangent to circles ω_A, ω_B, ω_C that is also tangent to the incircle of triangle ABC.

Chapter 19

Complete Quadrilaterals

We begin by defining what a complete quadrilateral is - you've already seen them numerous times in the configuration for Menelaus' Theorem and Brokard's Theorem!

Definition. A **complete quadrilateral** is the figure determined by four lines, no three of which are concurrent. The most common configuration in which one would see a complete quadrilateral is when there is a quadrilateral $ABCD$ and one takes the intersections $E = AB \cap CD$ and $F = DA \cap BC$.

Complete quadrilaterals have a number of amazing properties that we will explore in depth. But first, we'll talk about spiral similarity, the idea behind many of those properties.

Definition. Consider two similar and similarly oriented figures in the plane. The **spiral similarity** that sends one figure to the other is the composition of the rotation about a point and the homothety centered at that point that sends one figure to the other. This point is known as the **spiral center** of the two figures, and is unique.

Now, note that any two segments are similar - hence, there exists a spiral similarity taking any segment to any another segment. How do we find the spiral center?

Delta 19.1. Let AB and CD be two segments in the same plane. Let P be the intersection of lines AC and BD, and let Q be the second intersection of the circumcircles of triangles PAB and PCD. Prove that Q is the center of the spiral similarity taking segment AB to segment CD.

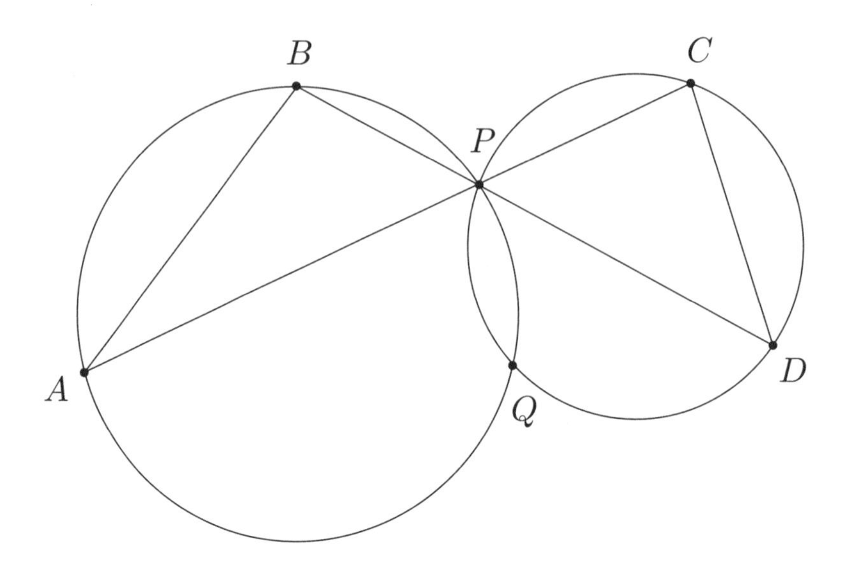

Proof. Assume the configuration shown above (other configurations can be handled similarly). Then

$$\angle QAB = 180^\circ - \angle QPB = \angle QPD = \angle QCD$$

and similarly

$$\angle QBA = \angle QDC$$

so triangles QAB and QCD are similar, hence Q is the desired spiral center. □

//If Q is the center of the spiral similarity that sends segment AB to segment CD, verify that it is also the center of the spiral similarity that sends segment AC to segment BD!

Delta 19.2. (USAMO 2006) Let $ABCD$ be a quadrilateral, and let E and F be points on sides AD and BC, respectively, such that $\frac{AE}{ED} = \frac{BF}{FC}$. Ray FE meets rays BA and CD at S and T, respectively. Prove that the circumcircles of triangles SAE, SBF, TCF, and TDE pass through a common point.

Proof. Assume the configuration shown above (other configurations can be handled similarly). Let P be the second intersection of the circumcircles of triangles SAE and SBF. We have $\angle APE = \angle ASE = \angle BPF$, and $\angle PAE = \angle PSE = \angle PBF$. Therefore triangles PAE and PBF are similar, and hence triangles PAB and PDC are similar as well. Therefore P is the center of the spiral similarity taking segment AD to segment BC.

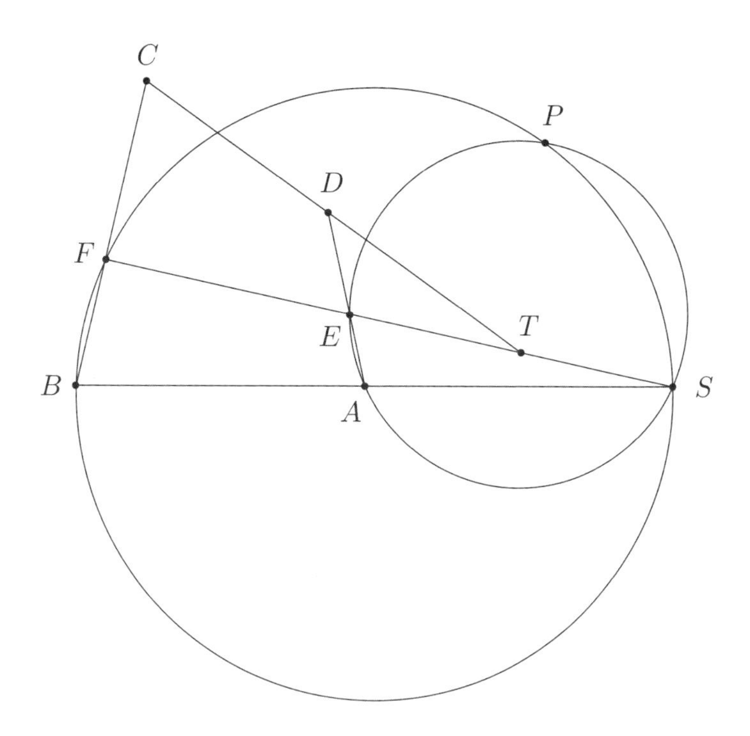

Now let Q be the second intersection of the circumcircles of triangles TCF and TDE. We analogously obtain that Q is the center of the spiral similarity taking segment AD to segment BC so $P = Q$ and the proof is complete. $\square$

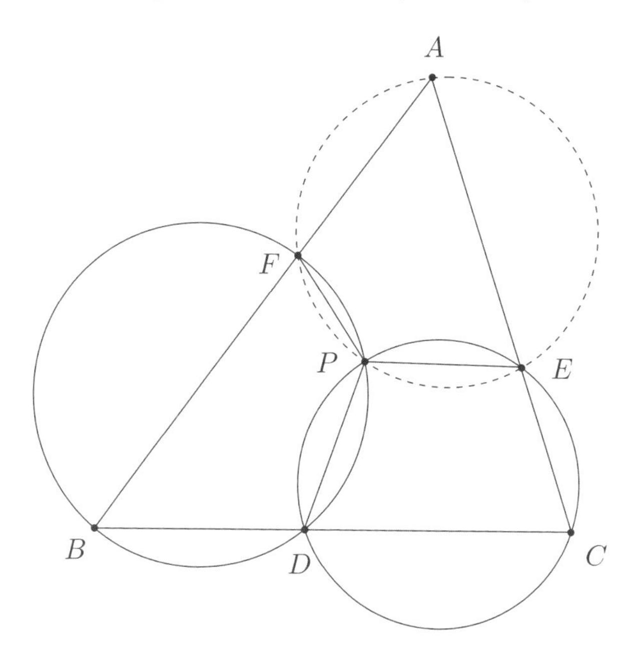

Now, we will tackle one of the most important theorems involving complete quadrilaterals - you already saw it in the claim in the proof of Sondat's Theorem (**Theorem 8.6**).

Theorem 19.1. (Miquel's Pivot Theorem) Let ABC be a triangle and let D, E, F be points lying on sides BC, CA, AB respectively. Then the circumcircles of triangles AEF, BFD, CDE concur.

Proof. First, stare at the picture on the previous page and try to get the proof on your own. The argument should normally go as follows. Let the circumcircles of triangles BFD and CDE intersect again at P. Then we have that

$$\angle EPF = 360^\circ - \angle FPD - \angle DPE = 360^\circ - (180^\circ - \angle B) - (180^\circ - \angle C) \\ = 180^\circ - \angle A$$

so quadrilateral $AEPF$ is cyclic. This completes the proof. □

Delta 19.3. (USAMO 2013) In triangle ABC, points P, Q, R lie on sides BC, CA, AB respectively. Let ω_A, ω_B, ω_C denote the circumcircles of triangles AQR, BRP, CPQ, respectively. Given the fact that segment AP intersects ω_A, ω_B, ω_C again at X, Y, Z, respectively, prove that

$$\frac{YX}{XZ} = \frac{BP}{PC}.$$

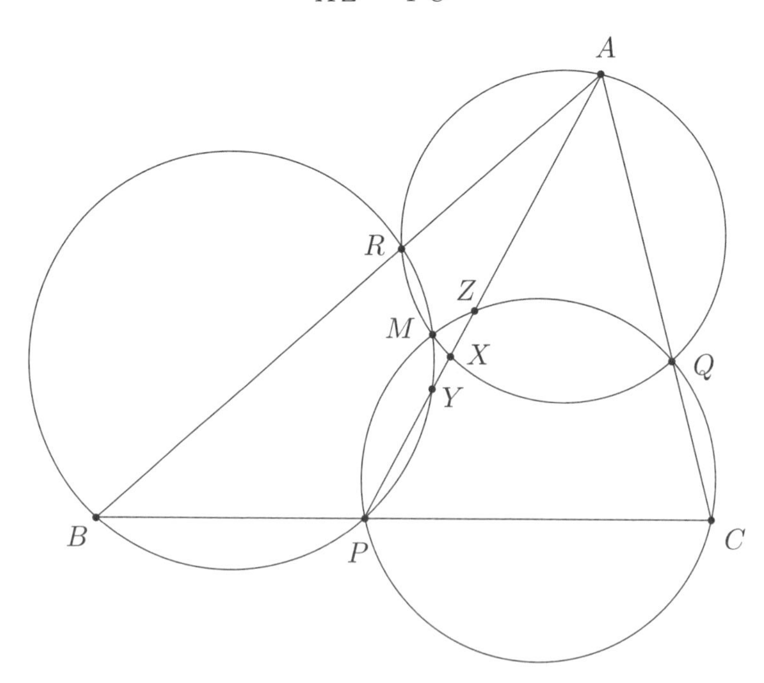

Proof. First note that by Miquel's Pivot Theorem, the circles ω_A, ω_B, ω_C concur at a point M. Since $P = YZ \cap BC$ and since ω_B is the circumcircle of triangle BPY and ω_C is the circumcircle of triangle CPZ, by **Delta 19.1** we have that M is the center of the spiral similarity that takes segment BY to segment CZ. Hence, M is also the center of the spiral similarity that takes segment YZ to segment BC. Also note that

$$\angle MXZ = \angle MQA = 180^\circ - \angle MQC = \angle MPC$$

so this spiral similarity also takes X to P. Therefore X and P are corresponding points on segments YZ and BC respectively, so we have the desired ratio equality. □

Delta 19.4. (IMO 2013) Let ABC be an acute triangle with orthocenter H, and let W be a point on the side BC, lying strictly between B and C. The points M and N are the feet of the altitudes from B and C, respectively. Denote by ω_1 is the circumcircle of triangle BWN, and let X be the point on ω_1 such that WX is a diameter of ω_1. Analogously, denote by ω_2 the circumcircle of triangle CWM, and let Y be the point such that WY is a diameter of ω_2. Prove that X, Y and H are collinear.

Proof. Look back at the proof of **Delta 2.13** and think about how Miquel's Pivot Theorem simplifies the argument! □

Theorem 19.2. (Miquel's Theorem) Let $ABCD$ be a quadrilateral and let $E = AB \cap CD$ and $F = DA \cap BC$. Then the circumcircles of triangles ABF, BCE, CDF, DAE concur at a point M, called the **Miquel Point** of complete quadrilateral $ABCDEF$.

First Proof. Let the circumcircles of triangles CDF and BCE intersect again at M. Then by **Delta 19.1** we have that M is the center of the spiral similarity that takes segment FD to segment BE. Thus, M is also the center of the spiral similarity that takes segment FB to segment DE and so it lies on the circumcircles of triangles DAE and ABF as desired. □

Second Proof. Applying Miquel's Pivot Theorem to triangle ABF with points C, D, E we have that the circumcircles of triangles DAE, BCE, and CDF concur. Applying Miquel's Pivot Theorem three more times in the same way then yields the desired result. □

Delta 19.5. (APMO 2015) Let ABC be a triangle, and let D be a point on side BC. A line through D intersects side AB at X and ray AC at Y. The circumcircle of triangle BXD intersects the circumcircle ω of triangle ABC again at point Z distinct from point B. The lines ZD and ZY intersect ω again at V and W respectively. Prove that $AB = VW$.

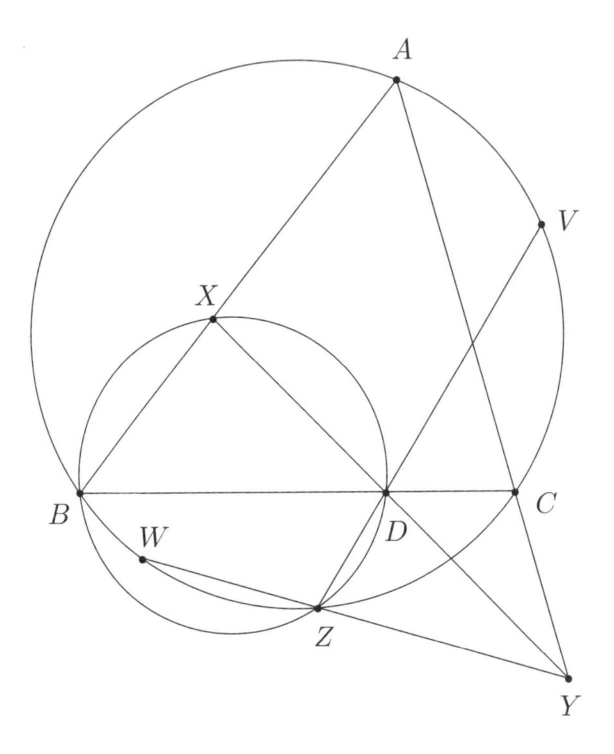

Proof. Assume the configuration shown above (other configurations can be handled similarly). Z is the Miquel point of complete quadrilateral $ACDXYB$ so Z lies on the circumcircle of triangle CDY. Therefore

$$\angle WZV = 180^\circ - \angle DZY = \angle DCY = 180^\circ - \angle ACB$$

so the chords AB and VW of ω subtend equal arcs. This implies that $AB = VW$ as desired. □

Delta 19.6. (APMO 2014) Circles ω and Ω meet at points A and B. Let M be the midpoint of the arc AB of circle ω (M lies inside Ω). A chord MP of circle ω intersects Ω at Q (Q lies inside ω). Let ℓ_P be the tangent line to ω at P, and let ℓ_Q be the tangent line to Ω at Q. Prove that the circumcircle of the triangle formed by the lines ℓ_P, ℓ_Q and AB is tangent to Ω.

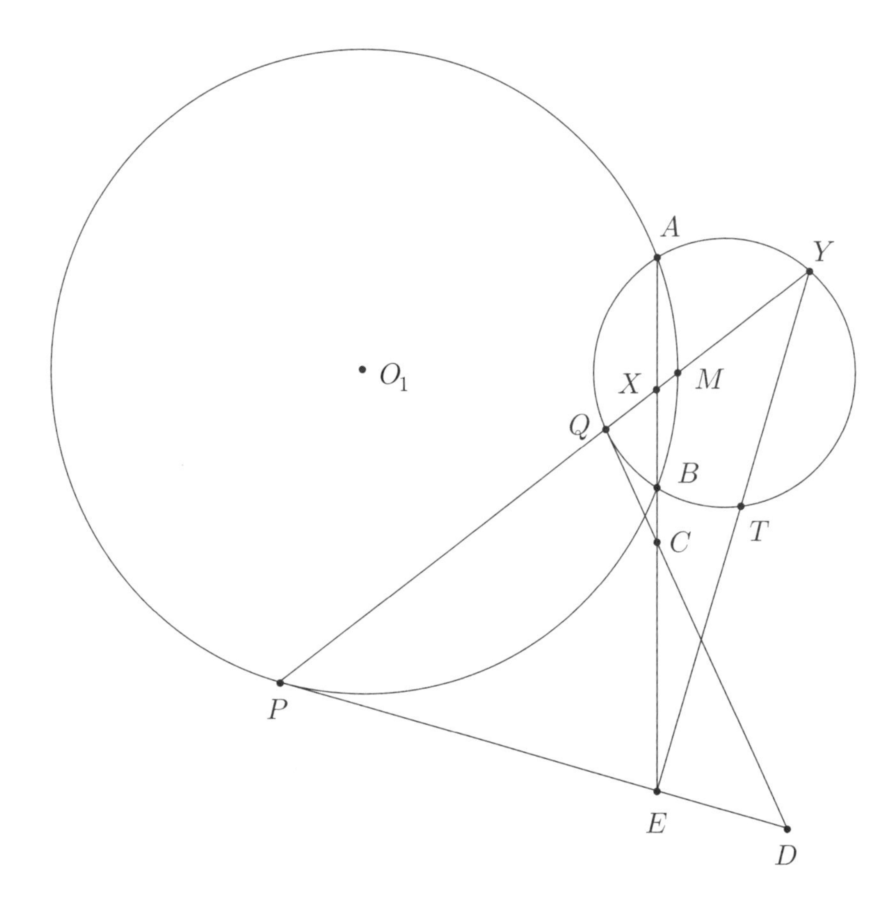

Proof. Let O_1 be the center of ω. Let $X = PM \cap AB$, $C = AB \cap \ell_Q$, $D = \ell_P \cap \ell_Q$, and $E = AB \cap \ell_P$. Notice that $\angle MPE = 90^\circ - \angle PMO_1 = \angle AXM = \angle PXE$ so $EP = EX$. This implies that $EX^2 = EP^2 = EB \cdot EA$. Now let Y be the second intersection of line PM with Ω and let T be the second intersection of line EY with Ω. By power of a point we have $EX^2 = EA \cdot EB = ET \cdot EY$ so line EX is tangent to the circumcircle of triangle YXT. Therefore $\angle TQC = \angle TYX = \angle TXC$ so quadrilateral $TCQX$ is cyclic. Similarly, we have $EP^2 = ET \cdot EY$ so line EP is tangent to the circumcircle of triangle YPT. Therefore $\angle EXT = \angle TYP = \angle EPT$ so quadrilateral $EPXT$ is cyclic as well. Therefore, T is Miquel point of the complete quadrilateral $ECQPXD$. Hence, T lies on circumcircle of triangle DEC. We now have $\angle EDT + \angle TYQ = \angle TCX + \angle TYQ = \angle TQX + \angle TYQ = \angle ETQ$. This implies the desired tangency, so we are done. □

Delta 19.7. Let $ABCD$ be a quadrilateral and let $E = AB \cap CD$ and $F = DA \cap BC$. Let M be the Miquel point of complete quadrilateral $ABCDEF$ and let O_1, O_2, O_3, O_4 be the circumcenters of triangles ABF, BCE, CDF, DAE respectively. Show that points M, O_1, O_2, O_3, O_4 are concyclic. The circle that contains them is called the **Steiner circle** of the complete quadrilateral $ABCDEF$.

The next exercise can be proved by a simple angle chase (try it yourself!). However, the proof we present is another one of those gems in Olympiad geometry - remember it!

Proof. Invert about a circle centered at M with arbitrary radius. The circumcircles of triangles ABF, BCE, CDF, DAE invert to four lines that form a complete quadrilateral $XYZTUV$ with Miquel point M. The circumcenters of these triangles invert to the reflections of M over lines XY, YZ, ZT, TX. Every three of these reflections are collinear as they lie on the Steiner lines of M with respect to the triangles XYV, YZU, ZTV, TXU. Hence, all four reflections are collinear which implies the desired result. □

Delta 19.8. Let $ABCD$ be a cyclic quadrilateral with circumcenter O and let $E = AB \cap CD$ and $F = DA \cap BC$. Let M be the Miquel point of complete quadrilateral $ABCDEF$. Show that M lies on line EF and that $OM \perp EF$.

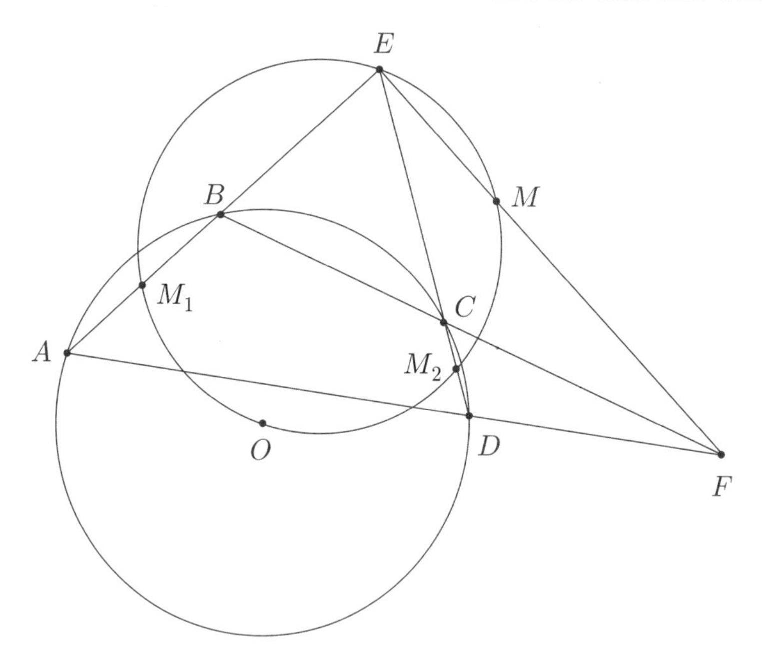

Proof. Assume the configuration shown above (other configurations can be handled similarly). Then from all the cyclic quadrilaterals we have

$$\angle EMA = \angle EDA = 180^\circ - \angle ABF = 180^\circ - \angle FMA$$

so M lies on line EF as desired. Now, let M_1, M_2 be the midpoints of segments AB, CD respectively. Since M is the center of the spiral similarity taking segment AB to segment DC we also have that M is the center of the spiral similarity taking segment AM_1 to segment DM_2. Therefore M lies on the circumcircle of triangle EM_1M_2. But since $\angle OM_1E = \angle OM_2E = 90°$, this circumcircle has diameter OE so $\angle OME = 90°$. This completes the proof. □

Theorem 19.3. (Gauss-Bodenmiller Theorem) Let $ABCD$ be a quadrilateral and let $E = AB \cap CD$ and $F = DA \cap BC$. Then the circles with diameters AC, BD, and EF (the diagonals of the complete quadrilateral $ABCDEF$) are coaxial and their common radical axis contains the orthocenters of triangles ABF, BCE, CDF, DAE.

Proof. We begin with the following claim:

Claim. Let ABC be a triangle and let M, N be points on sides CA and AB respectively. Then the orthocenter H of triangle ABC lies on the radical axis of the circles with diameters BM and CN.

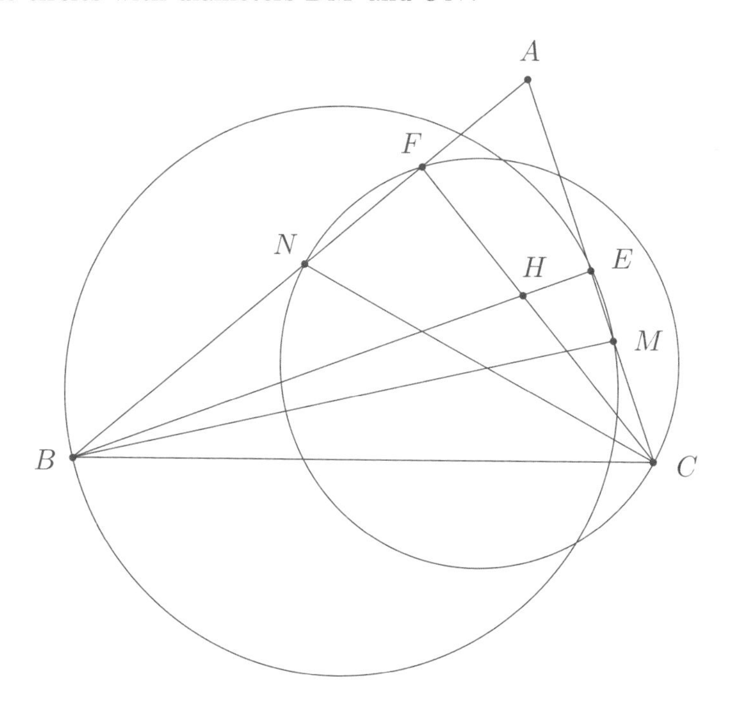

Proof. Let E, F be the feet of the B, C-altitudes in triangle ABC respectively. Then $\angle BEM = \angle CFN = 90°$ so E lies on the circle with diameter BM and F lies on the circle with diameter CN. Hence, it suffices to show that

$HB \cdot HE = HC \cdot HF$ (so that the power of H with respect to the two circles is equal). But since the reflections of H over the sidelines of the triangle ABC lie on the circumcircle of triangle ABC we have that $HB \cdot HE$ and $HC \cdot HF$ are both equal to half the power of H with respect to the circumcircle of triangle ABC. This completes the proof of the claim.

Let H_1, H_2, H_3, H_4 be the orthocenters of triangles ABF, BCE, CDF, DAE respectively. Returning to the problem, note that segments AC, BD, and FE are cevians in triangle ABF so from the claim we know that H_1 is the radical center of the circles with diameters AC, BD, EF. Similarly, points H_2, H_3, H_4 are also radical centers of these circles. Hence, either these circles are coaxial or the orthocenters of triangles ABF, BCE, CDF, DAE coincide. But the latter situation is clearly impossible, so this completes the proof. $\square$

This generalizes a beautiful result we first saw in **Section 5**!

Corollary 19.1. (Newton line) Let $ABCD$ be a quadrilateral and let $E = AB \cap CD$ and $F = DA \cap BC$. Then the midpoints of segments AC, BD, EF are collinear.

We end the section with a difficult problem from the 2009 IMO Shortlist.

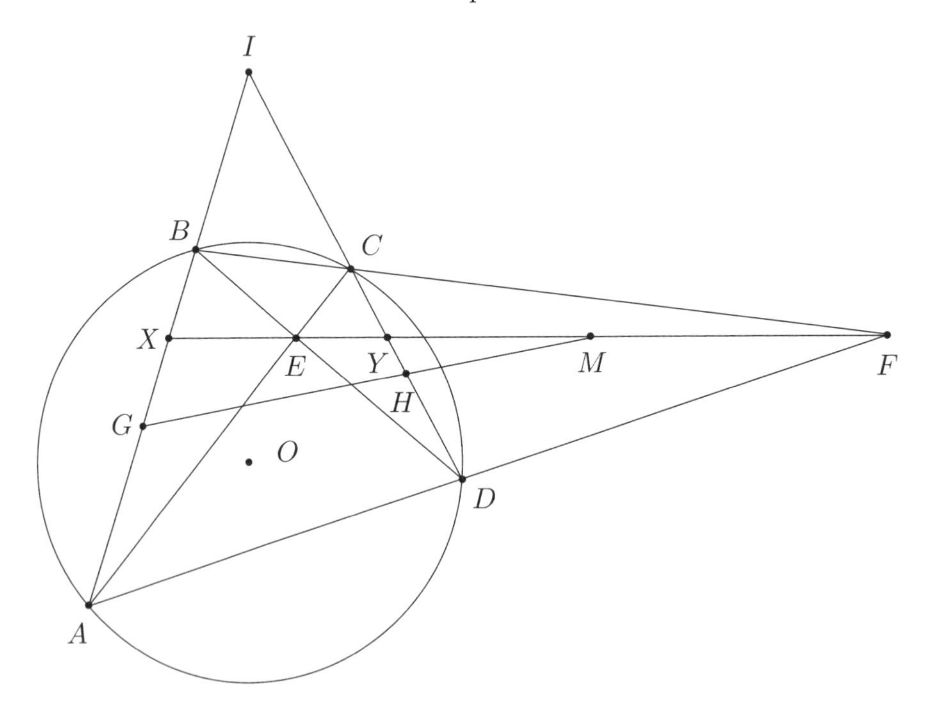

Delta 19.9. (IMO Shortlist 2009) Given a cyclic quadrilateral $ABCD$, let the diagonals AC and BD meet at E and the lines AD and BC meet at F. The

midpoints of AB and CD are G and H, respectively. Show that line EF is tangent to the circumcircle of triangle EGH.

Proof. Let M be the midpoint of segment EF and let O be the circumcenter of quadrilateral $ABCD$. Then points G, H, M all lie on the Newton line of complete quadrilateral $ACBDEF$, and hence are collinear. Let $I = AB \cap CD$. Points I, G, O, H all lie on the circle with diameter OI and thus are concyclic. Moreover, by Brokard's Theorem we know that $IO \perp EF$, hence line EF is an antiparallel to side GH in triangle GHI. Letting $X = EF \cap AB$ and $Y = EF \cap CD$, this implies that quadrilateral $GHYX$ is cyclic.

Now, since lines AC, BD, FX concur at E we have that $(A, B; X, I)$ is harmonic and since $(F, E; X, Y) \stackrel{D}{=} (A, B; X, I)$ this means that $(F, E; X, Y)$ is harmonic as well. Therefore $ME^2 = MX \cdot MY$ and since by power of a point we have that $MX \cdot MY = MG \cdot MH$, we find $ME^2 = MG \cdot MH$. This implies the desired tangency and completes the proof. □

Assigned Problems

Epsilon 19.1. (Morocco TST 2015) Let $ABA'B'$ be a convex quadrilateral, with $AA' \cap BB' = S$ and Let T be the intersection of the circumcircles of triangles ABS and $A'B'S$. Let C and C' be points on lines AB and $A'B'$ respectively such that B is between A and C, and B' is between A' and C'. and Let K and L be points on segments SB and SA respectively, such that points K, B, C, T are concyclic and points A', C', T, L are concyclic. Prove that points C, C', K, L are collinear if and only if

$$\frac{CA}{BC} = \frac{C'A'}{C'B'}$$

Epsilon 19.2. (IMO 1985) A circle with center O passes through the vertices A and C of a triangle ABC and intersects the segments AB and BC again at distinct points K and N respectively. Let M be the point of intersection of the circumcircles of triangles ABC and KBN (apart from B). Prove that $\angle OMB = 90°$.

Epsilon 19.3. (ELMO Shortlist 2014) Let ABC be a triangle with circumcenter O. Let P be a point inside triangle ABC, and let points D, E, F be on sides BC, AC, AB respectively such that the Miquel point of triangle DEF with respect to triangle ABC is P. Let the reflections of D, E, F over the midpoints of the sides of triangle ABC that they lie on be R, S, T respectively. Let Q be the Miquel point of triangle RST with respect to triangle ABC. Show that $OP = OQ$.

Epsilon 19.4. (IMO Shortlist 2006) Points A_1, B_1, C_1 are chosen on the sides BC, CA, AB of a triangle ABC respectively. The circumcircles of triangles AB_1C_1, BC_1A_1, CA_1B_1 intersect the circumcircle of triangle ABC again at points A_2, B_2, C_2 respectively ($A_2 \neq A, B_2 \neq B, C_2 \neq C$). Points A_3, B_3, C_3 are symmetric to A_1, B_1, C_1 with respect to the midpoints of the sides BC, CA, AB respectively. Prove that the triangles $A_2B_2C_2$ and $A_3B_3C_3$ are similar.

Epsilon 19.5. (USA TST 2009) Let ABC be an acute triangle. Point D lies on side BC. Let O_B, O_C be the circumcenters of triangles ABD and ACD, respectively. Suppose that the points B, C, O_B, O_C lies on a circle centered at X. Let H be the orthocenter of triangle ABC. Prove that $\angle DAX = \angle DAH$.

Epsilon 19.6. (Switzerland TST 2006) Let triangle ABC be an acute-angled triangle with $AB \neq AC$. Let H be the orthocenter of triangle ABC, and let M be the midpoint of the side BC. Let D be a point on the side AB and E a point on the side AC such that $AE = AD$ and the points D, H, E are on the same line. Prove that the line HM is perpendicular to the common chord of the circumscribed circles of triangle ABC and triangle ADE.

Epsilon 19.7. (Generalization of IMO 2011 Problem 6) Let ABC be a triangle and a point P. A line pass through P intersects the circumcircles of triangles PBC, PCA, PAB again at P_a, P_b, P_c respectively. Let ℓ_a, ℓ_b, ℓ_c, be the lines tangent to the circumcircles of triangles PBC, PCA, PAB at points P_a, P_b, P_c, respectively. Prove that the circumcircle of the triangle determined by the lines ℓ_a, ℓ_b, ℓ_c is tangent to the circumcircle of triangle ABC. (Hint: Invert about a circle centered at P and show that the tangency point is the Miquel point of a complete quadrilateral in the inverted diagram).

Chapter 20

Apollonian Circles and Isodynamic Points

This section will consist of information related to an important configuration - the Apollonian circle - surprisingly not often discussed in other Olympiad geometry texts. We begin with a definition.

Definition. Let AB be a segment and let k be a positive real number. Then the locus of points P satisfying $\frac{AP}{BP} = k$ is known as an **Apollonian circle**. Note that in the case $k = 1$, our circle is degenerate - namely, it coincides with the perpendicular bisector of segment AB.

But why is this locus a circle? Let R be the point inside segment AB such that $\frac{AR}{BR} = k$ and let S be the point on line AB outside of segment AB such that $\frac{AS}{BS} = k$. Then it's clear that $(A, B; R, S)$ is harmonic. Now, consider any point P satisfying $\frac{AP}{BP} = k$. Since $\frac{AP}{BP} = \frac{AR}{BR}$ by the Angle Bisector Theorem we have that line PR bisects angle $\angle APB$. Hence, $PR \perp PS$ and it's easy to see that the desired locus is the circle with diameter RS.

We proceed with the definition of the Apollonian circles of a triangle. In a triangle ABC, we denote the locus of points P such that $\frac{BP}{CP} = \frac{AB}{AC}$ as the A-Apollonian circle of triangle ABC. It is clear that every triangle ABC has precisely three Apollonian circles associated with it - namely its A, B, and C-Apollonian circles. Now, let's see some properties!

Delta 20.1. Show that the circumcircle of triangle ABC and the A-Apollonian circle of triangle ABC are orthogonal.

Proof. Let R and S be the feet of the A-interior angle bisector and A-exterior angle bisector of triangle ABC respectively and let M be the midpoint of segment RS (M is the center of the A-Apollonian circle of triangle ABC). Now assume without loss of generality that B lies between S and C.

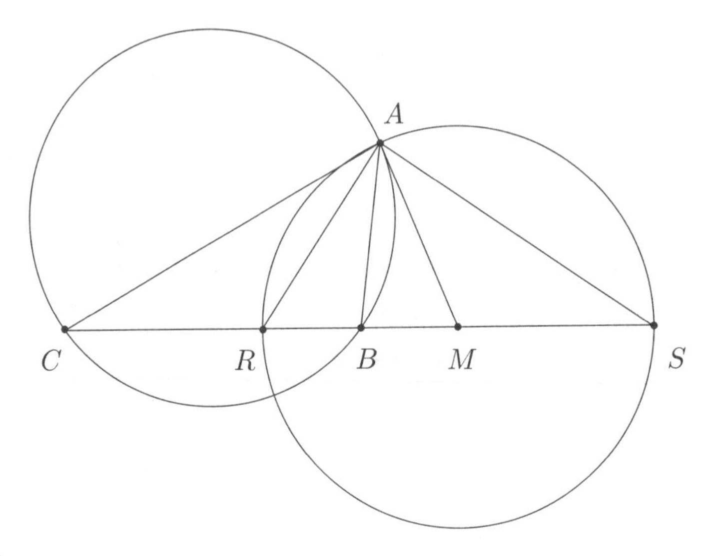

Note that

$$\begin{aligned}\angle MAB + \frac{\angle BAC}{2} &= \angle MAR = \angle MRA \\ &= \angle ACB + \frac{\angle BAC}{2};\end{aligned}$$

hence $\angle MAB = \angle ACB$. Therefore MA is tangent to the circumcircle of triangle ABC and this implies the desired result. $\square$

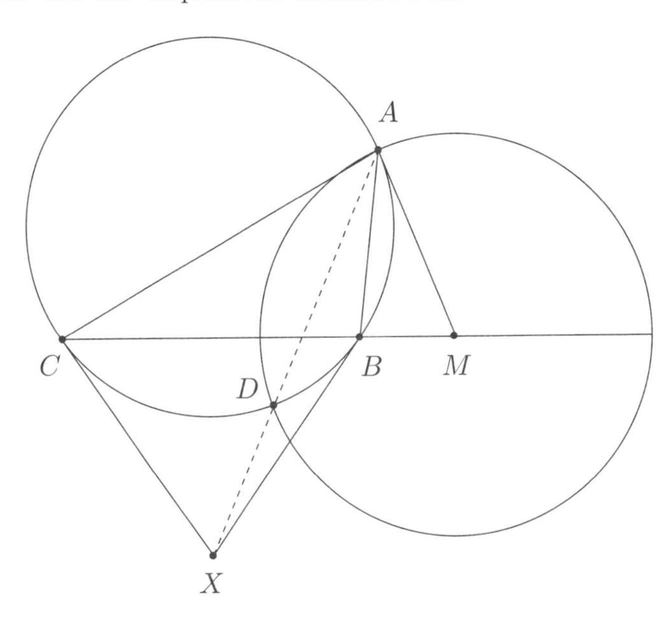

Delta 20.2. Let D be the second intersection of the circumcircle of triangle ABC and the A-Apollonian circle of triangle ABC. Then line AD is the A-symmedian of triangle ABC.

We give two proofs of the result - the first will help us in some later exercises, and the second is really short!

First proof. Let M be the center of the A-Apollonian circle of triangle ABC and let ω be the circumcircle of triangle ABC. Let the lines tangent to ω at B and C intersect at X. We know that line AX is the A-symmedian of triangle ABC so it suffices to show that X lies on line AD. By **Delta 20.1** we have that segments MA and MD are tangent to ω so line AD is the polar of M with respect to ω. Segments XB and XC are also tangent to ω so line BC is the polar of X with respect to ω. But M lies on line BC so by La Hire's Theorem X must lie on the polar of M - namely, line AD. This completes the proof. □

Second proof. By definition we have $\frac{DB}{DC} = \frac{AB}{AC}$. Hence, quadrilateral $ABDC$ is harmonic and so line AD is the A-symmedian of triangle ABC as desired. □

The next exercise is perhaps the most famous property of the Apollonian circles of a triangle.

Delta 20.3. Show that the three Apollonian circles of a non-equilateral triangle concur at exactly two points - one point is inside the triangle and is called the **First Isodynamic point** of the triangle and the other is outside of the triangle and is predictably called the **Second Isodynamic point** of the triangle.

Proof. Let J be an intersection of the B and C-Apollonian circles of triangle ABC (it's easy to see these circles do intersect). Then we know that $\frac{CJ}{AJ} = \frac{AB}{BC}$ and $\frac{AJ}{BJ} = \frac{BC}{CA}$ and after multiplying we obtain $\frac{BJ}{CJ} = \frac{AB}{CA}$, hence, J also lies on the A-Apollonian circle of triangle ABC. This completes the proof. □

Delta 20.4. (Serbia TST 2003) Let M and N be distinct points in the plane of triangle ABC that satisfy

$$AM : BM : CM = AN : BN : CN.$$

Show that line MN passes through the circumcenter of triangle ABC.

Proof. It's clear that N lies on the M-Apollonian circles of triangles MBC, MCA, and MAB and since three non-coinciding circles can all intersect in at most two points, we have that N is uniquely determined by M. Now, let O and R be the circumcenter and circumradius of triangle ABC respectively. Consider the inversion about the circumcircle of triangle ABC. Let M invert to a point M'. We have that

$$AM' = \frac{R^2}{OA \cdot OM} \cdot AM = \frac{R}{OM} \cdot AM$$

and similarly $BM' = \frac{R}{OM} \cdot BM$ and $CM' = \frac{R}{OM} \cdot CM$. Hence,

$$AM : BM : CM = AM' : BM' : CM'$$

so $N = M'$. This immediately implies M, N, O are collinear as desired. □

//One could also note that the circumcircle of triangle ABC is an Apollonian circle of segment MN, which would also imply the desired result.

Delta 20.5. (RMM 2009) Given four points A_1, A_2, A_3, A_4 in the plane, no three collinear, such that

$$A_1A_2 \cdot A_3A_4 = A_1A_3 \cdot A_2A_4 = A_1A_4 \cdot A_2A_3,$$

denote by O_i the circumcenter of triangle $A_jA_kA_l$ with $\{i, j, k, l\} = \{1, 2, 3, 4\}$. Assuming $\forall i\, A_i \neq O_i$, prove that the four lines A_iO_i are concurrent or parallel.

Proof. We work in the projective plane (so we can drop the "or parallel" condition). Denote by ω_{ij} the Apollonian circle of points A_k and A_l and ratio $\frac{A_iA_k}{A_iA_l} = \frac{A_jA_k}{A_jA_l}$ (which clearly passes through points A_i and A_j). By **Delta 20.1** and **Delta 20.3** we have that ω_{ij}, ω_{il} and ω_{ik} are coaxial and all orthogonal to the circumcircle of triangle $A_jA_lA_k$, so the radical axis r_i of these three circles passes through O_i. Hence r_i is actually the line A_iO_i. So lines A_iO_i, A_jO_j, A_kO_k are the radical axes of ω_{ij} and ω_{ik}, ω_{ij} and ω_{jk}, ω_{ik} and ω_{jk} respectively and therefore concur at the radical center of these three circles. Analogously lines A_iO_i, AjO_j, A_lO_l concur, so the proof is complete. □

Delta 20.6. Show that the circumcenter O, the Symmedian point K, the First Isodynamic point J, and the Second Isodynamic point J' of a triangle ABC are collinear.

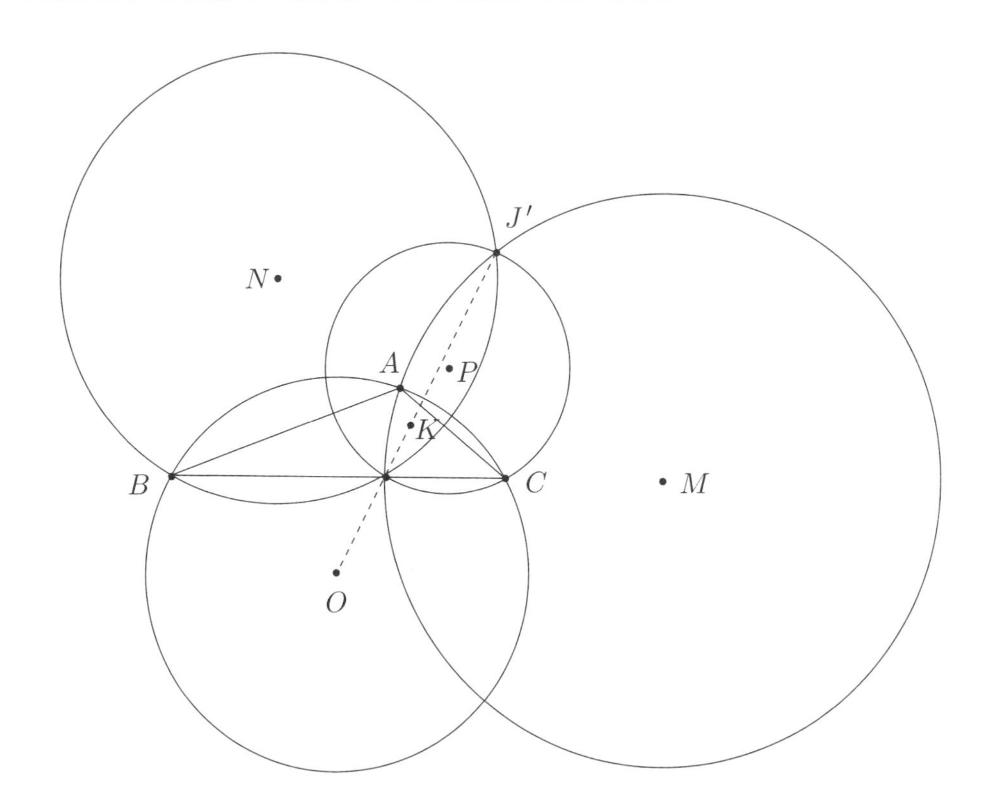

Proof. Let M, N, P be the centers of the A, B, C-Apollonian circles of triangles ABC respectively. We know from **Delta 20.1** that OB is tangent to the B-Apollonian circle of triangle ABC so the power of O with respect to the B-Apollonian circle of triangle ABC is OB^2. Similarly the power of O with respect to the C-Apollonian circle of triangle ABC is OC^2 but since by definition $OB = OC$, we have that O lies on the radical axis of the B and C-Apollonian circle of triangle ABC. Hence, O lies on line JJ'. Now, in the proof of **Delta 20.2** we showed that the A-symmedian of triangle ABC is the polar of M with respect to the circumcircle of triangle ABC. Similarly the B and C-symmedians of triangle ABC are the polars of N and P respectively with respect to the circumcircle of triangle ABC. Hence by La Hire's Theorem, K is the pole of the line determined by points M, N, P with respect to the circumcircle. This implies that line OK is perpendicular to the line determined by points M, N, P and since the radical axis of the Apollonian circles of triangle ABC is also perpendicular to this line, we have the desired collinearity. □

Delta 20.7. Show that the pedal triangle of the First Isodynamic point of triangle ABC with respect to triangle ABC is an equilateral triangle.

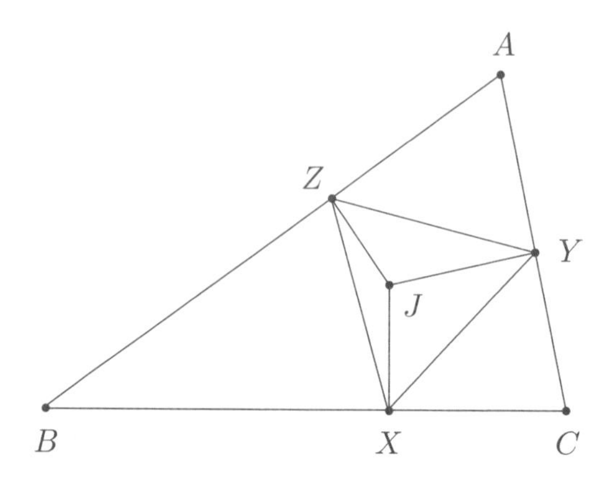

Proof. Let J be the First Isodynamic point of triangle ABC and let X, Y, Z be the projections of J on sides BC, CA, AB respectively. Since points J, Y, A, Z lie on a circle with diameter AJ we have that $YZ = AJ \sin A$. Similarly $ZX = BJ \sin B$ so

$$\frac{YZ}{ZX} = \frac{AJ}{BJ} \cdot \frac{\sin A}{\sin B} = \frac{AJ}{BJ} \cdot \frac{BC}{CA} = 1$$

thus $YZ = ZX$. Similarly we find that $ZX = XY$ so triangle XYZ is equilateral as desired. □

Delta 20.8. Show that among all equilateral triangles with vertices on each of the sides BC, CA, AB of triangle ABC, the pedal triangle of the First Isodynamic point of triangle ABC has the minimal area.

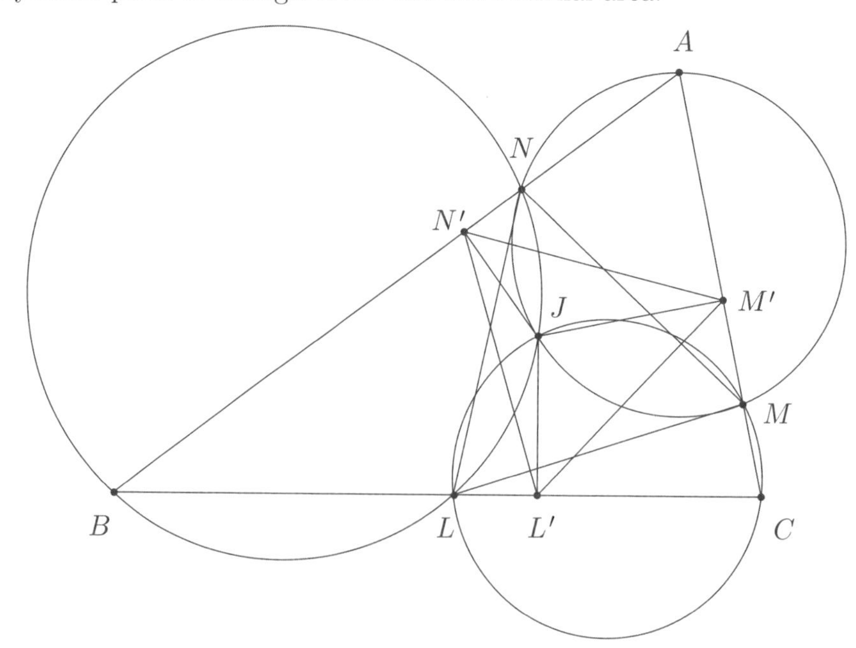

Proof. Let L, M, N be points on sides BC, CA, AB respectively such that triangle LMN is equilateral. Let J be the intersection of the circumcircles of triangles AMN, BNL, CLM (this point exists by Miquel's Pivot Theorem) and let L', M', N' be the projections of J onto sides BC, CA, AB respectively. Then angle chasing with cyclic quads yields $\angle JLM = \angle JCM = \angle JL'M'$ and similarly $\angle JLN = \angle JL'N'$. Hence $\angle M'L'N' = \angle JL'M' + \angle JL'N' = \angle JLM + \angle JLN = 60^\circ$ and doing the same thing for angle $\angle L'M'N'$ yields that triangle $L'M'N'$ is equilateral. Hence J is the First Isodynamic point of triangle ABC (now you see why we called it J). Moreover, J is the center of the spiral similarity with ratio $\frac{JL'}{JL}$ that takes triangle LMN to triangle $L'M'N'$. And because $\frac{JL'}{JL} \leq 1$, this implies the desired minimality. □

Now, we will discuss a pair of points you first saw in **Section 6** that happen to be closely connected to the Isodynamic points. For the following results, assume that the reference triangle ABC does not have an angle larger that 120°.

Definition. Let X, Y, Z be points in the plane of triangle ABC such that triangles BCX, CAY, ABZ are equilateral and don't intersect the interior of triangle ABC. Then the circumcircles of triangles BCX, CAY, ABZ intersect at F, the **First Fermat point** of triangle ABC. Let X', Y', Z' be points in the plane of triangle ABC such that triangles BCX', CAY', ABZ' are equilateral and all intersect the interior of triangle ABC. Then the circumcircles of triangles BCX', CAY', ABZ' intersect at F', the **Second Fermat point** of triangle ABC

How do we know these circumcircles intersect? Actually, the proof is similar to the proof of Miquel's Pivot Theorem. Let F be the second intersection of the circumcircles of triangles CAY and ABY. Then

$$\angle BFC = 360^\circ - \angle AFB - \angle CFA = 120^\circ = 180^\circ - \angle BXC$$

hence F lies on the circumcircle of triangle BCX as desired. Similar reasoning applies for the existence of the Second Fermat point.

Delta 20.9. Show that lines AX, BY, CZ concur at F and that

$$AX = BY = CZ.$$

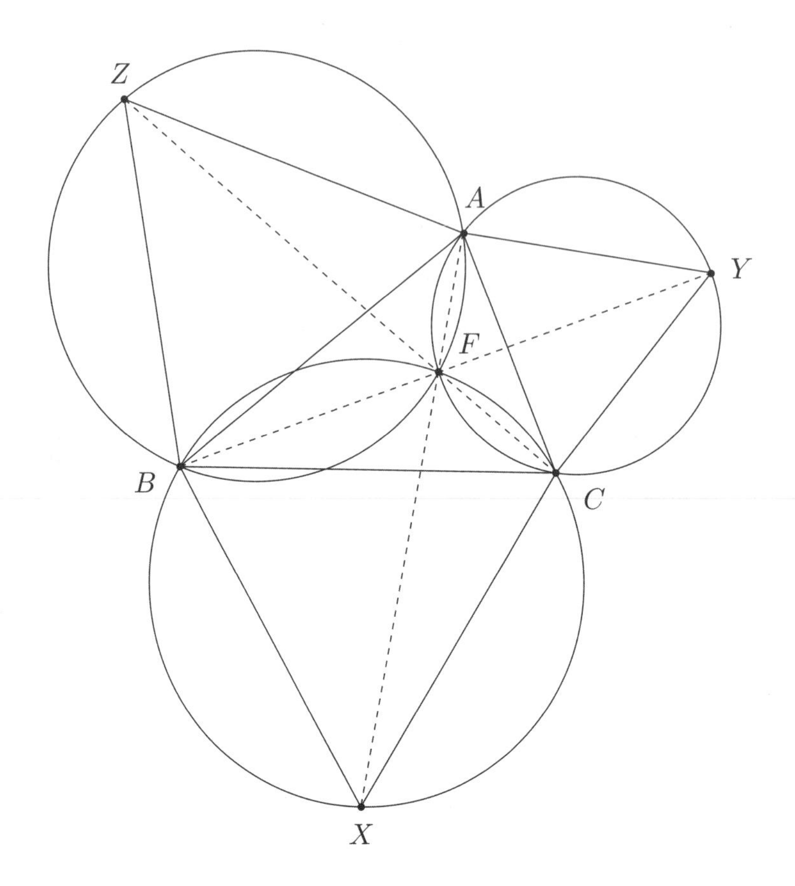

Proof. We have that $\angle XFB + \angle AFB = \angle XCB + 120^\circ = 180^\circ$ hence F lies on line AX. Similarly, F lies on lines BY and CZ. Now, by Ptolemy's Theorem on cyclic quadrilateral $XBFC$ we know that

$$FX \cdot BC = BX \cdot CF + CX \cdot BF \Longrightarrow FX = BF + CF$$

where we used the fact that triangle XBC is equilateral. Therefore

$$AX = AF + FX = AF + BF + CF$$

and similarly

$$BY = CZ = FA + FB + FC$$

so we are done. □

//Similarly, one can show that lines AX', BY', CZ' concur at F' and that $AX' = BY' = CZ'$.

Delta 20.10. Prove that the point P in the interior of triangle ABC that minimizes the sum $AP + BP + CP$ is the First Fermat Point of triangle ABC.

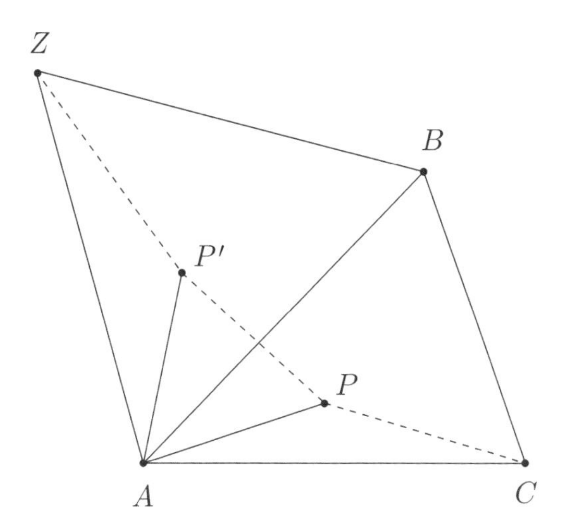

Proof. Let Z be the point in the plane of triangle ABC such that triangle ABZ is equilateral and does not intersect the interior of triangle ABC. Let P be a point inside triangle ABC and let P' be the point inside triangle ABZ such that triangle APP' is equilateral. Note that A is the center of the rotation that takes triangle APB to triangle $AP'Z$, hence $BP = ZP'$. Thus we have that

$$AP + BP + CP = PP' + ZP' + CP \geq CZ$$

with equality holding if and only if points C, P, P', Z are collinear. Since

$$\angle AP'P = \angle APP' = 60°$$

the collinearity holds if and only if

$$\angle CPA = 120°$$

and

$$\angle AP'Z = \angle APB = 120°,$$

which happens only when P is the First Fermat point of triangle ABC as desired. □

You might be asking yourself, why did we introduce Fermat points into the discussion of the Apollonian circles and Isodynamic points? That question is answered by the following result:

Delta 20.11. The First Fermat point F and the First Isodynamic point J of triangle ABC are isogonal conjugates with respect to triangle ABC.

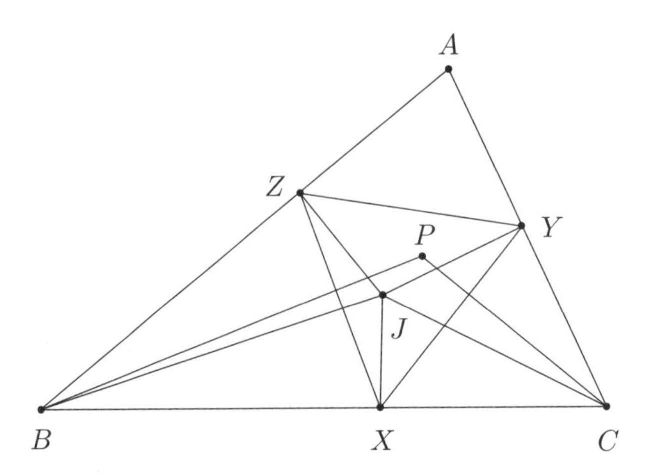

Proof. Let X, Y, Z be the projections from J onto sides BC, CA, AB respectively. We know from **Delta 20.5** that triangle XYZ is equilateral so

$$\begin{aligned}\angle BJC &= 180^\circ - \angle JBC - \angle JCB = 180^\circ - (\angle B - \angle JBA) - (\angle C - \angle JCA) \\ &= 180^\circ - (\angle B - \angle JXZ) - (\angle C - \angle JXY) = \angle A + 60^\circ\end{aligned}$$

and we know that $\angle BFC = 120^\circ$ so $\angle BFC + \angle BJC = 180^\circ + \angle A$. Similarly we have that $\angle CFA + \angle CJA = 180^\circ + \angle B$ and $\angle AFB + \angle AJB = 180^\circ + \angle C$.

Now, let P be the isogonal conjugate of J with respect to triangle ABC. We have that

$$\begin{aligned}\angle BJC + \angle BPC &= (180^\circ - \angle JBC - \angle JCB) + (l80^\circ - \angle PBC - \angle PCB) \\ &= 360^\circ - (\angle JBC + \angle PBC) - (\angle JCB + \angle PCB) \\ &= 360^\circ - \angle B - \angle C = 180^\circ + \angle A.\end{aligned}$$

Similarly we find $\angle CJA + \angle CPA = 180^\circ + \angle B$ and $\angle AJB + \angle APB = 180^\circ + \angle C$. This implies that $P = F$ as desired. □

//Similarly, one can show that the Second Fermat point and Second Isodynamic point of a triangle are isogonal conjugates with respect to that triangle.

We end the section with a beautiful problem that combines results about the First Isodynamic point with properties of pedal triangles and Menelaus' Theorem.

Delta 20.12. Let triangle XYZ be an equilateral triangle inscribed in a circle ω. Let P be a point in the interior of triangle XYZ and let A, B, C be the second intersections of lines XP, YP, ZP with ω. Let I, I_1, I_2, I_3 be the incenters of triangles ABC, PBC, PCA, PAB respectively. Show that lines AI_1, BI_2, CI_3, PI concur.

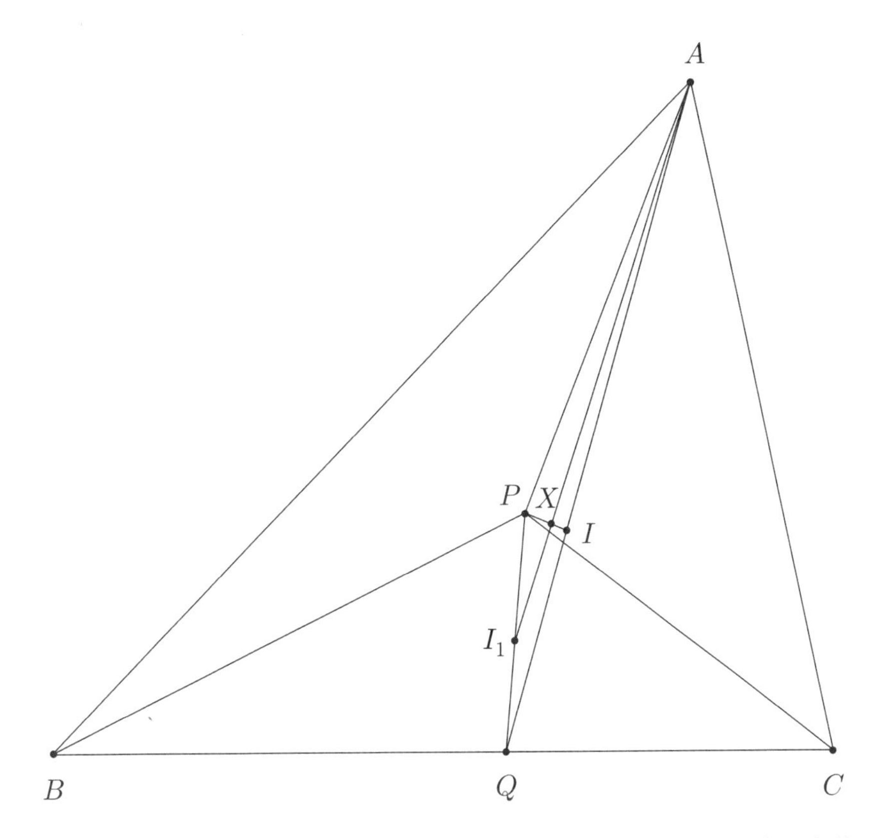

Proof. Note that triangle XYZ is the circumcevian triangle of P with respect to triangle ABC and so from the claim in the proof of **Theorem 7.4** we have that the pedal triangle of P with respect to triangle ABC is equilateral. Hence by the converse of **Delta 20.7**, P is an isodynamic point of triangle ABC. By definition we have $AP \cdot BC = BP \cdot CA = CP \cdot AB$. Alternatively, the internal angle bisectors of angles $\angle BPC$, $\angle CPA$, $\angle APB$ pass through the feet of the internal angle bisectors of angles $\angle BAC, \angle ABC, \angle BCA$ in triangle ABC. Thus, let Q be the common foot of the internal angle bisectors of angles $\angle BAC$ and $\angle BPC$ and let $X = PI \cap AI_1$. By Menelaus' Theorem for triangle PIQ with points A, X, I_1 we have

$$\frac{XI}{PX} = \frac{AI}{AQ} \cdot \frac{QI_1}{I_1P} = \frac{AB+AC}{AB+BC+CA} \cdot \frac{BC}{PB+PC} = \frac{AC}{PC} \cdot \frac{BC}{AB+BC+CA}.$$

Similarly, if $X' = PI \cap BI_2$ then by Menelaus' Theorem for the triangle formed by P, I, and the foot of the B-internal angle bisector in triangle ABC with points B, X', I_2 we get the relation

$$\frac{X'I}{PX'} = \frac{BC+AB}{AB+BC+CA} \cdot \frac{AC}{PA+PC} = \frac{AC}{PC} \cdot \frac{BC}{AB+BC+CA}.$$

Hence it follows that $X = X'$ and so lines AI_1, BI_2, PI concur at X. Analogously, we have that X also lies on line CI_3. This completes the proof. $\square$

Assigned Problems

Epsilon 20.1. (ELMO Shortlist 2014) Let $A_1A_2A_3 \cdots A_{2014}$ be a cyclic 2014-gon. Prove that for every point P not the circumcenter of the 2014-gon, there exists a point $Q \neq P$ such that $\frac{A_iP}{A_iQ}$ is constant for $i \in \{1, 2, 3, \cdots, 2014\}$.

Epsilon 20.2. Let J and J' be the First and Second Isodynamic points of triangle ABC respectively. Show that the inversion about the circumcircle of triangle ABC takes J to J'.

Epsilon 20.3. (USA TST 2008) Let P, Q, and R be the points on sides BC, CA, and AB of an acute triangle ABC such that triangle PQR is equilateral and has minimal area among all such equilateral triangles. Prove that the perpendiculars from A to line QR, from B to line RP, and from C to line PQ are concurrent.

Epsilon 20.4. Let ABC be a triangle and let D, E, F be the feet of the A, B, C-internal angle bisectors respectively. Let X be the intersection of line BC with the perpendicular bisector of segment AD, and define Y and Z similarly. Show that points X, Y, Z are collinear.

Epsilon 20.5. (Singapore TST 2004) Let D be a point in the interior of triangle ABC such that $AB = ab$, $AC = ac$, $AD = ad$, $BC = bc$, $BD = bd$ and $CD = cd$ for some positive real numbers a, b, c, d. Prove that $\angle ABD + \angle ACD = 60^\circ$.

Epsilon 20.6. (Vladimir Zajic) Let D, E, F be points on sides BC, CA, AB of triangle ABC respectively such that triangle DEF is equilateral. Show that

$$DE \geq \frac{2\sqrt{2}K}{\sqrt{a^2 + b^2 + c^2 + 4\sqrt{3}K}}$$

where K is the area of triangle ABC.

Epsilon 20.7. Let J be the First Isodynamic point of triangle ABC and let A', B', C' be the reflections of J over lines BC, CA, AB respectively. Show that lines AA', BB', CC' concur at the First Fermat point of triangle ABC.

Epsilon 20.8. Let F be the First Fermat point of triangle ABC. Show that the Euler lines of triangles FBC, FCA, FAB concur at the centroid of triangle ABC.

Chapter 21

The Erdös-Mordell Inequality

The next result is probably the most beautiful geometric inequality in triangle geometry.

Theorem 21.1. (The Erdős-Mordell Inequality) If from a point P inside a given triangle ABC perpendiculars PH_1, PH_2, PH_3 are drawn to its sides, then

$$PA + PB + PC \geq 2(PH_1 + PH_2 + PH_3)$$

with equality holding if and only if triangle ABC is equilateral and if P is its center.

This was conjectured by Paul Erdős in 1935, and first proved by Mordell in the same year. Several proofs of this inequality have been given, using Ptolemy's Theorem by André Avez, angular computations with similar triangles by Leon Bankoff, area inequality by V. Komornik, or using trigonometry by Mordell and Barrow. We will present three very different proofs here.

First proof. The unusually stringent equality condition should suggest that perhaps the proof proceeds in two stages, with different equality conditions. This is indeed the case.

We will first prove that

$$PA \geq \frac{AB}{BC} \cdot PH_2 + \frac{AC}{BC} \cdot PH_3.$$

As a matter of fact, this step (called **Mordell's Lemma**) is so important that practically every proof of the Erdős-Mordell Inequality uses it as a lemma. So, let's prove it! Rewrite the inequality as

$$PA \sin A \geq PH_2 \sin C + PH_3 \sin B,$$

and note that that $PA \sin A = H_2H_3$ (by the Law of Sines in triangle AH_2H_3). On the other hand, the right hand side of the above inequality is the length of the projection of H_2H_3 on BC, and therefore we have equality if and only if H_2H_3 is parallel to the side BC.

Now, adding the inequality

$$PA \geq \frac{AB}{BC} \cdot PH_2 + \frac{AC}{BC} \cdot PH_3$$

to its two analogues yields

$$PA+PB+PC \geq PH_1\left(\frac{CA}{AB}+\frac{AB}{CA}\right)+PH_2\left(\frac{AB}{BC}+\frac{BC}{AB}\right)+PH_3\left(\frac{BC}{CA}+\frac{CA}{AB}\right),$$

with equality occurring if and only if the triangles $H_1H_2H_3$ and ABC are homothetic - in other words, if and only if P is the circumcenter of triangle ABC. Now for the second step: we note that each of the terms in the parentheses is at least 2 by the AM-GM Inequality. This gives

$$PA + PB + PC \geq 2(PX + PY + PZ),$$

with equality if and only if $AB = BC = CA$, and so our proof is complete. □

Now, let's give a proof that is much more straightforward (albeit less pretty).

Second proof. [MB] We transform it into a trigonometric inequality. Let $h_1 = PH_1$, $h_2 = PH_2$ and $h_3 = PH_3$.

Apply the Law of Sines and then the Law of Cosines to obtain

$$\begin{aligned}
PA \sin A = H_2H_3 &= \sqrt{{h_2}^2 + {h_3}^2 - 2h_2h_3 \cos(180^\circ - A)}, \\
PB \sin B = H_3H_1 &= \sqrt{{h_3}^2 + {h_1}^2 - 2h_3h_1 \cos(180^\circ - B)}, \\
PC \sin C = H_1H_2 &= \sqrt{{h_1}^2 + {h_2}^2 - 2h_1h_2 \cos(180^\circ - C)}.
\end{aligned}$$

So, we need to prove that

$$\sum_{\text{cyclic}} \frac{1}{\sin A}\sqrt{{h_2}^2 + {h_3}^2 - 2h_2h_3 \cos(180^\circ - A)} \geq 2(h_1 + h_2 + h_3).$$

The main trouble is that the left hand side has heavy terms with square root expressions. Our strategy is to find a lower bound without square roots. To this end, we express the terms inside the square root as the sum of two squares.

$$\begin{aligned} H_2H_3{}^2 &= h_2{}^2 + h_3{}^2 - 2h_2h_3\cos(180^\circ - A) \\ &= h_2{}^2 + h_3{}^2 - 2h_2h_3\cos(B + C) \\ &= h_2{}^2 + h_3{}^2 - 2h_2h_3(\cos B\cos C - \sin B\sin C). \end{aligned}$$

Using $\cos^2 B + \sin^2 B = 1$ and $\cos^2 C + \sin^2 C = 1$, one finds that

$$H_2H_3{}^2 = (h_2\sin C + h_3\sin B)^2 + (h_2\cos C - h_3\cos B)^2.$$

Since $(h_2\cos C - h_3\cos B)^2$ is clearly nonnegative, we get

$$H_2H_3 \geq h_2\sin C + h_3\sin B.$$

Hence,

$$\begin{aligned} \sum_{\text{cyclic}} \frac{\sqrt{h_2{}^2 + h_3{}^2 - 2h_2h_3\cos(180^\circ - A)}}{\sin A} &\geq \sum_{\text{cyclic}} \frac{h_2\sin C + h_3\sin B}{\sin A} \\ &= \sum_{\text{cyclic}} \left(\frac{\sin B}{\sin C} + \frac{\sin C}{\sin B}\right) h_1 \\ &\geq \sum_{\text{cyclic}} 2\sqrt{\frac{\sin B}{\sin C}\cdot\frac{\sin C}{\sin B}} h_1 \\ &= 2h_1 + 2h_2 + 2h_3. \end{aligned}$$

as desired. □

The next proof is perhaps the simplest of the three.

Third proof. Let Q be the reflection of P over the A-internal angle bisector in triangle ABC and let D, E, F be the feet of the perpendiculars from Q to sides BC, CA, AB respectively. It's clear that $AQ + QD$ is greater than the length of the A-altitude in triangle ABC so

$$BC \cdot (AQ + QD) \geq 2[ABC] = BC \cdot QD + CA \cdot QE + AB \cdot QF.$$

But $QE = PH_3$ and $QF = PH_2$ and $AQ = AP$ so this implies that

$$AP \geq \frac{AB}{BC} \cdot PH_2 + \frac{CA}{BC} \cdot PH_3$$

and then we proceed as in the last two solutions. □

//The final proof (if one were to proceed using P and not Q) also implies the following inequality:

$$AP \geq \frac{CA}{BC} \cdot PH_2 + \frac{AB}{BC} \cdot PH_3$$

In fact, we can prove something even stronger than Erdős-Mordell:

Theorem 21.2. (Barrow's Inequality) Let P be an interior point of a triangle ABC and let U, V, W be the points where the internal bisectors of angles $\angle BPC$, $\angle CPA$, $\angle APB$ intersect the sides BC, CA, AB respectively. Then, we have

$$PA + PB + PC \geq 2(PU + PV + PW).$$

Proof. ([MB] and [AK]) We begin with a classic claim known as **Wolstenholme's Inequality**:

Claim: Let $x, y, z, \theta_1, \theta_2, \theta_3$ be real numbers with $\theta_1 + \theta_2 + \theta_3 = \pi$. Then, the following inequality holds:

$$x^2 + y^2 + z^2 \geq 2(yz \cos\theta_1 + zx \cos\theta_2 + xy \cos\theta_3).$$

Proof. Using $\theta_3 = 180° - (\theta_1 + \theta_2)$, we have the identity

$$x^2 + y^2 + z^2 - 2(yz \cos\theta_1 + zx \cos\theta_2 + xy \cos\theta_3) \quad = \\ [z - (x \cos\theta_2 + y \cos\theta_1)]^2 + [x \sin\theta_2 - y \sin\theta_1]^2 \geq 0$$

Returning to the problem, let $d_1 = PA$, $d_2 = PB$, $d_3 = PC$, $l_1 = PU$, $l_2 = PV$, $l_3 = PW$, $2\theta_1 = \angle BPC$, $2\theta_2 = \angle CPA$, and $2\theta_3 = \angle APB$. We need to show that $d_1 + d_2 + d_3 \geq 2(l_1 + l_2 + l_3)$. By the Angle Bisector Theorem on triangle BPC and then the Law of Cosines on triangles BPU and CPU we have that

$$d_2 \cdot CU = d_3 \cdot BU \Longrightarrow d_2^2(d_3^2 + l_1^2 - 2d_3 l_1 \cos\theta_1) = d_3^2(d_2^2 + l_1^2 - 2d_2 l_1 \cos\theta_1)$$

which yields $l_1 = \frac{2d_2d_3}{d_2+d_3} \cos\theta_1$. Analogously we deduce:

$$l_1 = \frac{2d_2d_3}{d_2 + d_3} \cos\theta_1, \; l_2 = \frac{2d_3d_1}{d_3 + d_1} \cos\theta_2, \;\; \text{and} \;\; l_3 = \frac{2d_1d_2}{d_1 + d_2} \cos\theta_3,$$

It now follows by the HM-GM inequality and Wolstenholme's Inequality that

$$l_1 + l_2 + l_3 \leq \sqrt{d_2 d_3}\cos\theta_1 + \sqrt{d_3 d_1}\cos\theta_2 + \sqrt{d_1 d_2}\cos\theta_3 \leq \frac{1}{2}\left(d_1 + d_2 + d_3\right).$$

This completes the proof. Note that the equality in both inequalities holds if and only if triangle ABC is equilateral and P is its center. □

As you can imagine, due to its beauty and importance, the Erdős-Mordell inequality inspired many mathematicians to find variations, extensions and even generalizations on this quintessential lemma in triangle geometry. We already saw Barrow's sharpening, but let's try and extend the result for polygons this time. To this end, let us begin with the following lemma.

Delta 21.1. Let $x_1, x_2, \ldots, x_n$ and $\theta_1, \theta_2, \ldots, \theta_n$ be two sets of positive real numbers such that

$$\theta_1 + \theta_2 + \ldots + \theta_n = \pi.$$

Then,

$$\sum_{i=1}^{n} x_i x_{i+1} \cos\theta_i \leq \cos\frac{\pi}{n}\sum_{i=1}^{n} x_i^2,$$

where the indices are taken modulo n.

Obviously, when $n = 3$, we recover Wolstenholme's inequality. The proof for the case when $n = 4$ was given by Florian [Flo] and the proof for general n was obtained by Lenhard [Len]. We won't include it here as it is too tedious, and will only minimize the beauty of the result itself. We now detonate the bomb!

Corollary 21.1. (Generalization of Barrow) If P is a point in the interior of a convex n-gon, then the sum of the distances from P to the sides of the polygon is at most $\cos\left(\frac{\pi}{n}\right)$ times the sum of its distances to the vertices.

Proof. Proceed exactly as in the proof of Barrow's Inequality.

An interesting corollary of this generalization is the following neat inequality that generalizes the Euler-Chappel Inequality.

Corollary 21.2. Given a bicentric polygon $\mathcal{P}$ (a polygon with both an inscribed and circumscribed circle) with vertices A_1, A_2, $\ldots$, A_n, we have that

$$\frac{R}{r} \geq \frac{1}{\cos\frac{\pi}{n}},$$

where R, r are the circumradius and inradius of $\mathcal{P}$, respectively.

//Clearly, for $n = 2$, this reduces to the extremely famous $R \geq 2r$.

Let's see some Erdös-Mordell applied to a few of Olympiad-style problems now. We begin with a famous IMO problem!

Delta 21.2. (IMO 1991) Let ABC be a triangle and P an interior point in ABC. Show that at least one of the angles $\angle PAB$, $\angle PBC$, $\angle PCA$ is less than or equal to $30°$.

Proof. Let D, E, F be the feet of the perpendiculars from P to sides BC, CA, AB respectively. Assume for the sake of contradiction that each of the angles $\angle PAB$, $\angle PBC$, $\angle PCA$ has measure greater than $30°$. Then $\frac{PD}{PB} = \sin PBC > \frac{1}{2}$ and similarly $\frac{PE}{PC} > \frac{1}{2}$ and $\frac{PF}{PA} > \frac{1}{2}$. This implies that $2(PD + PE + PF) > PA + PB + PC$ which contradicts the Erdős-Mordell Inequality. This completes the proof. □

Also, note that the generalization of Erdős-Mordell for polygons proves the more general version of **Delta 21.2**.

Delta 21.3. (Hojoo Lee and Cosmin Pohoata, Mathematical Reflections) Let $A_1A_2\ldots A_n$ be a convex polygon and let P be a point in its interior. Prove that

$$\min_{i\in\{1,2,\ldots,n\}} \angle PA_iA_{i+1} \leq \frac{\pi}{2} - \frac{\pi}{n}$$

where indices are taken modulo n.

Proof. Proceed exactly as in the proof of **Delta 21.2**.

Delta 21.4. (USA TST 2001) Let h_a, h_b, h_c be the lengths of the altitudes of a triangle ABC from A, B, C respectively. Let P be any point inside the triangle. Show that

$$\frac{PA}{h_b + h_c} + \frac{PB}{h_a + h_c} + \frac{PC}{h_a + h_b} \geq 1.$$

Proof. Let D, E, F be the feet of the perpendiculars from P to sides BC, CA, AB respectively. From the sidenote after the third proof of Erdős-Mordell, we have that

$$a \cdot PA \geq PE \cdot b + PF \cdot c \quad \text{and} \quad a \cdot PA \geq PE \cdot c + PF \cdot b.$$

By averaging the two inequalities we obtain

$$PA \geq \frac{(b+c)(PE+PF)}{2a}$$

and analogous inequalities for PB and PC. Thus, letting S be the area of triangle ABC, we may write

$$\sum_{cyc} \frac{PA}{h_b + h_c} = \sum_{cyc} \frac{PA}{\frac{2S}{b} + \frac{2S}{c}} = \sum_{cyc} \frac{bc \cdot PA}{2S(b+c)}.$$

And now by using the averaged inequalities we obtain

$$\begin{aligned} \sum_{cyc} \frac{bc \cdot PA}{2S(b+c)} &\geq \sum_{cyc} \frac{bc \cdot \frac{(b+c)(PE+PF)}{2a}}{2S(b+c)} \\ &= \sum_{cyc} \frac{bc(PE+PF)}{4aS} \\ &= \frac{\sum_{cyc} b^2c^2(PE+PF)}{4abcS} \\ &= \frac{\sum_{cyc} PD(a^2b^2 + a^2c^2)}{4abcS} \\ &\geq \frac{\sum_{cyc} 2a^2bc \cdot PD}{4abcS} \\ &= 1 \end{aligned}$$

where we used AM-GM for the last inequality and the fact that

$$a \cdot PD + b \cdot PE + c \cdot PF = 2S$$

for the final equality. This completes the proof. □

Delta 21.5. Let M be a point inside an arbitrary triangle ABC and let R_a, R_b, R_c be the circumradii of triangles MBC, MCA, MAB respectively. Prove that

$$\frac{1}{MA} + \frac{1}{MB} + \frac{1}{MC} \geq \frac{1}{R_a} + \frac{1}{R_b} + \frac{1}{R_c}$$

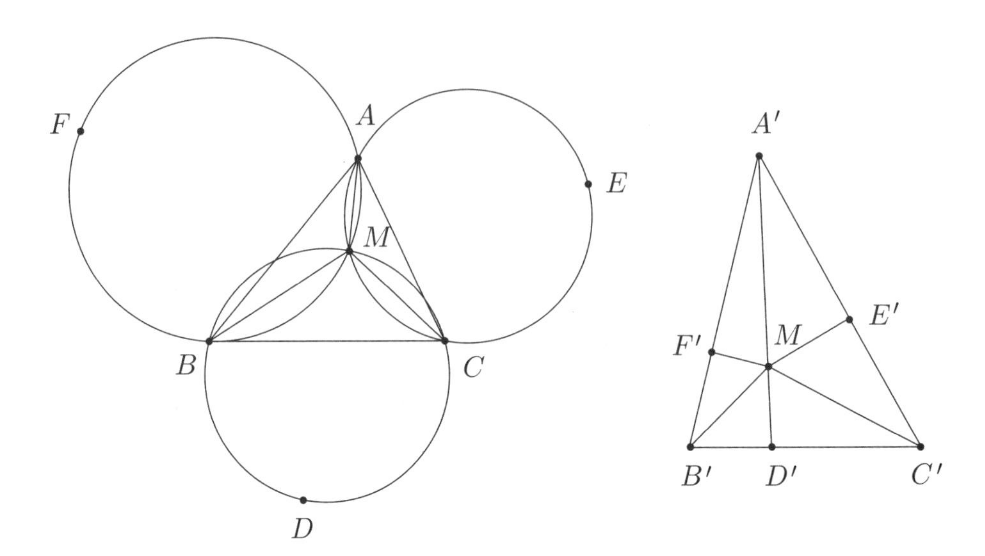

Proof. Let D, E, F be the points diametrically opposite from M on the circumcircles of triangles MBC, MCA, MAB respectively. Invert about the circle with center M and radius 1 - denote the inverses of points by adding an apostrophe. It's clear that lines BC, CA, AB invert to the circumcircles of triangles $MB'C', MC'A', MA'B'$ respectively and that the circumcircles of triangles MBC, MCA, MAB invert to lines $B'C', C'A', A'B'$ respectively. Moreover, it's easy to see that points D', E', F' are the feet of the perpendiculars from M to lines $B'C', C'A', A'B'$ respectively. Hence by the Erdős-Mordell Inequality we have that

$$MA' + MB' + MC' \geq 2(MD' + ME' + MF').$$

But we know that

$$MA' = \frac{1}{MA}$$

and analogous results for MB' and MC' and

$$MD' = \frac{1}{2R_a}$$

and analogous results for ME' and MF'. Substituting then completes the proof. □

Delta 21.6. Let ABC be a triangle with incenter I and let M, N, P be the midpoints of the arcs BC, CA, AB which do not contain the vertices of the triangle. Prove that

$$MI + NI + PI \geq AI + BI + CI.$$

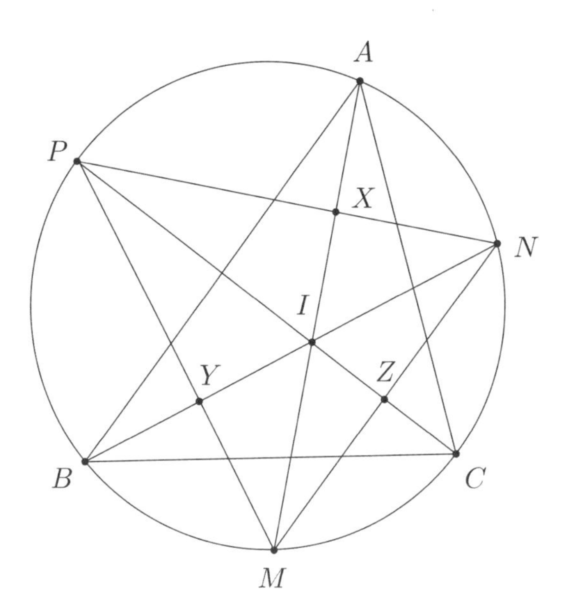

Proof. Let $X = AI \cap NP$ and $Y = BI \cap PM$ and $Z = CI \cap MN$. An easy angle chase shows that X, Y, Z are the feet of the perpendiculars from I to sides NP, PM, MN of triangle MNP respectively. Moreover, we know that $NA = NI$ and $PA = PI$ so quadrilateral $INAP$ is a kite. Therefore $AI = 2XI$ and similarly $BI = 2YI$ and $CI = 2ZI$. Applying the Erdős-Mordell Inequality to triangle MNP and point I then completes the proof. □

The next problem was number 5 in the 1996 IMO - however, it turned out to be the hardest problem!

Delta 21.7. (IMO 1996) Let $ABCDEF$ be a convex hexagon such that AB is parallel to DE, BC is parallel to EF, and CD is parallel to FA. Let R_A, R_C, R_E denote the circumradii of triangles FAB, BCD, DEF, respectively, and let P denote the perimeter of the hexagon. Prove that

$$R_A + R_C + R_E \geq \frac{P}{2}.$$

Proof. Let A', C', E' be the reflections of A, C, E over the midpoints of segments BF, DB, FB respectively. Let the line through F perpendicular to $E'F$ meet the lines through D perpendicular to $C'D$ and through B perpendicular to $A'B$ at B' and D' respectively. Also let the line through B perpendicular to $A'B$ meet the line through D perpendicular to $C'D$ at F'.

It's clear that $D'A' = 2R_A$ and $E'B' = 2R_B$ and $F'C' = 2R_C$ so it suffices to show that

$$D'A' + E'B' + F'C' \geq A'F + A'B + C'B + C'D + E'D + E'F$$

Now, angle chasing with the cyclic quads $B'DE'F$ and $D'FA'B$ and $F'BC'D$ yields that triangles $A'C'E'$ and $D'F'B'$ are similar so

$$A'B \cdot B'D' + A'F \cdot F'D' = C'B \cdot B'D' + E'F \cdot F'D'.$$

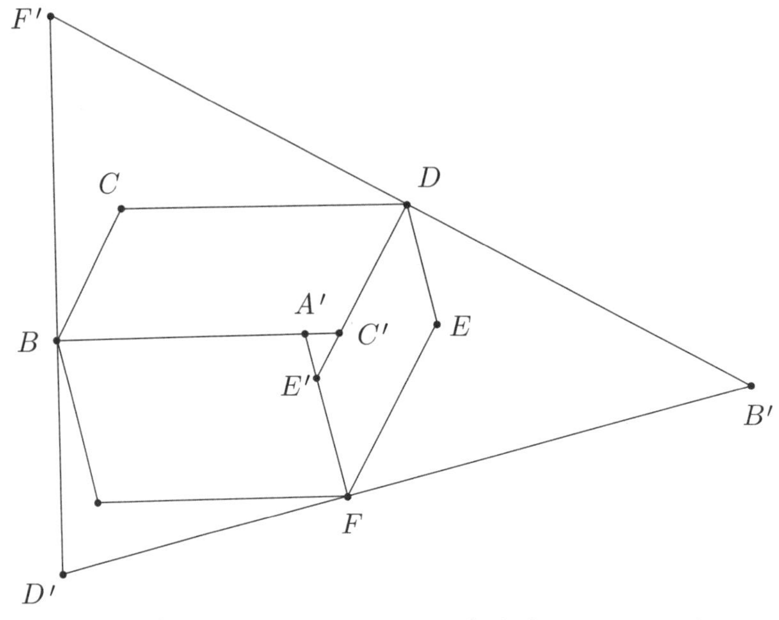

Therefore by Mordell's Lemma on triangle $B'D'F'$ and point A' we have that

$$A'D' \geq A'B \cdot \frac{B'D'}{B'F'} + A'F \cdot \frac{F'D'}{B'F'} = C'B \cdot \frac{B'D'}{B'F'} + E'F \cdot \frac{F'D'}{B'F'}$$

Obtaining five similar inequalities and summing we have

$$2(A'D' + B'E' + C'F') \geq \sum_{cyc}(A'B + C'B)\left(\frac{B'D'}{B'F'} + \frac{B'F'}{B'D'}\right)$$

and after some obvious applications of AM-GM we are done. □

We finish with a result given as the last problem of the USA Team Selection Test in 2000. This problem was extremely difficult for contestants, and remained unsolved on Art of Problem Solving forums for quite a while. However, with the tools of this chapter, it's a piece of cake!

Delta 21.8. (USA TST 2000) Let ABC be a triangle inscribed in a circle of radius R, and let P be a point in the interior of triangle ABC. Prove that

$$\frac{PA}{BC^2} + \frac{PB}{CA^2} + \frac{PC}{AB^2} \geq \frac{1}{R}.$$

Proof. We use Mordell's Lemma. Let D, E, F be the feet of the perpendiculars from P to sides BC, CA, AB respectively and let S be the area of triangle ABC. We obtain:

$$\begin{aligned}
\sum_{cyc} \frac{PA}{a^2} &\geq \sum_{cyc} \left(\frac{c \cdot PE}{a^3} + \frac{b \cdot PF}{a^3} \right) \\
&= \sum_{cyc} PD \left(\frac{b}{c^3} + \frac{c}{b^3} \right) \\
&\geq \sum_{cyc} \frac{2PD}{bc} \\
&= \frac{\sum_{cyc} 2a \cdot PD}{abc} \\
&= \frac{4S}{abc} \\
&= \frac{1}{R}
\end{aligned}$$

where we used AM-GM for the last inequality. This completes the proof. □

Assigned Problems

Epsilon 21.1. Let H and O be the orthocenter and circumcenter of triangle ABC respectively. Show that

$$HA + HB + HC \leq OA + OB + OC$$

Epsilon 21.2. Prove that in any acute-angled triangle ABC, we have that

$$\frac{3}{2} \geq \cos A + \cos B + \cos C \geq 2\cos B\cos C + 2\cos C\cos A + 2\cos A\cos B.$$

Epsilon 21.3. Let ABC be a triangle. Prove that

$$\sin\frac{A}{2} + \sin\frac{B}{2} + \sin\frac{C}{2} \geq \frac{r}{2R}\cdot\left(\frac{1}{\sin\frac{A}{2}} + \frac{1}{\sin\frac{B}{2}} + \frac{1}{\sin\frac{C}{2}}\right) \geq \frac{3r}{R},$$

with equality if and only if ABC is equilateral.

Epsilon 21.4. (Leonard Carlitz, AMM) Show that in an acute triangle,

$$h_1 + h_2 + h_3 \leq 3(R + r),$$

where the h_i are the lengths of the altitudes. Show that the equality case takes place if and only if the triangle is equilateral.

Epsilon 21.5. Let P be an interior point of triangle ABC. Denote by R_a , R_b , R_c the circumradii of the triangles PBC , PCA and PAB respectively. Prove that

$$R_a + R_b + R_c \geq PA + PB + PC.$$

Epsilon 21.6. (Moldova TST 2001) If P is a point lying on the segment OH of the acute-angled triangle ABC, where O and H denote the circumcenter, and the orthocenter, respectively, prove that

$$6r \leq PA + PB + PC \leq 3R,$$

where r and R denote the inradius, and the circumradius of ABC, respectively.

Epsilon 21.7. Let P be a point inside triangle ABC. Prove that

$$a\cdot\frac{PA}{d_a} + b\cdot\frac{PB}{d_b} + c\cdot\frac{PC}{d_c} \geq 2(a + b + c),$$

where d_a, d_b, d_c are the distances $\delta(P, BC)$, $\delta(P, CA)$, $\delta(P, AB)$ from P to the sides BC, CA, AB, respectively.

Epsilon 21.8. Let P be a point inside a triangle ABC. With the same notations as in the previous problem, prove that

$$PA \cdot d_a + PB \cdot d_b + PC \cdot d_c \leq \frac{PA^2 + PB^2 + PC^2}{2}.$$

Epsilon 21.9. (Razvan Satnoianu, AMM) Let P be a point in the interior of triangle ABC. Let r, s, t be the distances from P to the vertices A, B, C, respectively, and let x, y, z be the distances from P to the sides BC, CA, AB, respectively.

(a) Prove that $q^r + q^s + q^r + 3 \geq 2(q^x + q^y + q^z)$ for any $q \geq 1$.

(b) Prove that $q^{s+t} + q^{t+r} + q^{r+s} + 6 \geq q^{2x} + q^{2y} + q^{2z} + 2(q^x + q^y + q^z)$ for any $q \geq 1$.

Epsilon 21.10. The incircle k of triangle ABC touches the sides BC, CA, AB at points A', B', C', respectively. For any point K on k, let d be the sum of the distances from K to the sides of the triangle $A'B'C'$. Prove that

$$KA + KB + KC > 2d.$$

Epsilon 21.11. (Kazarinoff) Let P be a point inside tetrahedron $ABCD$. Let G, H, L, K be the feet of the perpendiculars from P to triangles BCD, ACD, ABD, ABC respectively. Show that

$$PA + PB + PC + PD > 2\sqrt{2}(PG + PH + PL + PK)$$

Chapter 22

Sondat's Theorem and the Neuberg Cubic

We begin with an extremely powerful theorem in modern geometry.

Theorem 22.1. (Sondat's Theorem) Let ABC and $A'B'C'$ be two triangles such that the perpendiculars from vertices A, B, C to sides $B'C'$, $C'A'$, $A'B'$ of triangle $A'B'C'$ are concurrent at some point O. Then

(a) Perpendiculars from vertices A', B', C' to sides BC, CA, AB of triangle ABC are concurrent at some point O'.

(b) If $O = O'$, then lines AA', BB', CC' are concurrent.

(c) If $O \neq O'$, but the lines AA', BB', CC' are still concurrent at some point P, then line OO' passes through P and is perpendicular to the perspectrix of triangles ABC and $A'B'C'$.

Proof. (a) This was **Corollary 2.2**. Recall that triangles ABC and $A'B'C'$ are called orthologic triangles and that O and O' are called the orthology centers of these two triangles.

(b) We begin with a beautiful collinearity theorem due to Nikolaos Dergiades.

Claim. Let ABC be a triangle and let Γ_a, Γ_b, Γ_c be circles having segments BC, CA, AB respectively as chords. Let D be the second intersection of Γ_b and Γ_c, E the second intersection of Γ_c and Γ_a, and F the second intersection of Γ_a and Γ_b. Furthermore, let the perpendicular line to AD passing through D intersect the line BC at X. Similarly, define Y and Z. Then points X, Y, Z are collinear.

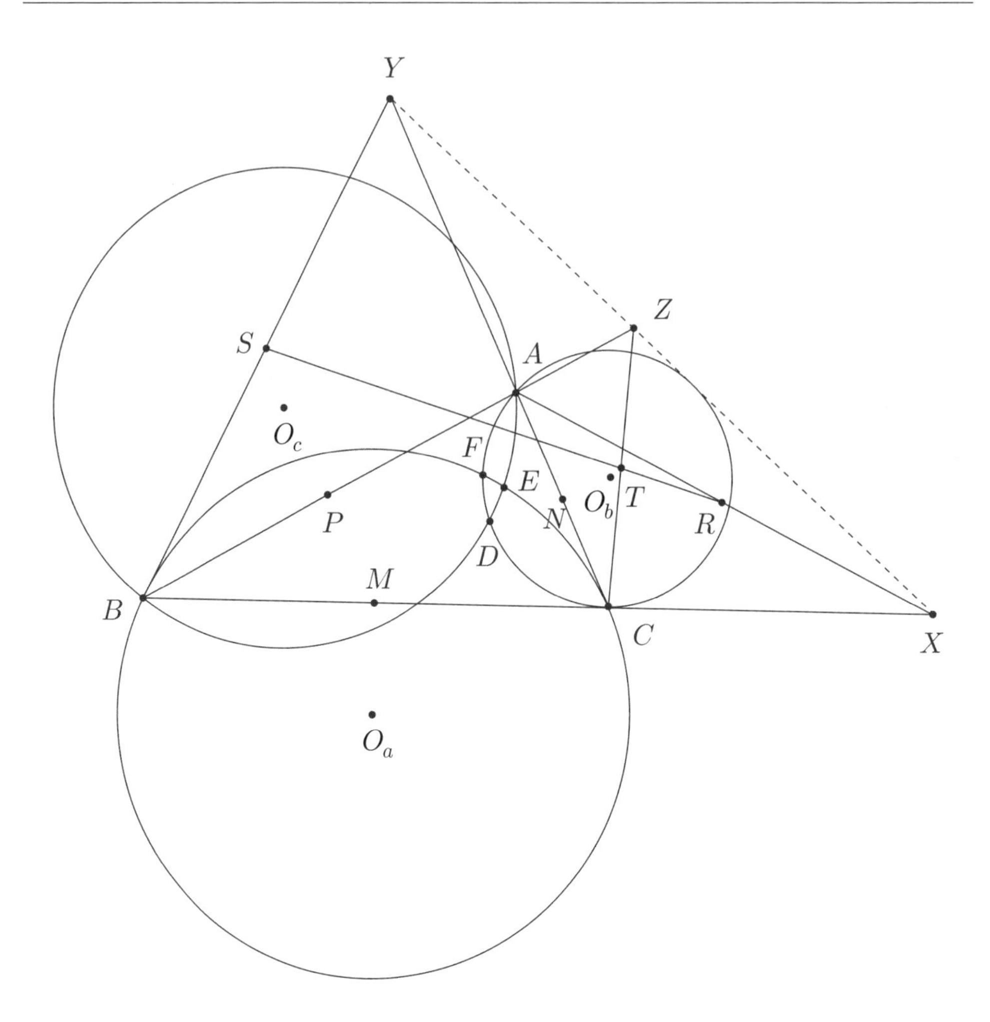

Proof. Let O_a, O_b, O_c be the centers of circles Γ_a, Γ_b, Γ_c respectively, let M, N, P be the midpoints of sides BC, CA, AB respectively, and let R, S, T be the midpoints of segments AX, BY, CZ respectively. Line O_bO_c is the perpendicular bisector of segment AD so $O_bO_c \parallel DX$, which implies that $R \in O_bO_c$. Similarly, we get that $S \in O_cO_a$ and $T \in O_aO_b$. On the other hand, points R, S, T also lie on the midlines NP, PM, and MN respectively, so $R = O_bO_c \cap NP$, $S = O_cO_a \cap PM$, $T = O_aO_b \cap MN$. However, the lines O_aM, O_bN, O_cP are the perpendicular bisectors of sides BC, CA, AB respectively so they concur at the circumcenter O of triangle ABC. Desargues' Theorem on triangles $O_aO_bO_c$ and MNP then yields that points R, S, T are collinear. Now let $X' = BC \cap YZ$ and consider the complete quadrilateral $BCYZX'A$. Letting R' be the midpoint of segment AX', we have that R', S, T all lie on the Newton-Gauss line of this complete quadrilateral and hence are collinear. But this means that $R' = R$ which implies that $X' = X$ as desired.

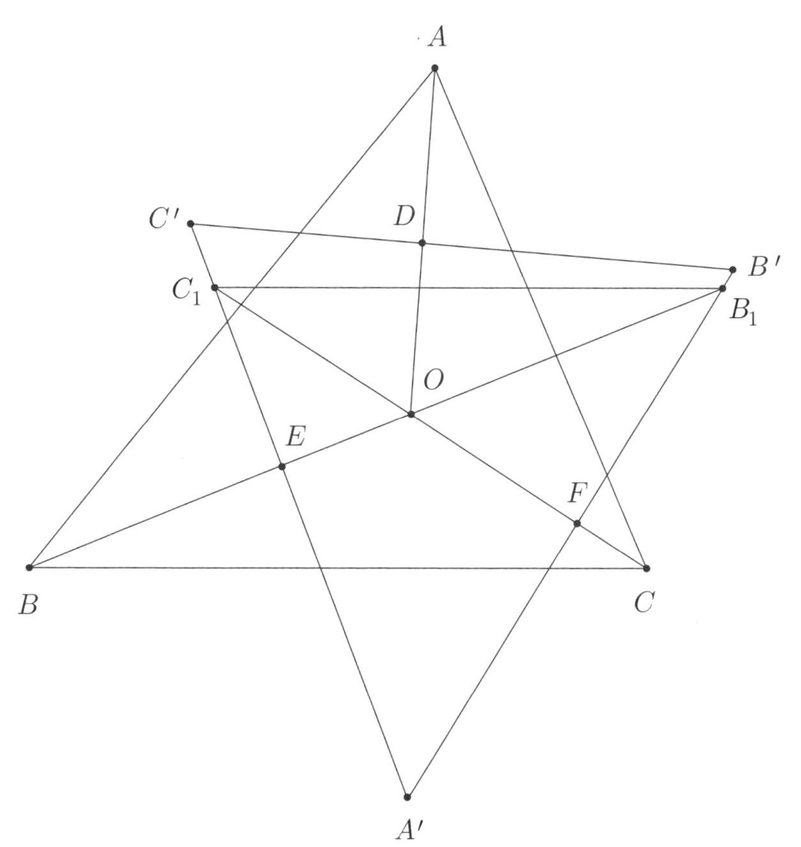

Returning to the problem (and majorly abusing notation), let D, E, F be the intersection points of lines AO, BO, CO with lines $B'C'$, $C'A'$, and $A'B'$ respectively. Furthermore, let $B_1 = BE \cap A'B'$ and $C_1 = CF \cap A'C'$. Point O is the orthocenter of triangle $A'B_1C_1$, so B_1, C_1, E, F all lie on the circle with diameter B_1C_1. In particular, B_1C_1 is an antiparallel to line EF in triangle B_1OC_1. On the other hand, $B_1C_1 \perp A'O$ and by definition $A'O \perp BC$, hence $B_1C_1 \| BC$. Thus, BC is an antiparallel to EF in triangle B_1OC_1 as well, which means that quadrilateral $BCEF$ is cyclic. Similarly, we get that quadrilaterals $CAFD$ and $ABDE$ are also cyclic, and so by the claim, the intersections $X = B'C' \cap BC$, $Y = C'A' \cap CA$, $Z = A'B' \cap AB$ are collinear. But then Desargues' Theorem yields that lines AA', BB', and CC' are concurrent, so the proof for part (b) is complete.

(c) We begin with three claims this time!

Claim 1. Let ABC and $A_1B_1C_1$ be two triangles that are perspective at P. Let $A_2B_2C_2$ be a triangle that is homothetic to triangle $A_1B_1C_1$, with homothety center P. Let $X = BC \cap B_1C_1$, $Y = CA \cap C_1A_1$, $Z = AB \cap A_1B_1$, and let $X' = BC \cap B_2C_2$, $Y' = CA \cap C_2A_2$, $Z' = AB \cap A_2B_2$. Then, the lines determined by points X, Y, Z and X', Y', Z' are parallel.

Proof. Since lines A_1A_2, YY', ZZ' concur at A by Desargues' Theorem we have that the intersections $A_1Y \cap A_2Y'$, $A_1Z \cap A_2Z'$, $YZ \cap Y'Z'$ are collinear. But since triangles $A_1B_1C_1$ and $A_2B_2C_2$ are homothetic we have $A_1Y \parallel A_2Y'$ and $A_1Z \parallel A_2Z'$ so $A_1Y \cap A_2Y'$ and $A_1Z \cap A_2Z'$ are points at infinity. Therefore $YZ \cap Y'Z'$ is also a point at infinity which implies that $YZ \parallel Y'Z'$ as desired.

Claim 2. Keeping the notation from part (b), suppose ABC and $A'B'C'$ are orthologic triangles with $O = O'$ (so that lines AA', BB', CC' are concurrent). Let X be the intersection of BC with $B'C'$. Then, lines OX and AA' are perpendicular.

Proof. As in the proof of part (b), let D be the intersection of AO with $B'C'$, E the intersection of BO with $C'A'$, and F the intersection of CO with $A'B'$. By the proof of (b), points B, C, E, F are concyclic and points A, C, F, D are concyclic. Similarly, if D' is the intersection of $A'O$ with BC and E' is the intersection of $B'O$ with CA, points A', B', D', E' are concyclic. And finally, it's clear that points B', E', C, F lie on the circle with diameter $B'C$ and hence are also concyclic. Power of a Point on O then yields

$$OA' \cdot OD' = OB' \cdot OE' = OC \cdot OF = OA \cdot OD.$$

Hence, $OA' \cdot OD' = OA \cdot OD$, i.e. points A, A', D, and D' are concyclic. Therefore line AA' is an antiparallel to side DD' in triangle DOD' and since OX is clearly a diameter of the circumcircle of triangle DOD' this implies that $AA' \perp OX$ as desired.

Claim 3. Using the same notations and hypothesis from Claim 2, furthermore let P be the common point of lines AA', BB', CC', let Y be the intersection of CA with $C'A'$, and let Z be the intersection of AB with $A'B'$. Then, OP is perpendicular to the line passing through X, Y, and Z.

Proof. Let D' be the intersection of $A'O$ with BC and let X' be the intersection of OX with AA'. Since Claim 2 implies $OX \perp AA'$, the circle with diameter $A'X$ passes through D' and X' so

$$OX \cdot OX' = OA' \cdot OD'.$$

Let $B'O$ meet CA at E'. By the proof of (b), points A', B', D', E' are concyclic; therefore,

$$OA' \cdot OD' = OB' \cdot OE'.$$

But the circle of diameter $B'Y$ passes through E' and the intersection Y' of OY and BB'. Hence,

$$OB' \cdot OE' = OY' \cdot OY.$$

Combining the three Power of a Point identities above, we get that $OX \cdot OX' = OY \cdot OY'$, so points X, X', Y, Y' are concyclic. Therefore line XY is an antiparallel to side $X'Y'$ in triangle $X'OY'$ and since OP is clearly a diameter of the circumcircle of triangle $X'OY'$ this implies that $XY \perp OP$. Analogously we have $ZX \perp OP$ and so the proof is complete.

We are finally ready to see what happens when $O \neq O'$! We maintain the notation from Claim 3 above. Furthermore, let A_1 be the intersection of the perpendicular from O to BC with the line AA', let B_1 be the intersection of the parallel to $A'B'$ through A_1 with BB', and let C_1 be the intersection of the parallel to $B'C'$ through B_1 with CC'. By Desargues' Theorem on triangles $A_1B_1C_1$ and $A'B'C'$, which are perspective at P, the lines A_1C_1 and $A'C'$ are parallel; thus, triangles $A_1B_1C_1$ and $A'B'C'$ are homothetic, with homothety center P. In particular, since $AO \perp B'C'$ and $B'C' \parallel B_1C_1$, it follows that $AO \perp B_1C_1$ and similarly $BO \perp C_1A_1$ and $CO \perp A_1B_1$. In other words, triangles ABC and $A_1B_1C_1$ are orthologic. But magic happened! Unlike with triangles ABC and $A'B'C'$, their orthology centers coincide at O. Since triangles ABC and $A_1B_1C_1$ are also perspective at P, it follows from Claim 3 that the points $R = B_1C_1 \cap BC$, $S = C_1A_1 \cap CA$, $T = A_1B_1 \cap AB$ determine a line perpendicular to OP.

We are practically done, for we can sense that Claim 1 is around the corner. As in the last paragraph, we can define A_2 to be the intersection of AA' with the perpendicular from O' to $B'C'$, B_2 to be the intersection of BB' with the parallel to AB through A_2, and C_2 to be the intersection of CC' with the parallel to BC through B_2. Similarly, we get that triangles ABC and $A_2B_2C_2$ are homothetic, with homothety center P, and consequently that triangles $A'B'C'$ and $A_2B_2C_2$ are orthologic with coincident orthology center O' while also being perspective at P. By Claim 3, it again follows that the points $R' = B_2C_2 \cap B'C'$, $S' = C_2A_2 \cap C'A'$, $T' = A_2B_2 \cap A'B'$ determine a line perpendicular to $O'P$. But Claim 1 tells us that $RS \parallel R'S'$. It follows that $OP \parallel OP'$. In particular, this means that points O, O', and P are collinear on a line that is perpendicular to line RS. This completes the proof of Sondat's Theorem. □

Now, let's see some examples in which this result is incredibly useful.

Delta 22.1. (Romanian TST 2010) Let ABC be a scalene triangle. Let ω be a circle that intersects side BC at points A_1 and A_2, intersects side CA at B_1

and B_2, and intersects side AB at C_1 and C_2. Let the lines tangent to ω at A_1 and A_2 meet at A', and define B' and C' similarly. Prove that the lines AA', BB' and CC' concur.

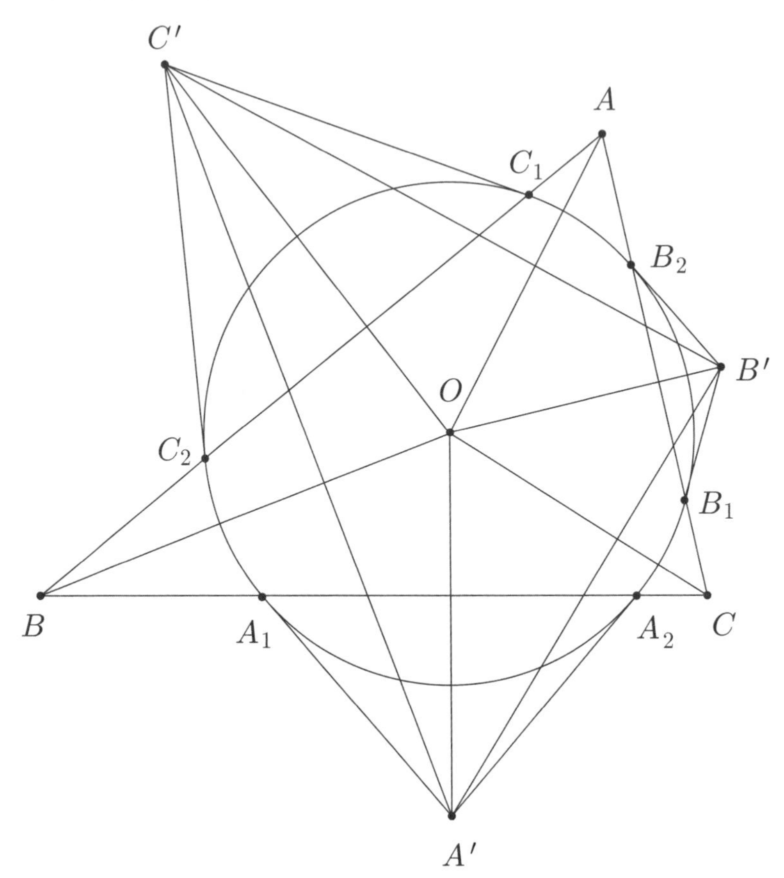

Proof. Let O be the center of ω - it's clear that $OA' \perp BC$ and $OB' \perp CA$ and $OC' \perp AB$. Now line CA is the polar of B' with respect to ω and line AB is the polar of C' with respect to ω so by La Hire's Theorem line $B'C'$ is the polar of A with respect to ω. Therefore $OA \perp B'C'$ and similarly $OB \perp C'A'$ and $OC \perp A'B'$. Hence triangles ABC and $A'B'C'$ are orthologic and their orthology centers coincide at O, so by Sondat's Theorem we have that lines AA', BB', CC' concur as desired. □

Delta 22.2. Let ABC be a non-isosceles triangle and let O, I, N denote its circumcenter, incenter, and nine-point center respectively. Let A', B', C' be the orthogonal projections of point O on lines AI, BI, CI respectively. Let ℓ_a be a line through A perpendicular to $B'C'$ and define lines ℓ_b, ℓ_c similarly. Prove that lines ℓ_a, ℓ_b, ℓ_c concur at a point on line IN.

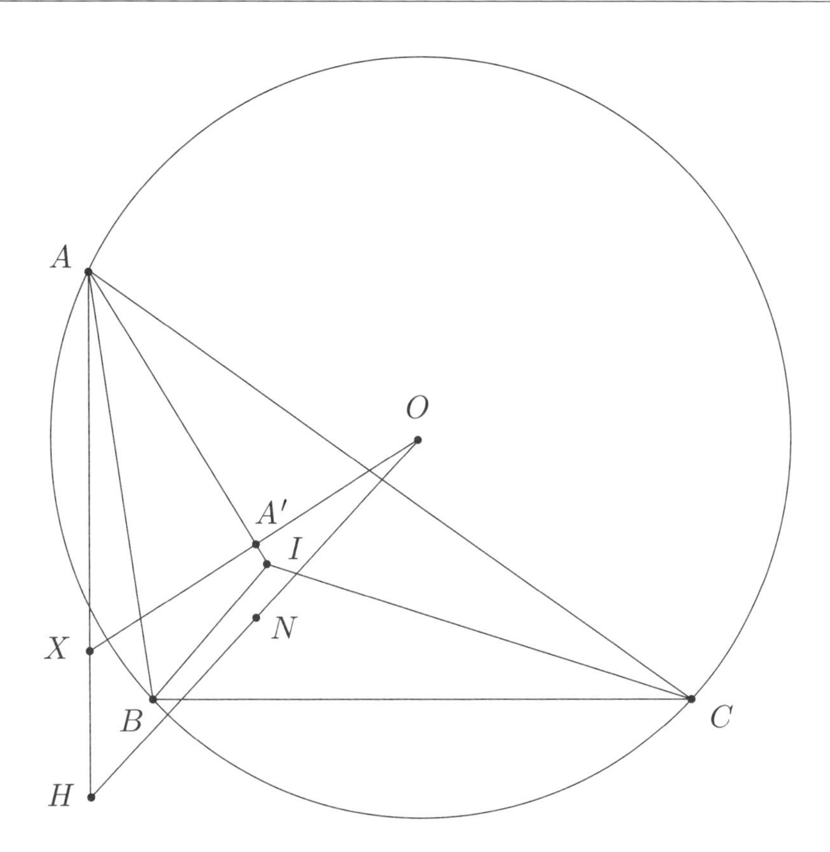

Proof. Let H be the orthocenter of triangle ABC and let X be the reflection of O over line AI. Since O is the isogonal conjugate of H with respect to triangle ABC we have that line AX is the A-altitude of triangle ABC. Hence A' lies directly between the lines through O and H perpendicular to BC. Since N is the midpoint of segment OH, this implies that $NA' \perp BC$. Similarly $NB' \perp CA$ and $NC' \perp AB$. Therefore triangles ABC and $A'B'C'$ are orthologic with N as an orthology center. But these triangles are also clearly perspective at I, so by Sondat's Theorem their other orthology center lies on line IN. But this other orthology center is precisely the concurrency point of lines ℓ_a, ℓ_b, ℓ_c, so we are done. $\square$

Delta 22.3. (Romanian TST 2004) The incircle of a non-isosceles triangle ABC has center I and touches the sides BC, CA and AB in A', B' and C', respectively. The lines AA' and BB' intersect in P, the lines AC and $A'C'$ intersect in M, and the lines BC and $B'C'$ intersect in N. Prove that the lines IP and MN are perpendicular.

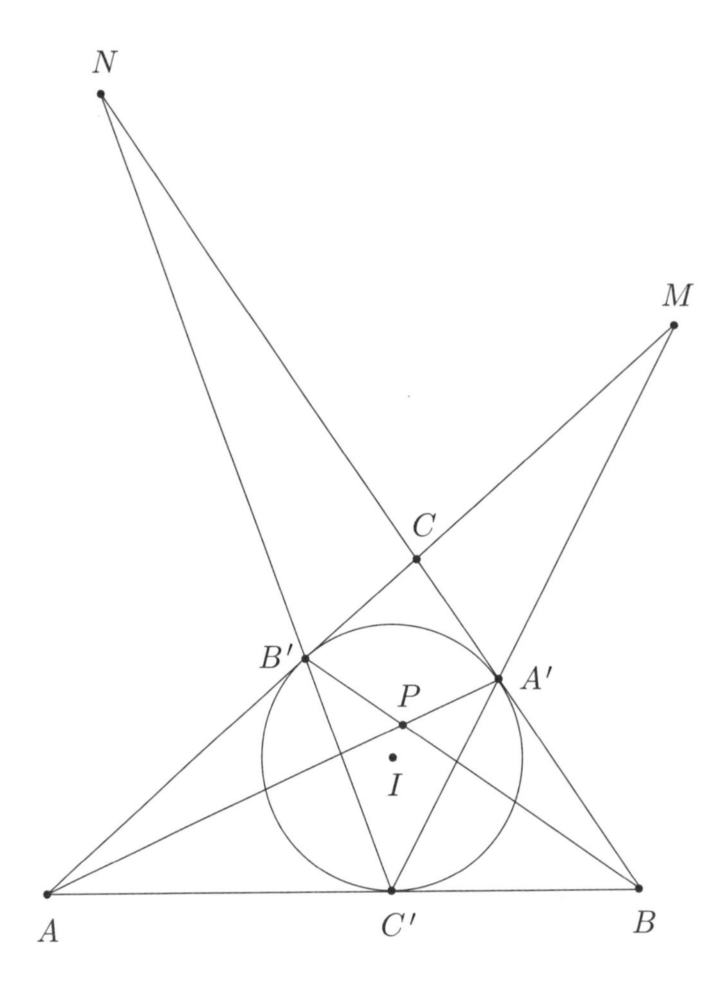

Proof. Since trivially $IA' \perp BC$ and $IB' \perp CA$ and $IC' \perp AB$ and $IA \perp B'C'$ and $IB \perp C'A'$ and $IC \perp A'B'$ we have that triangles ABC and $A'B'C'$ are orthologic and that their orthology centers coincide at I. It's clear that these triangles are also perspective at P (the Gergonne point of triangle ABC) and that their perspectrix is line MN so by Claim 3 in the proof of part (c) of Sondat's Theorem we have that $IP \perp MN$ as desired. □

For the next few problems, make sure to recall the various properties of isogonal conjugates and pedal triangles!

Delta 22.4. (ELMO Shortlist 2014) We are given triangles ABC and DEF such that $D \in BC$, $E \in CA$, $F \in AB$ and $AD \perp EF$, $BE \perp FD$, $CF \perp DE$. Let O be the circumcenter of triangle DEF, and let the circumcircle of triangle DEF intersect BC, CA, AB again at R, S, T respectively. Prove that the perpendiculars to BC, CA, AB that pass through D, E, F respectively intersect at a point X, and the lines AR, BS, CT intersect at a point Y, such that O, X, Y are collinear.

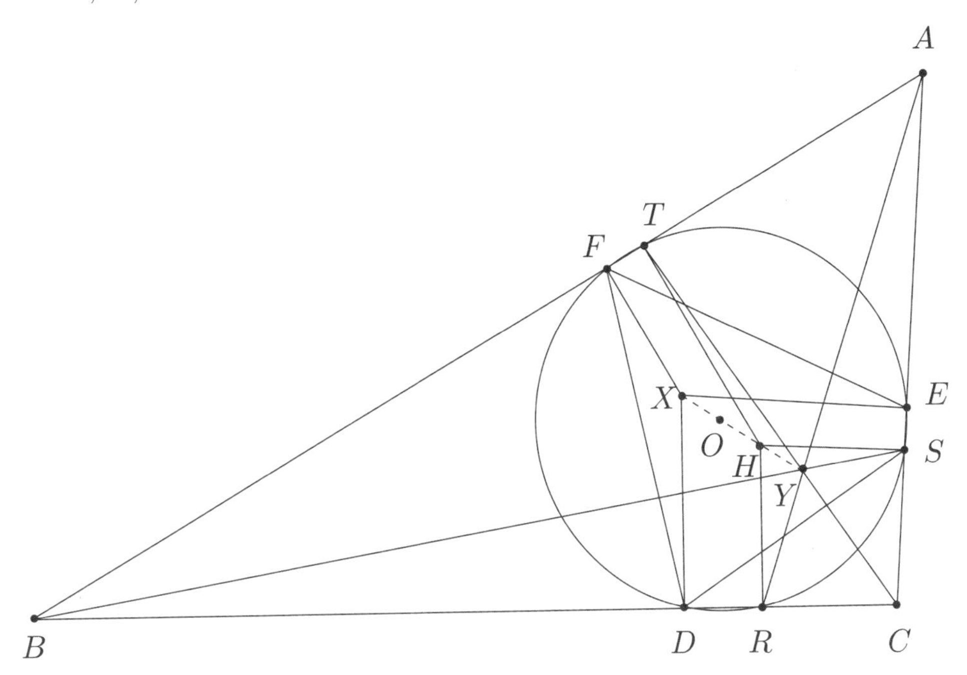

Proof. It's clear that the perpendiculars to EF, FD, DE that pass through A, B, C respectively concur at the orthocenter H of triangle DEF. Hence triangles ABC and DEF are orthologic and so the perpendiculars to BC, CA, AB through D, E, F respectively concur at a point X. Note that triangle DEF is the pedal triangle of X with respect to triangle ABC, so H is the isogonal conjugate of X with respect to triangle ABC and triangle RST is the pedal triangle of H with respect to triangle ABC. Note that triangles ABC and RST are orthologic and have orthology centers H and X. But as we proved in **Delta 3.2**, since lines AD, BE, CF concur at H we that lines AR, BS, CT concur at a point Y. Then by Sondat's Theorem on triangles ABC and RST we have that points H, X, Y are collinear. But we know that O is the midpoint of segment HX, so the proof is complete. □

Delta 22.5. (Iran 2001) Suppose that triangle ABC has circumcenter O and nine-point center N. Let N' be the isogonal conjugate of N with respect to triangle ABC. Suppose the perpendicular bisector of segment OA meets BC at A_1. Define B_1 and C_1 similarly. Prove that the points A_1, B_1, C_1 determine a line perpendicular to ON'.

Proof. We begin with a claim about N'.

Claim. Let O_a, O_b, O_c be the circumcenters of triangles BOC, COA, AOB respectively. Then lines AO_a, BO_b, CO_c concur at N'.

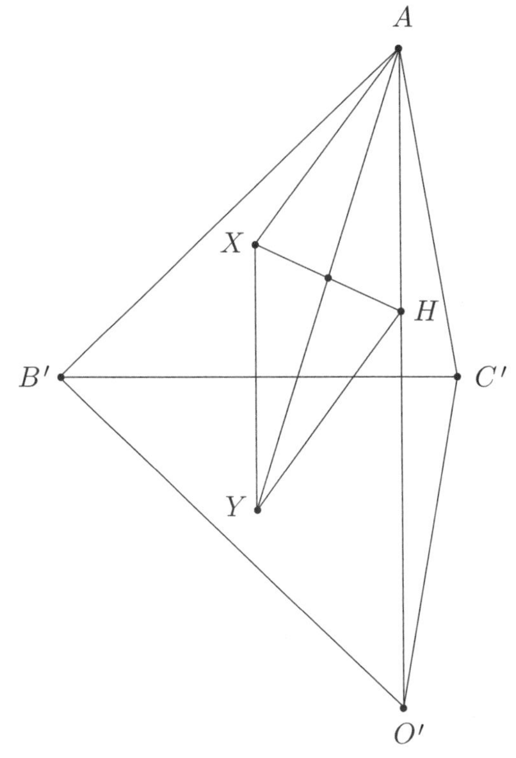

Proof. It clearly suffices to show that N' lies on line AO_a, because then by symmetry it will lie on the other two lines as well. Invert about a circle centered at A with arbitrary radius. Let points B and C invert to points B' and C' respectively. Then line AN' is the line determined by A and the nine-point center of triangle $AB'C'$. The circumcircle of triangle ABC inverts to line $B'C'$ so O inverts to reflection of A over line $B'C'$, which we shall call O'. Now, let Y be the circumcenter of triangle $B'O'C'$. It's clear that O_a inverts to a point on line AY so it suffices to show that line AY passes through the nine-point center of triangle $AB'C'$. Let X and H be the circumcenter and orthocenter of triangle $AB'C'$ respectively. Since Y is the reflection of X over line $B'C'$ we have $AH = XY$ and that $AH \parallel XY$ (as both lines are perpendicular to line $B'C'$). Therefore quadrilateral $AHXY$ is a parallelogram and so line AX passes through the midpoint of segment HX, which is precisely the nine-point center of triangle $AB'C'$. This completes the proof of the claim.

Returning to the problem, the claim shows that triangles ABC and $O_aO_bO_c$ are perspective at N'. Moreover, it's easy to see that lines OO_a, OO_b, OO_c are the perpendicular bisectors of segments BC, CA, AB respectively (and so concur at O). Also, lines O_bO_c, O_cO_a, O_aO_b are the perpendicular bisectors of segments OA, OB, OC respectively. Therefore triangles ABC and $O_aO_bO_c$ are also orthologic with their orthology centers coinciding at O. Also we have that $B_1 = CA \cap O_cO_a$ and $C_1 = AB \cap O_aO_b$ so line B_1C_1 is the perspectrix of triangles ABC and $O_aO_bO_c$. Hence, by Claim 3 in the proof of part (c) of Sondat's Theorem, $ON' \perp B_1C_1$ as desired. □

Delta 22.6. Let P be a point inside triangle ABC satisfying

$$\angle PBC + \angle PCA + \angle PAB = 90^\circ.$$

If P' is the isogonal conjugate of P with respect to triangle ABC, show that line PP' passes through the circumcenter of triangle ABC.

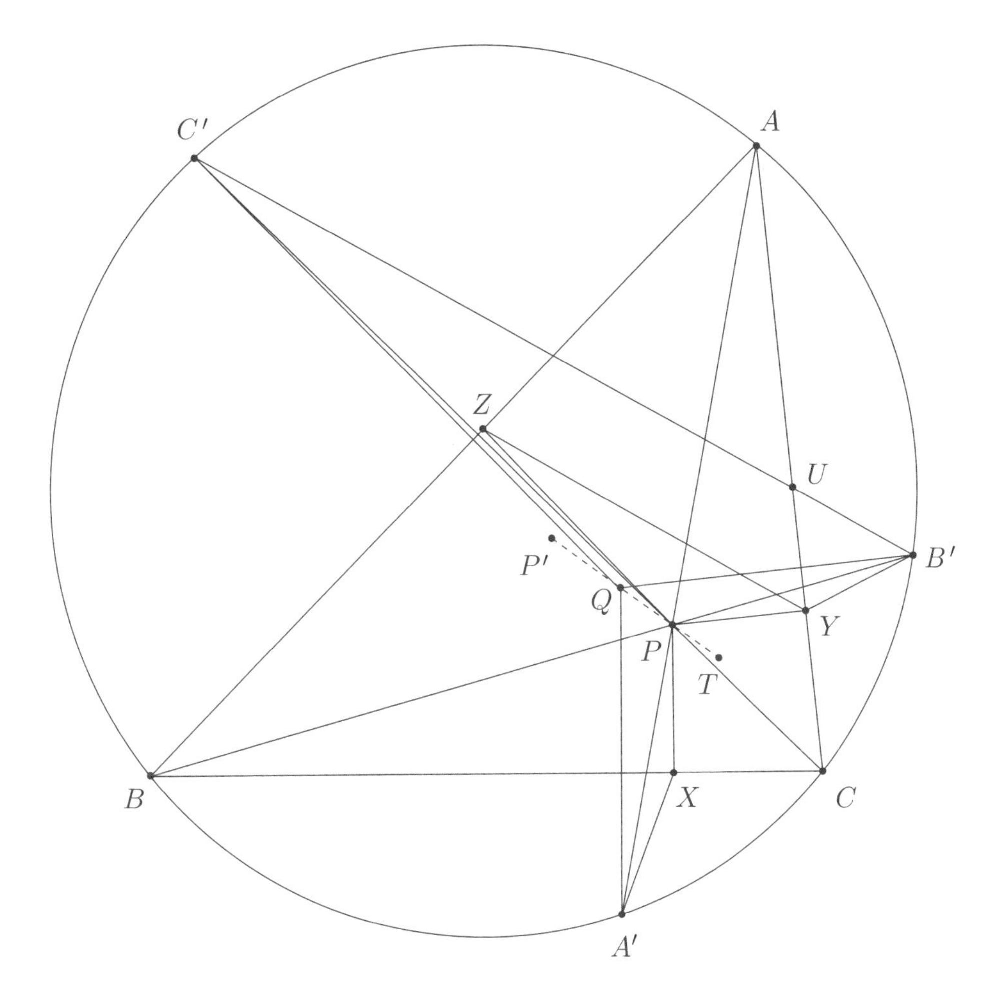

Proof. Let X, Y, Z be the projections of P on sides BC, CA, AB of triangle ABC respectively and let A', B', C' be the second intersections of lines

AP, BY, CZ with the circumcircle of triangle ABC. Let $U = CA \cap B'C'$. Then

$$\begin{aligned} \angle AUC' &= \angle UAB' + \angle UB'A \\ &= \angle PBC + \angle PCA \\ &= 90^\circ - \angle PAB. \end{aligned}$$

and since quadrilateral $AYPZ$ is cyclic we also have

$$\angle AYZ = \angle APZ = 90^\circ - \angle PAB.$$

Therefore $B'C' \parallel YZ$ and similarly $C'A' \parallel ZX$ and $A'B' \parallel XY$. This implies that triangles $A'B'C'$ and XYZ are homothetic, so let T be their homothety center. We know that the perpendiculars to YZ, ZX, XY that pass through A, B, C respectively intersect at P'. Hence triangles ABC and $A'B'C'$ are orthologic and one of their orthology centers is P'. Let Q be their other orthology center. It's clear that $QA' \parallel PX$ and $QB' \parallel PY$ and $QC' \parallel PZ$ so by homothety points P, Q, T are collinear. Moreover, by Sondat's Theorem on triangles ABC and $A'B'C'$ we have that points P, P', Q are collinear; thus, points T, P, P' are collinear. Now let O be the circumcenter of triangle XYZ. We know that O is the midpoint of segment PP' and that by homothety, the circumcenter of triangle $A'B'C'$ lies on line OT. But the circumcenter of triangle $A'B'C'$ obviously coincides with the circumcenter of triangle ABC and we've shown that line OT coincides with line PP', so this completes the proof. □

Delta 22.7. (Sam Korsky) Let I_a, I_b, I_c be the A, B, C-excenters respectively of triangle ABC. Let H be the orthocenter of triangle ABC and let X be the circumcenter of triangle $I_aI_bI_c$. Let M_a, M_b, M_c be the midpoints of BC, CA, AB respectively. Show that lines I_aM_a, I_bM_b, I_cM_c concur on line XH.

Proof. Let M be the midpoint of segment XH. Noting that triangle ABC is the orthic triangle of triangle $I_aI_bI_c$ and applying the result from the second proof of **Delta 2.5** we have that the perpendiculars from M_a, M_b, M_c to lines I_bI_c, I_cI_a, I_aI_b respectively concur at M. Moreover, since $M_bM_c \parallel BC$ and since triangle ABC is the orthic triangle of triangle $I_aI_bI_c$ we have that the lines through I_a perpendicular to M_bM_c passes through X. Obtaining analogous results for I_b and I_c we find that triangles $M_aM_bM_c$ and $I_aI_bI_c$ are orthologic and that their two orthology centers are X and M.

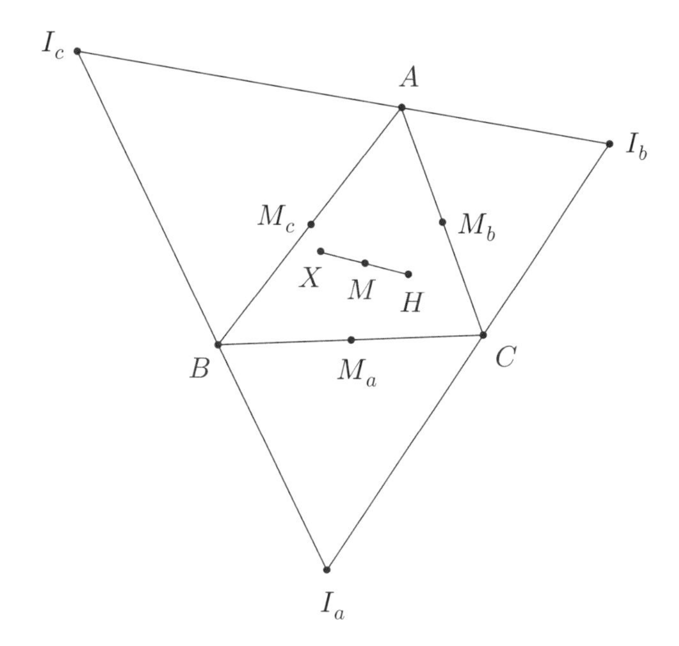

Now since lines AM_a, BM_b, CM_c concur at the centroid of triangle ABC and lines AI_a, BI_b, CI_c concur at the incenter of triangle ABC, by the Cevian Nest Theorem we have that lines I_aM_a, I_bM_b, I_cM_c concur. Hence by Sondat's Theorem on triangles $M_aM_bM_c$ and $I_aI_bI_c$ we obtain the desired result. □

We continue with one of our favorite problems.

Delta 22.8. Let P be a point in the plane of triangle ABC such that the Euler lines of triangles PBC, PCA, PAB concur. Show that this concurrency point lies on the Euler line of triangle ABC.

Proof. Let O and G be the circumcenter and centroid of triangle ABC respectively. Let O_a, O_b, O_c be the circumcenters of triangles PBC, PCA, PAB respectively and let G_a, G_b, G_c be the centroids of triangles PBC, PCA, PAB respectively. Suppose that lines O_aG_a, O_bG_b, O_cG_c concur at some point X. Note that the homothety centered at P with ratio $\frac{3}{2}$ takes segment G_bG_c to the A-midline of triangle ABC, so $OO_a \perp G_bG_c$ and similarly $OO_b \perp G_cG_a$ and $OO_c \perp G_aG_b$. Now let M be the midpoint of segment BC. The homothety centered at M with ratio 3 sends segment GG_a to segment AP so $GG_a \parallel AP$. But line O_bO_c is the perpendicular bisector of segment AP so $GG_a \perp O_bO_c$ and similarly $GG_b \perp O_cO_a$ and $GG_c \perp O_aO_b$. Hence triangles $O_aO_bO_c$ and $G_aG_bG_c$ are orthologic and that their two orthology centers are O and G.

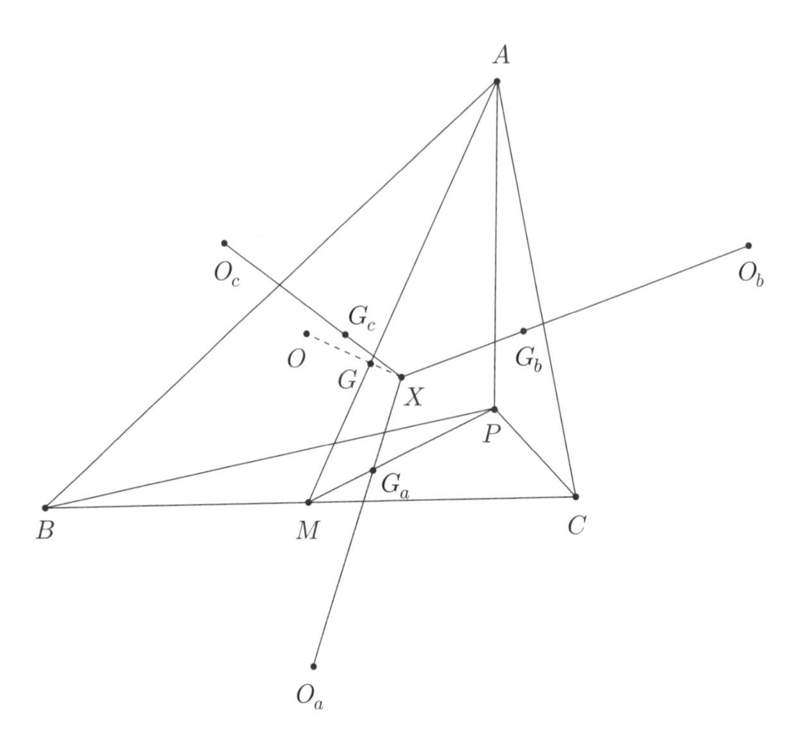

Therefore by Sondat's Theorem on triangles $O_aO_bO_c$ and $G_aG_bG_c$, their perspector X lies on line OG as desired. □

That last problem was not chosen randomly - it provides a setting for a remarkable locus of points with respect to a given triangle.

Definition. The **Neuberg Cubic** of a triangle ABC is the locus of points P not on the circumcircle of triangle ABC and not on the line at infinity satisfying any of the following equivalent conditions:

1. The Euler lines of triangles PBC, PCA, PAB concur.
2. If A', B', C' are the reflections of P over lines BC, CA, AB respectively. Then lines AA', BB', CC' concur.
3. If O_a, O_b, O_c are the circumcenters of triangles PBC, PCA, PAB respectively then lines AO_a, BO_b, CO_c concur.
4. Let X, Y, Z be the second intersection of lines AP, BP, CP with the circumcircles of triangles PBC, PCA, PAB respectively and let H_a, H_b, H_c be the orthocenters of triangles BCX, CAY, ABZ respectively. Then lines AH_a, BH_b, CH_c concur.
5. If Q is the isogonal conjugate of P with respect to triangle ABC then line PQ is parallel to the Euler line of triangle ABC

Here we'll prove that properties (1), (2), (3), (4) are equivalent - the other implications are beyond the scope of this book.

Delta 22.9. Let P be a point not on the circumcircle of triangle ABC and not on the line at infinity and let O_a, O_b, O_c are the circumcenters of triangles PBC, PCA, PAB respectively. Show that lines AO_a, BO_b, CO_c concur if and only if the Euler lines of triangles PBC, PCA, PAB concur.

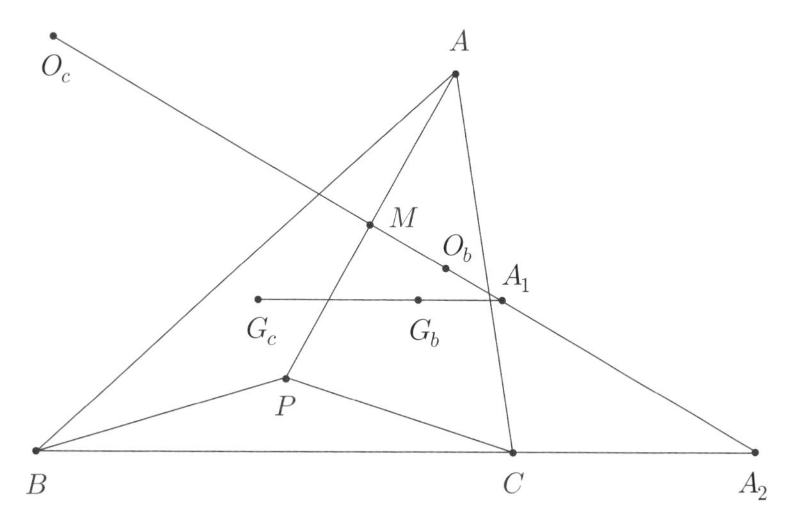

Proof. Let G_a, G_b, G_c be the centroids of triangles PBC, PCA, PAB respectively. Let $A_1 = O_bO_c \cap G_bG_c$ and $A_2 = BC \cap O_bO_c$, and define B_1, C_1, B_2, C_2 similarly. Let M be the midpoint of segment AP - it's clear that the homothety centered at M with ratio 3 takes segment G_bG_c to segment CB and hence also takes A_1 to A_2. Therefore

$$\frac{G_bA_1}{G_cA_1} = \frac{CA_2}{BA_2}$$

and multiplying cyclically implies

$$\frac{G_bA_1}{G_cA_1} \cdot \frac{G_cB_1}{G_aB_1} \cdot \frac{G_aC_1}{G_bC_1} = 1 \iff \frac{CA_2}{BA_2} \cdot \frac{AB_2}{CB_2} \cdot \frac{BC_2}{AC_2} = 1$$

so by Menelaus' Theorem triangles $G_aG_bG_c$ and $O_aO_bO_c$ are perspective if and only if triangles $O_aO_bO_c$ and ABC are perspective. Desargues' Theorem then implies the desired result. □

Corollary 22.1. Show that if P is on the Neuberg Cubic of triangle ABC then A is on the Neuberg Cubic of triangle BCP.

Proof. Since P is on the Neuberg Cubic of triangle ABC we have that the Euler lines of triangles PBC, PCA, PAB concur. But by **Delta 22.8** this concurrency point lies on the Euler line of triangle ABC, so the Euler lines of triangles ACP, APB, ABC concur. Hence A is on the Neuberg Cubic of triangle BCP as desired.

Delta 22.10. Let P be a point not on the circumcircle of triangle ABC and not on the line at infinity and let O_a, O_b, O_c are the circumcenters of triangles PBC, PCA, PAB respectively. Let X, Y, Z be the second intersection of lines AP, BP, CP with the circumcircles of triangles PBC, PCA, PAB respectively. Let H_a, H_b, H_c be the orthocenters of triangles BCX, CAY, ABZ respectively. Then lines AO_a, BO_b, CO_c concur if and only if the lines AH_a, BH_b, CH_c concur.

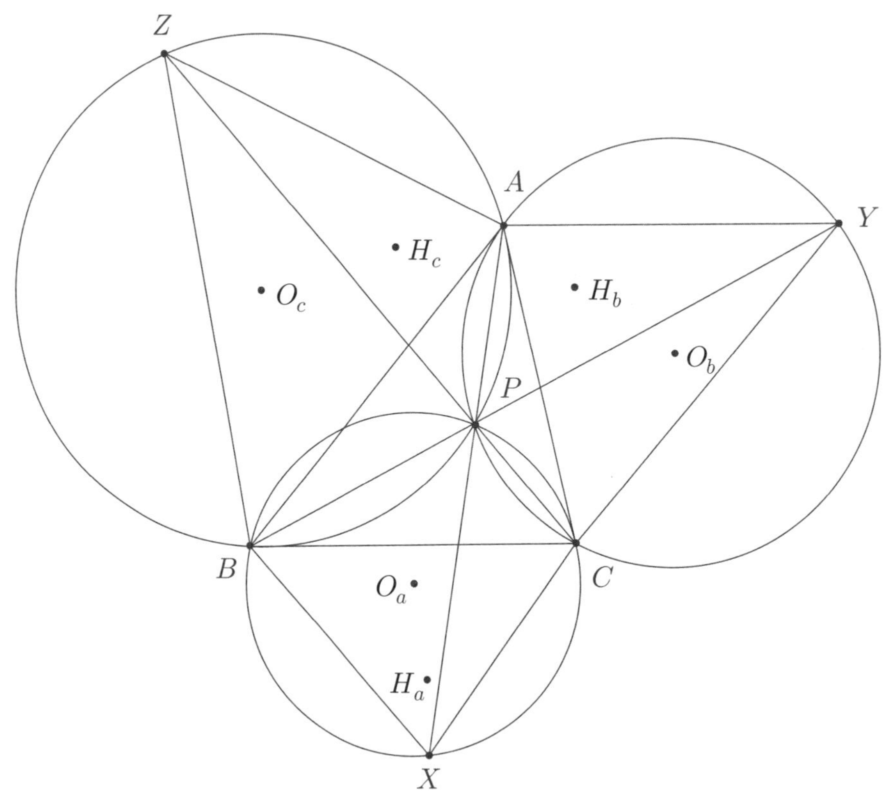

Proof. First we consider the "only if" direction, so assume that lines AO_a, BO_b, CO_c concur at a point Q. Assume that P is inside triangle ABC. Note that

$$\angle BCO_a = \angle BPC - 90^\circ$$

and

$$\angle ACH_b = 90^\circ - \angle YAC = 90^\circ - \angle YPC = \angle BPC - 90^\circ$$

so lines CO_a and CH_b are isogonal with respect to angle $\angle ACB$. Now let $A' = BH_c \cap CH_b$ and define B', C' similarly. A', B', C' are the isogonal conjugates of O_a, O_b, O_c respectively with respect to triangle ABC so lines AA', BB', CC' concur at the isogonal conjugate of Q with respect to triangle ABC. Thus, applying the converse of Brianchon's Theorem on degenerate hexagon $ABCA'B'C'$ we find that there exists a conic touching each of the lines AH_b, AH_c, BH_c, BH_a, CH_a, CH_b so applying Brianchon's Theorem to hexagon $ABCH_aH_bH_c$ yields that lines AH_a, BH_b, CH_c concur as desired. These steps are reversible, so the proof is complete. □

Delta 22.11. Let P be a point in triangle ABC and let A', B', C' be the reflections of P over lines BC, CA, AB respectively. Let O_a, O_b, O_c be the circumcenters of triangles PBC, PCA, PAB respectively. Show that lines AA', BB', CC' concur if and only if lines AO_a, BO_b, CO_c concur.

Proof. First we consider the "only if" direction, so assume that lines AA', BB', CC' concur. We begin with a claim.

Claim. Let P be a point in the plane of triangle ABC and let A_1, B_1, C_1 be the reflections of P over sides BC, CA, AB of triangle ABC respectively. Let P' be the isogonal conjugate of P with respect to triangle ABC and let A_2, B_2, C_2 be the reflections of P' over sides BC, CA, AB respectively. Suppose that lines AA_1, BB_1, CC_1 concur. Then lines AA_2, BB_2, CC_2 concur as well.

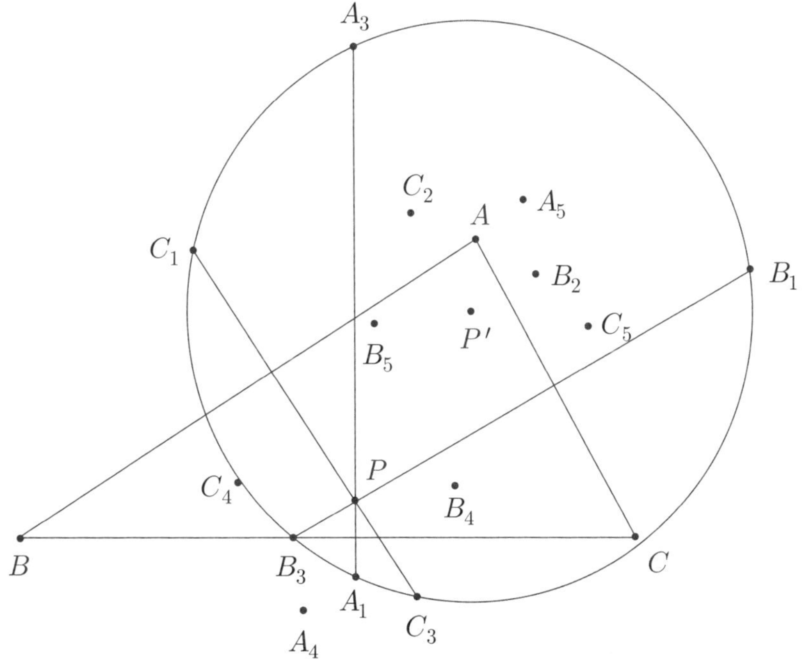

Proof. Let A_3, B_3, C_3 be the second intersections of lines A_1P, B_1P, C_1P with the circumcircle of triangle $A_1B_1C_1$. Let A_4, B_4, C_4 be the reflections of P over lines B_3C_3, C_3A_3, A_3B_3 respectively and let A_5, B_5, C_5 be the reflections of P' over lines B_2C_2, C_2A_2, A_2B_2 respectively. Now, consider the composition of inversion about the circle centered at P with radius $\sqrt{PA_1 \cdot PA_3}$ and reflection over P. This maps A_1, B_1, C_1 to A_3, B_3, C_3 respectively and since A is the circumcenter of triangle PB_1C_1, it maps A to A_4, and similarly maps B to B_4 and C to C_4. Since lines AA_1, BB_1, CC_1 are concurrent, the inversion yields that the circumcircles of triangles PA_3A_4, PB_3B_4, PC_3C_4 are coaxial. Now, a quick angle chase shows that

$$\angle PA_3C_3 = \angle PC_1A_1 = \angle PBC = \angle P'BA = \angle P'A_2C_2.$$

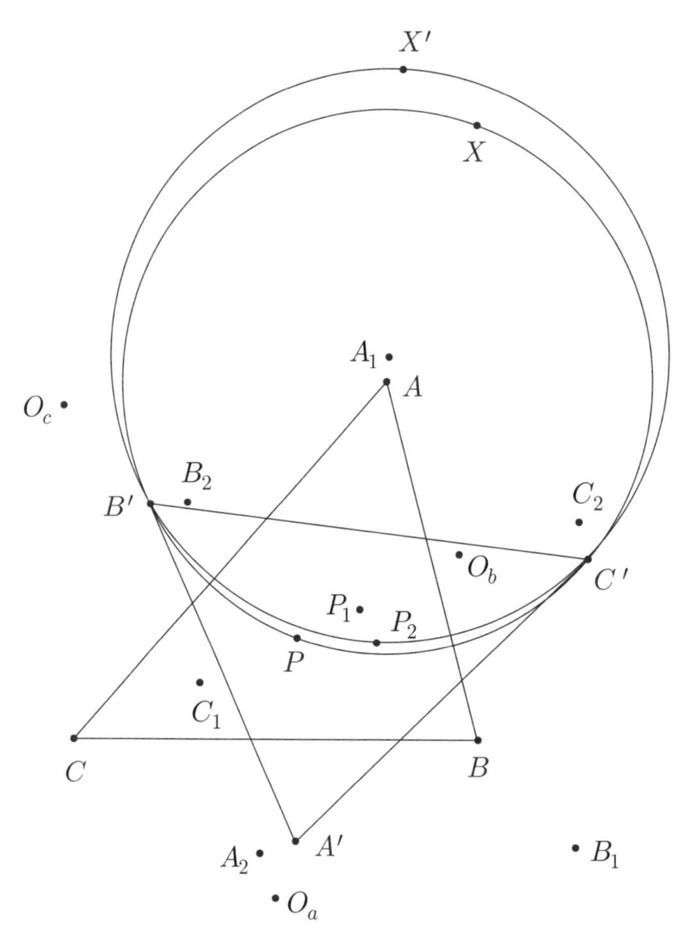

This and analogous angle equalities show that the figures $A_3B_3C_3P$ and $A_2B_2C_2P'$ are similar. Therefore the circumcircles of triangles $P'A_2A_5$, $P'B_2B_5$, $P'C_2C_5$ are coaxial and hence their centers are collinear. But it's clear that the centers of these circles are precisely the points $BC \cap B_2C_2$,

$CA \cap C_2A_2$, $AB \cap A_2B_2$ so by Desargues' Theorem, we have the desired concurrency.

Returning to the problem, let P_1, P_2 be the isogonal conjugates of P with respect to triangles ABC and $A'B'C'$ respectively. Let A_1, B_1, C_1 be the circumcenters of triangles $P_2B'C', P_2C'A', P_2A'B'$ respectively and let A_2, B_2, C_2 be the reflections of P_1 over lines B_1C_1, C_1A_1, A_1B_1 respectively. Note that A, B, C are the circumcenters of triangles $PB'C', PC'A', PA'B'$ respectively. Now, let X be the point diametrically opposite to P on the circumcircle of triangle $PB'C'$ and let X' be the point diametrically opposite to P_2 on the circumcircle of triangle $P_2B'C'$. Note that

$$\angle C'B'X = 90^\circ - \angle PB'C'$$
$$\angle A'B'X' = 90^\circ + \angle P_2B'A'$$

so this and it's analog for angles $\angle B'C'X$ and $\angle A'C'X'$ imply

$$\angle C'B'X = 180^\circ - \angle A'B'X'$$
$$\angle B'C'X = 180^\circ - \angle A'C'X'$$

so X and X' are isogonal conjugates with respect to triangle $A'B'C'$. Hence $\angle PA'X = \angle P_2A'X'$. Moreover, assuming without loss of generality the configuration shown above, we have that

$$\begin{aligned}\angle A'P_2X' = \angle A'P_2B' + \angle B'P_2X' &= 180^\circ - \angle P_2A'B' - \angle P_2B'A' + \angle B'C'X' \\ &= 360^\circ - \angle PA'C' - \angle PB'C' - \angle A'C'X \\ &= 360^\circ - \angle A'PB' + \angle A'C'B' - \angle A'C'X \\ &= 360^\circ - \angle A'PB' - \angle B'C'X \\ &= 360^\circ - \angle A'PB' - \angle B'PX \\ &= \angle A'PX\end{aligned}$$

so triangles $A'PX$ ad $A'P_2X'$ are similar. This implies that triangles $A'PA$ and $A'P_2A_1$ are similar - hence, lines $A'A$ and $A'A_1$ are isogonal conjugates with respect to angle $\angle B'A'C'$. Thus, since lines AA', BB', CC' concur at some point T, lines $A'A_1, B'B_1, C'C_1$ concur at the isogonal conjugate of T with respect to triangle $A'B'C'$. Now since A', B', C' are the reflections of P_2 over lines B_1C_1, C_1A_1, A_1B_1 respectively and since P_1 is the circumcenter of triangle $A'B'C'$ we have that P_1 and P_2 are isogonal conjugates with respect to triangle $A_1B_1C_1$. Since lines A_1A', B_1B', C_1C' concur, by the claim we have that lines A_1A_2, B_1B_2, C_1C_2 concur. Now another angle chase yields

$$\begin{aligned}\angle P_1B_1C_1 = \angle P_2B_1A_1 &= \frac{\angle P_2B_1C'}{2} = \angle P_2A'C' = \angle PA'B' = \angle PCA \\ &= \frac{\angle PO_bA}{2} = \angle PO_bO_c.\end{aligned}$$

This and analogous angle equalities yield that figures $A_1B_1C_1P_1$ and $O_aO_bO_cP$ are similar and since A, B, C are the reflections of P over lines O_bO_c, O_cO_a, O_aO_b respectively we have that lines AO_a, BO_b, CO_c concur as desired. These steps are reversible, so the proof is complete. □

We finish the section with a a difficult Olympiad problem that is actually just a simple application of the properties of the Neuberg Cubic!

Delta 22.12. (Romanian TST 2012) Let $ABCD$ be a cyclic quadrilateral such that the triangles BCD and CDA are not equilateral. Prove that if the Simson line of A with respect to triangle BCD is perpendicular to the Euler line of triangle BCD, then the Simson line of B with respect to triangle ACD is perpendicular to the Euler line of triangle ACD.

Proof. The Simson line of A with respect to triangle BCD is perpendicular to AA' where A' is the isogonal conjugate of A with respect to triangle BCD (recall also that A' is a point at infinity) so AA' is parallel to the Euler line of triangle BCD. Therefore A lies on the Neuberg Cubic of triangle BCD so by **Corollary 22.1** we have that B lies on the Neuberg Cubic of triangle ACD. Then, proceeding as we did with A, the Simson line of B with respect to triangle ACD is perpendicular to the Euler line of triangle ACD as desired. □

Assigned Problems

Epsilon 22.1. Let I be the incenter of triangle ABC and let A', B', C' be the reflections of I over lines BC, CA, AB respectively. Let P be the point at which lines AA', BB', CC' concur (this point exists by a simple application of Jacobi's Theorem). Show that IP is parallel to the Euler line of triangle ABC.

Epsilon 22.2. Show that the orthocenter of triangle ABC is on the Neuberg Cubic of triangle ABC.

Epsilon 22.3. Show that the circumcenter of triangle ABC is on the Neuberg Cubic of triangle ABC.

Epsilon 22.4. Show that the incenter of triangle ABC is on the Neuberg Cubic of triangle ABC.

Epsilon 22.5. Show that the Fermat points of triangle ABC are on the Neuberg Cubic of triangle ABC.

Epsilon 22.6. Show that the Isodynamic points of triangle ABC are on the Neuberg Cubic of triangle ABC. (Hint: if a point is on the Neuberg Cubic, why must its isogonal conjugate be as well?)

Epsilon 22.7. (RMM 2009) Given four points A_1, A_2, A_3, A_4 in the plane, no three collinear, such that

$$A_1A_2 \cdot A_3A_4 = A_1A_3 \cdot A_2A_4 = A_1A_4 \cdot A_2A_3,$$

denote by O_i the circumcenter of triangle $A_jA_kA_l$ with $\{i, j, k, l\} = \{1, 2, 3, 4\}$. Assuming $\forall i\, A_i \neq O_i$, prove that the four lines A_iO_i are concurrent or parallel. (Hint: use the previous problem)

Epsilon 22.8. Let P be a point on the Neuberg Cubic of triangle ABC and let X, Y, Z be the projections of P onto lines BC, CA, AB respectively. Show that P lies on the Neuberg Cubic of triangle XYZ.

Chapter 23

Introduction to Complex Numbers

Sometimes, it's better to approach Olympiad geometry problems with algebra rather than geometry. The most powerful method to do so is by interpreting geometric configurations in the complex plane.

As you may recall from Pre-Calculus, a complex number z is a number of the form $z = a + bi$ for real values of a and b (where i satisfies $i^2 = -1$). In this section, the complex coordinates of points will be denoted by a lowercase letter (e.g, point A has complex coordinate a) unless specified otherwise. The **conjugate** of a complex number $z = a + bi$ will be denoted by $\overline{z}$ and satisfies $\overline{z} = a - bi$. On the complex plane, the conjugate of a complex number is its reflection over the real line. The distance between two complex numbers z_1 and z_2 on the complex plane will be denoted by $|z_1 - z_2|$ and trivially satisfies $(z_1 - z_2)(\overline{z_1} - \overline{z_2}) = |z_1 - z_2|^2$. Similarly the **magnitude** of a complex number $z = a + bi$ (its distance from the origin) will be denoted by $|z| = \sqrt{a^2 + b^2}$.

How do we interpret a complex number? In one sense, it provides the location of a point in the complex plane. On the other hand, one can also interpret a complex number as the composition of a counter-clockwise rotation and a homothety on the complex plane. Given a complex number z, we can write it in **polar form** as $z = re^{i\theta}$ where r is the magnitude of z and $0^\circ \leq \theta < 360^\circ$ is the clockwise angle line OZ makes with the positive real axis, where O is the origin. Then multiplication by z is equivalent to the composition of a counter-clockwise rotation by θ centered at O and a homothety with ratio r centered at O. θ is called the **argument** of z and is denoted by $\arg(z)$. Note that for this section, we will be using directed angles.

Why are complex numbers useful in Olympiad geometry? Their power comes from a very simple fact - if complex number z is on the unit circle, then $\overline{z} = \frac{1}{z}$. This means that for points on the unit circle, we never have to worry about their conjugates (which often make computations tedious). The essence of "complex bashing" is to express the points and lines in a diagram in terms of the complex coordinates of points on what you designate to be the unit circle.

Now, let's interpret some geometric properties with complex numbers! The reader is encouraged to read the next few pages, but after we prove these identities we will provide a list of them without proof. When we get to Olympiad problems, we will use these properties without citing them.

Delta 23.1. (The Angle Between Two Lines) The angle θ formed by the two lines determined by segments AB and CD (from AB to CD in the clockwise direction) satisfies

$$\frac{a-b}{|a-b|} = e^{i\theta}\frac{c-d}{|c-d|}$$

Proof. Translations preserve the angle between two lines so translate segments AB and CD to segments OA' and OC' where O is the origin. It's clear that $a' = a - b$ and $c' = c - d$. Write a' and c' in polar form so that $a' = a - c = r_1 e^{i\theta_1}$ and $c' = c - d = r_2 e^{i\theta_2}$. It's easy to see that $\theta = \arg(a') - \arg(c') = \theta_1 - \theta_2$. Therefore we have that

$$\frac{\frac{a-b}{|a-b|}}{\frac{c-d}{|c-d|}} = \frac{e^{i\theta_1}}{e^{i\theta_2}} = e^{i\theta}$$

which is exactly what we wanted. □

//By squaring both sides, this equation can also be written in the more useful form

$$\frac{a-b}{\overline{a}-\overline{b}} = e^{2i\theta}\frac{c-d}{\overline{c}-\overline{d}}.$$

Verify this!

Corollary 23.1. (Parallel Lines) Segments AB and CD are parallel if and only if

$$\frac{a-b}{\overline{a}-\overline{b}} = \frac{c-d}{\overline{c}-\overline{d}}$$

Proof. Apply **Delta 23.1** with $\theta = 0°$ and square both sides. □

Corollary 23.2. (Collinearity) Points A, B, C are collinear if and only if

$$\frac{a-b}{\overline{a}-\overline{b}} = \frac{c-b}{\overline{c}-\overline{b}}$$

Proof. Note that points A, B, C are collinear if and only if $AB \parallel BC$ and apply **Corollary 23.1**. □

Corollary 23.3. (The Equation of a Line) The line through points A and B has equation

$$\frac{z-a}{\overline{z}-\overline{a}} = \frac{a-b}{\overline{a}-\overline{b}}$$

in z.

Proof. Replace c with z in **Corollary 23.2**. □

//In general, this means that the equations of lines can be written as $z = r\overline{z} + s$ for complex numbers r and s.

Corollary 23.4. (Perpendicularity) Segments AB and CD are perpendicular if and only if

$$\frac{a-b}{\overline{a}-\overline{b}} = -\frac{c-d}{\overline{c}-\overline{d}}$$

Proof. Apply **Delta 23.1** with $\theta = 90°$ and square both sides. □

Delta 23.2. (Similar Triangles) Triangles ABC and DEF are similar (and similarly oriented) if and only if

$$\frac{a-b}{a-c} = \frac{d-e}{d-f}$$

Proof. Note that

$$\frac{a-b}{a-c} = \frac{d-e}{d-f} \Longleftrightarrow \frac{|a-b|}{|a-c|} = \frac{|d-e|}{|d-f|} \Longleftrightarrow \frac{AB}{AC} = \frac{DE}{DF}$$

and by **Delta 23.1** that

$$\frac{a-b}{a-c} = \frac{d-e}{d-f} \Leftrightarrow \frac{\frac{a-b}{\overline{a}-\overline{b}}}{\frac{a-c}{\overline{a}-\overline{c}}} = \frac{\frac{d-e}{\overline{d}-\overline{e}}}{\frac{d-f}{\overline{d}-\overline{f}}} \Leftrightarrow e^{2i\angle BAC} = e^{2i\angle EDF} \Leftrightarrow \angle BAC = \angle EDF$$

which implies the desired result. □

//This can also be written symmetrically as

$$a(e-f)+b(f-d)+c(d-e)=0$$

Corollary 23.5. (Spiral Similarity) The center P of the spiral similarity taking segment AB to segment CD has complex coordinate

$$p=\frac{ad-bc}{a+d-b-c}$$

Proof. P is the unique point such that triangles PAB and PCD are similar and similarly oriented. Hence, we have by **Delta 23.2** that

$$\frac{p-a}{p-b}=\frac{p-c}{p-d}\Longrightarrow p=\frac{ad-bc}{a+d-b-c}$$

as desired. □

Delta 23.3. (Cyclicity) Points A, B, C, D are concyclic if and only if

$$\frac{(a-c)(b-d)}{(\overline{a}-\overline{c})(\overline{b}-\overline{d})}=\frac{(a-d)(b-c)}{(\overline{a}-\overline{d})(\overline{b}-\overline{c})}$$

Proof. Note that by **Delta 23.1**

$$\angle ACB=\angle ADB \Longleftrightarrow e^{2i\angle ACB}=e^{2i\angle ADB}\Longleftrightarrow \frac{\frac{c-b}{\overline{c}-\overline{b}}}{\frac{c-a}{\overline{c}-\overline{a}}}=\frac{\frac{d-b}{\overline{d}-\overline{b}}}{\frac{d-a}{\overline{d}-\overline{a}}}$$

which is equivalent to what we wanted to prove. □

Delta 23.4. (The Area of a Triangle) The area of triangle ABC is given by

$$\frac{i}{4}\begin{vmatrix} a & \overline{a} & 1 \\ b & \overline{b} & 1 \\ c & \overline{c} & 1 \end{vmatrix}$$

Proof. This can be bashed with Cartesian coordinates... we leave it as an exercise to the reader.

Delta 23.5. (Reflections) The reflection P' of point P over the line determined by segment AB has complex coordinate

$$p' = \frac{(a-b)\overline{p} + \overline{a}b - a\overline{b}}{\overline{a} - \overline{b}}$$

Proof. Consider the transformation that acts on every complex number z as follows:

$$z \mapsto \frac{z-a}{b-a}$$

This is a linear transformation and so preserves reflections. Note that $a \mapsto 0$ and $b \mapsto 1$ and $p \mapsto \frac{p-a}{b-a}$ and $p' \mapsto \frac{p'-a}{b-a}$ so since line AB is mapped to the real line we have

$$\frac{p'-a}{b-a} = \frac{\overline{p} - \overline{a}}{\overline{b} - \overline{a}} \Longrightarrow p' = \frac{(a-b)\overline{p} + \overline{a}b - a\overline{b}}{\overline{a} - \overline{b}}$$

as desired. □

Corollary 23.6. (Projections) The foot of the perpendicular Z of point P on the line determined by segment AB has complex coordinate

$$z = \frac{(\overline{a} - \overline{b})p + (a-b)\overline{p} + \overline{a}b - a\overline{b}}{2(\overline{a} - \overline{b})}$$

Proof. Using the same notation as in the proof of **Delta 23.5** we know Z is the midpoint of segment PP' so

$$z = \frac{p+p'}{2} = \frac{(\overline{a} - \overline{b})p + (a-b)\overline{p} + \overline{a}b - a\overline{b}}{2(\overline{a} - \overline{b})}$$

as desired. □

Delta 23.6. (Properties of the Orthocenter and Circumcenter) Let H and O be the orthocenter and circumcenter of triangle ABC respectively. Show that

$$h + 2o = a + b + c$$

Proof. Let G be the centroid of triangle ABC. Then since by known properties of the Euler line G lies on segment OH and satisfies $GH = 2OG$ we have that

$$h + 2o = 3g = a + b + c$$

as desired. $\square$

//For a triangle inscribed in the unit circle, **Delta 23.6** implies that the complex coordinate of its orthocenter is the sum of the complex coordinates of its vertices.

Delta 23.7. (The Circumcenter of an Arbitrary Triangle) Show that the circumcenter X of triangle ABC has complex coordinate

$$x = \frac{\begin{vmatrix} a & a\overline{a} & 1 \\ b & b\overline{b} & 1 \\ c & c\overline{c} & 1 \end{vmatrix}}{\begin{vmatrix} a & \overline{a} & 1 \\ b & \overline{b} & 1 \\ c & \overline{c} & 1 \end{vmatrix}}$$

Proof. If r is the radius of the circumcircle then x satisfies

$$|x-a|^2 = r^2$$
$$|x-b|^2 = r^2$$
$$|x-c|^2 = r^2$$

which after expanding can be written as

$$a\overline{x} + \overline{a}x + r^2 - x\overline{x} = a\overline{a}$$
$$b\overline{x} + \overline{b}x + r^2 - x\overline{x} = b\overline{b}$$
$$c\overline{x} + \overline{c}x + r^2 - x\overline{x} = c\overline{c}$$

and by treating this as a system of three linear equations with variables $x, \overline{x}$, and $r^2 - x\overline{x}$ by Cramer's Rule we have the desired result. $\square$

Corollary 23.7. (Circumcenters and Orthocenters of Triangles with a Vertex at the Origin) Let O be the origin. Show that the circumcenter X and orthocenter H of triangle AOB have complex coordinates

$$x = \frac{ab(\overline{a} - \overline{b})}{\overline{a}b - a\overline{b}}$$

and

$$h = \frac{(\overline{a}b + a\overline{b})(a-b)}{a\overline{b} - \overline{a}b}$$

Proof. By letting $c = 0$ in **Delta 23.7** we immediately obtain

$$x = \frac{ab(\overline{a} - \overline{b})}{\overline{a}b - a\overline{b}}$$

and since by **Delta 23.6** we have $h = a + b - 2x$ we also obtain

$$h = \frac{(\overline{a}b + a\overline{b})(a - b)}{a\overline{b} - \overline{a}b}$$

as desired. □

Finally, we can start discussing properties of points on the unit circle. Let's see just how much easier computations become!

Delta 23.8. (Equation of a Chord) If AB is a chord of the unit circle then the equation of line AB is given by

$$z = a + b - ab\overline{z}$$

Proof. We know from **Corollary 23.3** that the line has equation

$$\frac{z - a}{\overline{z} - \overline{a}} = \frac{a - b}{\overline{a} - \overline{b}} \Longrightarrow \frac{z - a}{\overline{z} - \frac{1}{a}} = \frac{a - b}{\frac{1}{a} - \frac{1}{b}} = -ab$$

and upon cross-multiplying we immediately obtain the desired result. □

Corollary 23.8. (Chord Intersection) Let AB and CD be two chords on the unit circle. If $P = AB \cap CD$ then

$$p = \frac{ab(c + d) - cd(a + b)}{ab - cd}$$

Proof. From **Delta 23.8** we have that p satisfies

$$p = a + b - ab\overline{p}$$
$$p = c + d - cd\overline{p}$$

so

$$\overline{p} = \frac{a + b - c - d}{ab - cd} \Longrightarrow p = \frac{\overline{a} + \overline{b} - \overline{c} - \overline{d}}{\overline{a}\overline{b} - \overline{c}\overline{d}} = \frac{\frac{1}{a} + \frac{1}{b} - \frac{1}{c} - \frac{1}{d}}{\frac{1}{ab} - \frac{1}{cd}}$$

and upon simplifying we obtain the desired result. □

Corollary 23.9. (Tangent Intersection) If the lines tangent to the unit circle at A and B intersect at P then

$$p = \frac{2ab}{a+b}$$

Proof. Just consider the "chords" AA and BB and use **Corollary 23.8.** □

Corollary 23.10. (Equation of a Tangent Line) The line tangent to the unit circle at point A has equation

$$z = 2a - a^2\overline{z}$$

Proof. Just substitute a in for b in **Delta 23.8**. □

Delta 23.9. (Reflection Over a Chord) Let AB be a chord of the unit circle and let P be a point in the plane. Then the reflection P' of P over line AB has complex coordinate

$$p' = a + b - ab\overline{p}$$

Proof. From **Delta 23.4** we have

$$p' = \frac{(a-b)\overline{p} + \overline{a}b - a\overline{b}}{\overline{a} - \overline{b}} = \frac{(a-b)\overline{p} + \frac{b}{a} - \frac{a}{b}}{\frac{1}{a} - \frac{1}{b}}$$

and upon simplifying we immediately obtain the desired result. □

Delta 23.10. (Properties of the Incircle) Let ABC be a triangle whose circumcircle is the unit circle. Let the complex coordinates of A, B, C be a^2, b^2, c^2 for complex numbers a, b, c respectively. Let $A_1, A_2, B_1, B_2, C_1, C_2, X, X_a, X_b, X_c$ be the points with complex coordinates $-bc$, $-ca$, $-ab$, bc, ca, ab, $-bc-ca-ab$, $ca+ab-bc$, $ab+bc-ca$, $bc+ca-ab$ respectively. Then points A_1, B_1, C_1 are the midpoints of arcs BC, CA, AB of the unit circle respectively not containing vertices of triangle ABC, points A_2, B_2, C_2 are the midpoints of arcs BAC, CBA, ACB of the unit circle respectively, X is the incenter of triangle ABC, and X_a, X_b, X_c are the A, B, C-excenters of triangle ABC respectively.

Proof. Note that $|-bc| = 1$ so A_1 lies on the unit circle. Moreover by **Delta 23.1** we have

$$e^{2i\angle A_1AB} = \frac{\frac{b^2-a^2}{\overline{b}^2-\overline{a}^2}}{\frac{-bc-a^2}{-\overline{b}\overline{c}-\overline{a}^2}} = \frac{-a^2b^2}{a^2bc} = -\frac{b}{c}$$

and

$$e^{2i\angle A_1AC} = \frac{\frac{c^2-a^2}{\overline{c}^2-\overline{a}^2}}{\frac{-bc-a^2}{-\overline{b}\overline{c}-\overline{a}^2}} = \frac{-a^2c^2}{a^2bc} = -\frac{c}{b}$$

so $\angle A_1AB = -\angle A_1AC$ which implies that A_1 lies on the A-internal angle bisector of triangle ABC. Hence, A_1 is the midpoint of arc BC of the unit circle not containing A. Now since $a_1 = -bc$ and $a_2 = bc$ we have that $a_1 + a_2 = 0$ - hence, A_2 is the point diametrically opposite to A_1 on the unit circle and so is as claimed the midpoint of arc BAC of the unit circle. We obtain analogous results for points B_1, B_2, C_1, C_2. Now we have that $x = -bc - ca - = a_1 + b_1 + c_1$ so by **Delta 23.6** X is the orthocenter of triangle $A_1B_1C_1$. Hence, X is the incenter of triangle ABC as desired. Similarly $x_a = ca + ab - bc = a_1 + b_2 + c_2$ so X_a is the orthocenter of triangle $A_1B_2C_2$ and thus is the A-excenter of triangle ABC as desired. Obtaining analogous results for X_b and X_c then completes the proof. □

//Note how this new information immediately gives us that A_1 is the center of a circle containing points B, C, X, X_a!

For those readers who skipped the tedious calculations of the last few pages, here is a list of all the identities one needs to know to successfully complex bash Olympiad problems:

- The clockwise angle θ from line AB to line CD satisfies

$$\frac{a-b}{|a-b|} = e^{i\theta}\frac{c-d}{|c-d|}$$

and

$$\frac{a-b}{\overline{a}-\overline{b}} = e^{2i\theta}\frac{c-d}{\overline{c}-\overline{d}}$$

- $AB \parallel CD$ if and only if

$$\frac{a-b}{\overline{a}-\overline{b}} = \frac{c-d}{\overline{c}-\overline{d}}$$

- A, B, C are collinear if and only if

$$\frac{a-b}{\overline{a}-\overline{b}} = \frac{c-b}{\overline{c}-\overline{b}}$$

- Line AB has equation

$$\frac{z-a}{\overline{z}-\overline{a}} = \frac{a-b}{\overline{a}-\overline{b}}$$

- $AB \perp CD$ if and only if

$$\frac{a-b}{\overline{a}-\overline{b}} = -\frac{c-d}{\overline{c}-\overline{d}}$$

- Triangles ABC and DEF are similar if and only if

$$\frac{a-b}{a-c} = \frac{d-e}{d-f}$$

which can be written symmetrically as

$$a(e-f) + b(f-d) + c(d-e) = 0$$

- If P is the center of the spiral similarity taking AB to CD then

$$p = \frac{ad-bc}{a+d-b-c}$$

- A, B, C, D are concyclic if and only if

$$\frac{(a-c)(b-d)}{(\overline{a}-\overline{c})(\overline{b}-\overline{d})} = \frac{(a-d)(b-c)}{(\overline{a}-\overline{d})(\overline{b}-\overline{c})}$$

- The area of triangle ABC is

$$\frac{i}{4}\begin{vmatrix} a & \overline{a} & 1 \\ b & \overline{b} & 1 \\ c & \overline{c} & 1 \end{vmatrix}$$

- The reflection of P over AB is

$$\frac{(a-b)\overline{p} + \overline{a}b - a\overline{b}}{\overline{a}-\overline{b}}$$

- The foot of the perpendicular from P on AB is

$$\frac{(\overline{a}-\overline{b})p + (a-b)\overline{p} + \overline{a}b - a\overline{b}}{2(\overline{a}-\overline{b})}$$

- In a triangle ABC with circumcenter O and orthocenter H we have

$$h + 2o = a + b + c$$

- The circumcenter of triangle ABC is

$$\frac{\begin{vmatrix} a & a\overline{a} & 1 \\ b & b\overline{b} & 1 \\ c & c\overline{c} & 1 \end{vmatrix}}{\begin{vmatrix} a & \overline{a} & 1 \\ b & \overline{b} & 1 \\ c & \overline{c} & 1 \end{vmatrix}}$$

- The circumcenter of triangle AOB where O is the origin is

$$\frac{ab(\overline{a}-\overline{b})}{\overline{a}b-a\overline{b}}$$

- The orthocenter of triangle AOB where O is the origin is

$$\frac{(\overline{a}b+a\overline{b})(a-b)}{a\overline{b}-\overline{a}b}$$

- The equation of chord AB of the unit circle is

$$z = a + b - ab\overline{z}$$

- The intersection of the lines determined by chords AB and CD of the unit circle is

$$\frac{ab(c+d)-cd(a+b)}{ab-cd}$$

- The intersection of the lines tangent to the unit circle at A and B is

$$\frac{2ab}{a+b}$$

- The equation of the line tangent to the unit circle at A is

$$z = 2a - a^2\overline{z}$$

- The reflection of P over chord AB of the unit circle is

$$a + b - ab\overline{p}$$

- If triangle ABC is inscribed in the unit circle and vertices A, B, C have complex coordinates a^2, b^2, c^2 for some $a, b, c \in \mathbb{C}$ then the midpoint of arc BC of the unit circle not containing A is $-bc$, the midpoint of arc BAC of the unit circle is bc, the incenter of triangle ABC is $-bc-ca-ab$, and the A-excenter of the unit circle is $ca + ab - bc$

Now, let's prove some theorems! We start with a somewhat tedious proof of Pascal's Theorem - however, despite the heavy computation, the proof's utility is that it avoids the numerous configuration issues in Menelaus-based proofs!

Theorem 23.1. (Pascal's Theorem Revisited) Let A, B, C, D, E, F be points on a circle (not necessarily in that order) and let $X = AB \cap DE$ and $Y = BC \cap EF$ and $Z = CD \cap FA$. Then points X, Y, Z are collinear.

Proof. Assume without loss of generality that the circumcircle of $ABCDEF$ is the unit circle. Then we have that

$$x = \frac{ab(d+e) - de(a+b)}{ab - de}$$
$$y = \frac{bc(e+f) - ef(b+c)}{bc - ef}$$
$$z = \frac{cd(f+a) - fa(c+d)}{cd - fa}$$

We immediately obtain

$$\overline{x} - \overline{y} = \frac{a+b-d-e}{ab-de} - \frac{b+c-e-f}{bc-ef} = \frac{(b-e)(bc+de+fa-ab-cd-ef)}{(ab-de)(bc-ef)}$$

and similarly

$$\overline{y} - \overline{z} = \frac{(c-f)(ab+cd+ef-bc-de-fa)}{(bc-ef)(cd-fa)}$$

so

$$\frac{\overline{x}-\overline{y}}{\overline{y}-\overline{z}} = -\frac{(b-e)(cd-fa)}{(c-f)(ab-de)}$$

and upon conjugating we see that

$$\frac{\overline{x}-\overline{y}}{\overline{y}-\overline{z}} = \frac{x-y}{y-z}$$

which implies the desired collinearity. □

Theorem 23.2. (Newton's Theorem Revisited) Let quadrilateral $ABCD$ have an inscribed circle ω and let ω touch sides AB, BC, CD, DA at M, N, P, Q respectively. Then lines AC, BD, MP, NQ concur.

Proof. Let $Z = MP \cap NQ$. It clearly suffices to show that Z lies on line AC, because then by symmetry it will also lie on line BD. Assume without loss of generality that ω is the unit circle. Then we have

$$z = \frac{mp(n+q) - nq(m+p)}{mp - nq}$$

and

$$a = \frac{2mq}{m+q}$$
$$c = \frac{2np}{n+p}$$

Note that

$$\overline{a} - \overline{c} = \frac{2}{m+q} - \frac{2}{n+p} = \frac{2(n+p-m-q)}{(m+q)(n+p)}$$

and

$$\overline{z} - \overline{a} = \frac{n+q-m-p}{nq-mp} - \frac{2}{m+q} = \frac{(m-q)(n+p-m-q)}{(m+q)(nq-mp)}$$

so

$$\frac{\overline{a} - \overline{c}}{\overline{z} - \overline{a}} = \frac{nq - mp}{2(m-q)(n+p)}$$

and upon conjugation we easily find

$$\frac{\overline{a} - \overline{c}}{\overline{z} - \overline{a}} = \frac{a - c}{z - a}$$

which implies the desired collinearity. □

The next theorem had a complicated synthetic proof - however with complex numbers, the proof becomes trivial!

Theorem 23.3. (The Steiner Line Revisited) Let ABC be a triangle and let P be a point on its circumcircle. Let X, Y, Z be the reflections of P over lines BC, CA, AB respectively. Then points X, Y, Z, H are collinear where H is the orthocenter of triangle ABC.

Proof. Assume without loss of generality that the circumcircle of triangle ABC is the unit circle. Then we have that

$$h = a + b + c$$
$$x = b + c - \frac{bc}{p}$$
$$y = c + a - ca\frac{ca}{p}$$

It clearly suffices to show that H lies on line XY, because by symmetry H will also then lie on line YZ. We have that

$$x - y = \frac{(a-b)(c-p)}{p} \Longrightarrow \frac{x-y}{\overline{x}-\overline{y}} = \frac{\frac{(a-b)(c-p)}{p}}{\frac{(a-b)(c-p)}{abc}} = \frac{abc}{p}.$$

Also

$$h - x = \frac{ap+bc}{p} \Longrightarrow \frac{h-x}{\overline{h}-\overline{x}} = \frac{\frac{ap+bc}{p}}{\frac{ap+bc}{abc}} = \frac{abc}{p}$$

so

$$\frac{x-y}{\overline{x}-\overline{y}} = \frac{h-x}{\overline{h}-\overline{x}}$$

which implies the desired collinearity. □

Delta 23.11. (The Anticenter) Let $ABCD$ be a cyclic quadrilateral and let ℓ_A be the Simson line of A with respect to triangle BCD. Define ℓ_B, ℓ_C, ℓ_D similarly. Show that lines $\ell_A, \ell_B, \ell_C, \ell_D$ concur (This concurrency point is called the **anticenter** of quadrilateral $ABCD$).

Proof. Assume without loss of generality that the circumcircle of quadrilateral $ABCD$ is the unit circle. Let H be the orthocenter of triangle BCD. We have that

$$h = b + c + d$$

We know that ℓ_A passes through the midpoint of AH so it passes through the point with complex coordinate

$$\frac{a+h}{2} = \frac{a+b+c+d}{2}.$$

Since the coordinates of this point are symmetric in a, b, c, d it's clear that lines ℓ_B, ℓ_C, ℓ_D also pass through it so we are done. □

Theorem 23.4. (Feuerbach's Theorem Revisited) The incircle of triangle ABC is tangent to the nine-point circle of triangle ABC.

Let O, H, N, X be the circumcenter, orthocenter, nine-point center, and incenter of triangle ABC respectively. Let R and r be the inradius of triangle ABC. Since the nine-point circle of triangle ABC has radius $\frac{R}{2}$ it clearly suffices to show that

$$R - 2r = 2XN$$

because if this equation holds then the distance between the centers of the incircle and nine-point circle will equal the difference of their radii and imply the desired tangency. But it's well-known that $OX^2 = R(R-2r)$ so it suffices to show that

$$OX^2 = 2R \cdot XN.$$

Now, assume without loss of generality that the circumcircle of triangle ABC is the unit circle. We know that

$$n = \frac{h+o}{2} = \frac{a^2+b^2+c^2}{2}$$
$$x = -bc - ca - ab$$

so

$$2R \cdot XN = 2|n - x| = |a+b+c|^2.$$

Moreover we have

$$OX^2 = |x - o|^2 = |bc + ca + ab|^2$$

so it suffices to show that $|a+b+c| = |bc+ca+ab|$. But note that

$$|a+b+c| = |\overline{a} + \overline{b} + \overline{c}| = \left|\frac{bc+ca+ab}{abc}\right| = |bc+ca+ab|$$

so we are done. □

Theorem 23.5. (Napoleon's Theorem) Let X, Y, Z be points in the plane of triangle ABC such that triangles BCX, CAY, ABZ are equilateral and do not intersect the interior of triangle ABC. Let R, S, T be the centers of triangles BCX, CAY, ABZ respectively. Then triangle RST is equilateral.

Proof. The rotation centered at B by 60° takes C to X so

$$x - b = \omega(c-b) \Longrightarrow x = \omega(c-b) + b$$

where ω is an appropriate primitive sixth root of unity. Similarly

$$y = \omega(a-c) + c$$
$$z = \omega(b-a) + a$$

Hence we have

$$r = \frac{b+c+x}{3} = \frac{2b+c+\omega(c-b)}{3}$$
$$s = \frac{c+a+y}{3} = \frac{2c+a+\omega(a-c)}{3}$$
$$t = \frac{a+b+z}{3} = \frac{2a+b+\omega(b-a)}{3}$$

Now a quick computation yields

$$s-r = \frac{c+a-2b+\omega(a+b-2c)}{3} = \omega\left(\frac{2a-b-c+\omega(2b-a-c)}{3}\right) = \omega(t-r)$$

where we used the fact that $\omega^2 = \omega - 1$. This implies that the rotation centered at R by 60° takes T to S, hence triangle RST is equilateral as desired. □

//In the above proof, since $r+s+t = a+b+c$ we also have that the center of triangle RST coincides with the centroid of triangle ABC.

Theorem 23.6. (Newton's Second Theorem) Let quadrilateral $ABCD$ have an an inscribed circle ω with center O and denote by E, F the midpoints of segments AC, BD respectively. Then points O, E, F are collinear.

Proof. Let ω touch sides AB, BC, CD, DA at M, N, P, Q respectively. Assume without loss of generality that ω is the unit circle. Then we have

$$a = \frac{2mq}{m+q}$$
$$c = \frac{2np}{n+p}$$

so

$$e = \frac{a+c}{2} = \frac{mq}{m+q} + \frac{np}{n+p} = \frac{npq+mpq+mnq+mnp}{(m+q)(n+p)}$$

hence

$$\frac{e-o}{\overline{e}-\overline{o}} = \frac{e}{\overline{e}} = \frac{\frac{npq+mpq+mnq+mnp}{(m+q)(n+p)}}{\frac{m+p+n+p}{(m+q)(n+p)}} = \frac{npq+mpq+mnq+mnp}{m+n+p+q}.$$

Since this is symmetric in m, n, p, q we obtain

$$\frac{e-o}{\overline{e}-\overline{o}} = \frac{f-o}{\overline{f}-\overline{o}}$$

which implies the desired collinearity. □

We end with a cute application of complex numbers.

Theorem 23.7. (Ptolemy's Inequality) For convex quadrilaterals $ABCD$ we have $AB \cdot CD + DA \cdot BC \geq AC \cdot BD$ with equality holding if and only if quadrilateral $ABCD$ is cyclic.

Proof. Since

$$(a-b)(c-d)+(a-d)(b-c)=(a-c)(b-d)$$

by the Triangle Inequality we have

$$|a-b||c-d|+|a-d||b-c|\geq|a-c||b-d|$$

with equality holding if and only if the line determined by $(a-b)(c-d)$ and $(a-d)(b-c)$ passes through the origin - in other words, if and only if

$$\frac{(a-b)(c-d)}{(a-d)(b-c)}\in\mathbb{R}$$

which is equivalent to points A, B, C, D lying on a circle in that order. This completes the proof. $\square$

Assigned Problems

Epsilon 23.1. Show that the nine-point circle exists.

Epsilon 23.2. Find the coordinates of the symmedian point of a triangle inscribed in the unit circle whose vertices have coordinates a, b, c.

Epsilon 23.3. Find the coordinates of the First Fermat point of a triangle inscribed in the unit circle whose vertices have coordinates a, b, c.

Epsilon 23.4. Show that if three complex numbers a, b, c satisfy

$$a^2 + b^2 + c^2 = bc + ca + ab$$

then they are the vertices of an equilateral triangle in the complex plane.

Epsilon 23.5. Let $A_0A_1A_2A_3A_4A_5A_6$ be a regular heptagon. Show that

$$\frac{1}{A_0A_1} = \frac{1}{A_0A_2} + \frac{1}{A_0A_3}$$

Epsilon 23.6. Let $A_0A_1A_2\ldots A_{n-1}$ be a regular n-gon inscribed in a circle with center O and radius R. Show that

$$\sum_{i=0}^{n-1} PA_i^2 = n(R^2 + PO^2)$$

for any point P in the plane of the n-gon.

Epsilon 23.7. Prove **Delta 23.4.**

Epsilon 23.8. (Mongolia 1996) Let O be the circumcenter of acute triangle ABC, and let M be a point on the circumcircle of triangle ABC. Let X, Y, and Z be the projections of M onto OA, OB, and OC, respectively. Prove that the incenter of triangle XYZ lies on the Simson line of M with respect to triangle ABC.

Epsilon 23.9. (MOP 2006) Let H be the orthocenter of triangle ABC. Points D, E, F lie on the circumcircle of triangle ABC such that $AD \parallel BE \parallel CF$. Let S, T, U be the reflections of D, E, F over lines BC, CA, AB respectively. Show that points S, T, U, H are concyclic.

Chapter 24

Complex Numbers in Olympiad Geometry

The authors believe that the best way to learn a new method, especially a "bash", is through seeing lots and lots of example problems. In this section, we'll provide them!

We'll start with two warm-up problem:

Delta 24.1. (Yugoslavia 1990) Let O and H be the circumcenter and orthocenter of triangle ABC respectively. Let P be the reflection of H over O. If G_1, G_2, G_3 are the centroids of triangles BCP, CAP, ABP respectively show that

$$AG_1 = BG_2 = CG_3 = \frac{4R}{3}$$

where R is the circumradius of triangle ABC.

Proof. Assume without loss of generality that the circumcircle of triangle ABC is the unit circle. Then since $h = a + b + c$ we have $p = -a - b - c$ so

$$g_1 = \frac{b + c + p}{3} = -\frac{a}{3}$$

hence

$$AG_1 = |a - g_1| = \left|\frac{4a}{3}\right| = \frac{4}{3}$$

and by symmetry we also have $BG_2 = CG_3 = \frac{4}{3}$ so the proof is complete. $\square$

Delta 24.2. Let AC be a diameter of circle ω and let B and D be point on ω on opposite sides of line AC. If lines AB and CD intersect at M and the lines tangent to ω at B and D intersect at N, show that $MN \perp AC$.

Proof. Assume without loss of generality that ω is the unit circle and let $a = -1$ and $c = 1$ (note how this captures the information that AC is a diameter of ω). Then we have

$$m = \frac{ab(c+d) - cd(a+b)}{ab - cd} = \frac{-b(1+d) - d(-1+b)}{-b-d} = \frac{2bd + b - d}{b+d}$$

$$n = \frac{2bd}{b+d}$$

so

$$\frac{m-n}{\overline{m} - \overline{n}} = \frac{\frac{b-d}{b+d}}{\frac{d-b}{b+d}} = -1.$$

But we also know

$$\frac{a-c}{\overline{a} - \overline{c}} = \frac{-2}{-2} = 1$$

so

$$\frac{m-n}{\overline{m} - \overline{n}} = -\frac{a-c}{\overline{a} - \overline{c}}$$

which implies the desired perpendicularity. □

Now we move toward more serious problems.

Delta 24.3. (IMO Shortlist 1998) Let O, H be the circumcenter and orthocenter of triangle ABC respectively. Let D, E, F be the reflections of A, B, C over lines BC, CA, AB respectively. Show that points D, E, F are collinear if and only if $OH = 2R$ where R is the circumradius of triangle ABC.

Proof. Assume without loss of generality that the circumcircle of triangle ABC is the unit circle. Then we have

$$d = b + c - \frac{bc}{a}$$

$$e = c + a - \frac{ca}{b}$$

$$f = a + b - \frac{ab}{c}$$

which means that

$$d - e = \frac{(a-b)(bc+ca-ab)}{ab} \Longrightarrow \frac{d-e}{\overline{d} - \overline{e}} = \frac{\frac{(a-b)(bc+ca-ab)}{ab}}{\frac{(b-a)(a+b-c)}{abc}} = -\frac{c(bc+ca-ab)}{a+b-c}.$$

Similarly

$$\frac{e-f}{\overline{e}-\overline{f}} = -\frac{a(ca+ab-bc)}{b+c-a}$$

so points D, E, F are collinear if and only if

$$\frac{d-e}{\overline{d}-\overline{e}} = \frac{e-f}{\overline{e}-\overline{f}} \iff a(ca+ab-bc)(a+b-c) = c(bc+ca-ab)(b+c-a)$$

which can be factored as

$$(a-c)(a^2b+a^2c+b^2c+b^2a+c^2a+c^2b-abc) = 0.$$

Now, since $h = a+b+c$ note that

$$\begin{aligned} OH^2 - 4R^2 = |a+b+c|^2 - 4 &= (a+b+c)\left(\frac{1}{a}+\frac{1}{b}+\frac{1}{c}\right) - 4 \\ &= \frac{(a+b+c)(bc+ca+ab)}{abc} - 4 \\ &= \frac{a^2b+a^2c+b^2c+b^2a+c^2a+c^2b-abc}{abc} \end{aligned}$$

so since $abc(a-c) \neq 0$ we have

$$OH = 2R \iff \sum_{cyc} a^2b + a^2c = abc \iff D, E, F \text{ collinear}$$

as desired. □

Delta 24.4. Let ABC be a triangle with circumcircle ω. Let M, N, P be the midpoints of sides BC, CA, AB respectively and let the line tangent to ω at A intersect line NP at A_1. Define B_1 and C_1 similarly. Show that points A_1, B_1, C_1 are collinear and that the line they determine is perpendicular to the Euler line of triangle ABC.

Proof. Assume without loss of generality that ω is the unit circle. Let O and H be the circumcenter and orthocenter of triangle ABC respectively. The goal will be to show

$$\frac{a_1-b_1}{\overline{a_1}-\overline{b_1}} = -\frac{h-o}{\overline{h}-\overline{o}}$$

because this will imply that $A_1B_1 \perp OH$ and since by symmetry we would also have $C_1A_1 \perp OH$, this would complete the proof. Since $h = a+b+c$, we start by calculating

$$\frac{h-o}{\overline{h}-\overline{o}} = \frac{a+b+c}{\frac{1}{a}+\frac{1}{b}+\frac{1}{c}} = \frac{abc(a+b+c)}{bc+ca+ab}.$$

Now, it's easy to see that $n = \frac{a+c}{2}$ and $p = \frac{a+b}{2}$ so line NP has equation

$$\frac{z-p}{\overline{z}-\overline{p}} = \frac{p-n}{\overline{p}-\overline{n}} \Longrightarrow \frac{z-\frac{a+b}{2}}{\overline{z}-\frac{a+b}{2ab}} = \frac{\frac{b-c}{2}}{\frac{c-b}{2bc}} = -bc$$

which simplifies to

$$z = \frac{(a+b)(a+c)}{2a} - bc\overline{z}.$$

We also know the equation of the line tangent to ω at A has equation

$$z = 2a - a^2\overline{z}$$

so

$$\frac{(a+b)(a+c)}{2a} - bc\overline{a_1} = 2a - a^2\overline{a_1} \Longrightarrow \overline{a_1} = \frac{3a^2 - bc - ca - ab}{2a(a^2 - bc)}.$$

Similarly

$$\overline{b_1} = \frac{3b^2 - bc - ca - ab}{2b(b^2 - ca)}$$

so

$$\overline{a_1} - \overline{b_1} = \frac{b(b^2 - ca)(3a^2 - bc - ca - ab) - a(a^2 - bc)(3b^2 - bc - ca - ab)}{2ab(a^2 - bc)(b^2 - ca)}.$$

How can we easily factor the expression

$$b(b^2 - ca)(3a^2 - bc - ca - ab) - a(a^2 - bc)(3b^2 - bc - ca - ab)?$$

Well remembering that we want

$$\frac{a_1 - b_1}{\overline{a_1} - \overline{b_1}} = -\frac{h-o}{\overline{h}-\overline{o}} = -\frac{abc(a+b+c)}{bc+ca+ab}$$

it seems likely that $\overline{a_1} - \overline{b_1}$ should have $bc + ca + ab$ as one of its factors. With this in mind, we write

$$\begin{aligned}
&b(b^2 - ca)(3a^2 - bc - ca - ab) - a(a^2 - bc)(3b^2 - bc - ca - ab) \\
&= (bc + ca + ab)(a(a^2 - bc) - b(b^2 - ac)) + 3ab(a(b^2 - ca) - b(a^2 - bc)) \\
&= (bc + ca + ab)(a^3 - b^3) + 3ab(b - a)(bc + ca + ab) \\
&= (a - b)^3(bc + ca + ab)
\end{aligned}$$

so

$$\overline{a_1} - \overline{b_1} = \frac{(a-b)^3(bc + ca + ab)}{2ab(a^2 - bc)(b^2 - ca)}$$

and an easy conjugation yields

$$\frac{a_1 - b_1}{\overline{a_1} - \overline{b_1}} = \frac{\frac{c(b-a)^3(a+b+c)}{2(a^2-bc)(b^2-ca)}}{\frac{(a-b)^3(bc+ca+ab)}{2ab(a^2-bc)(b^2-ca)}} = -\frac{abc(a+b+c)}{bc+ca+ab} = -\frac{h-o}{\overline{h}-\overline{o}}$$

as desired. □

Delta 24.5. Let $ABCDEF$ be a convex hexagon such that $\angle B + \angle D + \angle F = 360^\circ$ and $AB \cdot CD \cdot EF = BC \cdot DE \cdot FA$. Show that

$$BC \cdot AE \cdot FD = CA \cdot EF \cdot DB$$

Proof. Note that

$$\frac{c-b}{|c-b|} = e^{i\angle B}\frac{a-b}{|a-b|}$$
$$\frac{e-d}{|e-d|} = e^{i\angle D}\frac{c-d}{|c-d|}$$
$$\frac{a-f}{|a-f|} = e^{i\angle F}\frac{e-f}{|e-f|}$$

After multiplying these equations and noting that

$$\angle B + \angle D + \angle F = 360^\circ \implies e^{i(\angle B + \angle D + \angle F)} = 1$$
$$AB \cdot CD \cdot EF = BC \cdot DE \cdot FA \implies |c-b||e-d||a-f| = |a-b||c-d||e-f|$$

we find that

$$(b-c)(d-e)(f-a) = (a-b)(c-d)(e-f)$$

and upon expanding and rearranging we find

$$(b-c)(a-e)(f-d) = (c-a)(e-f)(d-b).$$

Taking the magnitude of both sides then completes the proof. □

Delta 24.6. (IMO 2000) Let AH_1, BH_2, CH_3 be the altitudes of an acute angled triangle ABC. Its incircle touches the sides BC, AC and AB at T_1, T_2, and T_3 respectively. Consider the reflections of the lines H_1H_2, H_2H_3 and H_3H_1 with respect to the lines T_1T_2, T_2T_3, and T_3T_1 respectively. Prove that these reflections determine a triangle whose vertices lie on the incircle of triangle ABC.

Proof. Let ω be the incircle of triangle ABC and without loss of generality let ω be the unit circle. Now, we have that

$$a = \frac{2t_2t_3}{t_2+t_3}$$
$$b = \frac{2t_1t_3}{t_1+t_3}$$
$$c = \frac{2t_1t_2}{t_1+t_2}$$

Since H_2 is the projection of B onto the line tangent to ω at T_2 (which we can treat as "chord" T_2T_2 of ω), we have that

$$h_2 = \frac{1}{2}\left(b + 2t_2 - t_2^2\overline{b}\right) = \frac{t_2t_3 + t_3t_1 + t_1t_2 - t_2^2}{t_1+t_3}.$$

Now, let P_2 be the reflection of H_2 over T_2T_3. Then

$$p_2 = t_2 + t_2 - t_2t_3\overline{h_2} = \frac{t_1(t_2^2+t_3^2)}{t_2(t_1+t_3)}.$$

Letting P_3 be the reflection of H_3 over T_2T_3 we similarly obtain

$$p_3 = \frac{t_1(t_2^2+t_3^2)}{t_3(t_1+t_2)}.$$

Therefore

$$p_2 - p_3 = \frac{t_1^2(t_3-t_2)(t_2^2+t_3^2)}{t_2t_3(t_1+t_3)(t_1+t_2)}$$

and so we can compute

$$\frac{p_2-p_3}{\overline{p_2}-\overline{p_3}} = \frac{\frac{t_1^2(t_3-t_2)(t_2^2+t_3^2)}{t_2t_3(t_1+t_3)(t_1+t_2)}}{\frac{(t_2-t_3)(t_2^2+t_3^2)}{t_2t_3(t_1+t_3)(t_1+t_2)}} = -t_1^2.$$

Now let Z be an intersection of line P_2P_3 with ω. Since Z lies on line P_2P_3 we have that

$$\frac{z-p_2}{\overline{z}-\overline{p_2}} = \frac{p_2-p_3}{\overline{p_2}-\overline{p_3}} = -t_1^2.$$

Moreover, since Z lies on ω we have that $\overline{z} = \frac{1}{z}$. Therefore we have that

$$\frac{z-p_2}{\frac{1}{z}-\overline{p_2}} = -t_1^2 \Longrightarrow z^2 - (p_2 + t_1^2\overline{p_2})z + t_1^2 = 0.$$

We can compute that

$$p_2 + t_1^2\overline{p_2} = \frac{t_1(t_2^2+t_3^2)}{t_2t_3}$$

so a quick application of the quadratic formula yields that z could be $\frac{t_1t_2}{t_3}$ or $\frac{t_1t_3}{t_2}$. By symmetry, this implies that the vertices of the triangle formed by the lines P_1P_2, P_2P_3, P_3P_1 (where P_1 is defined analogously to P_2 and P_3) have complex coordinates $\frac{t_1t_2}{t_3}, \frac{t_2t_3}{t_1}, \frac{t_3t_1}{t_2}$ all of which clearly lie on ω as desired. $\square$

Delta 24.7. (Cosmin Pohoata, USAMO 2014) Let ABC be a triangle with orthocenter H and let P be the second intersection of the circumcircle of triangle AHC with the internal bisector of the angle $\angle BAC$. Let X be the circumcenter of triangle APB and Y the orthocenter of triangle APC. Prove that the length of segment XY is equal to the circumradius of triangle ABC.

Proof. Let ω be the circumcircle of triangle ABC and assume without loss of generality that ω is the unit circle. Let A, B, C have complex coordinates a^2, b^2, c^2 respectively for some complex numbers a, b, c. We know that the reflection of H over line AC lies on ω so ω is the reflection of the circumcircle of triangle AHC over line AC. Hence the reflection P' of P over line AC lies on ω. Now, let M be the midpoint of arc BC of ω not containing A and let M' be the reflection of M over line AC. We know that

$$m = -bc$$

so

$$m' = a^2 + c^2 + \frac{a^2c}{b}.$$

Now since P' lies on line AM' we have

$$\frac{p' - a^2}{\overline{p'} - \overline{a}^2} = \frac{m' - a^2}{\overline{m'} - \overline{a}^2}.$$

But it's easy to compute that

$$\frac{m' - a^2}{\overline{m'} - \overline{a}^2} = \frac{\frac{c(bc+a^2)}{b}}{\frac{bc+a^2}{a^2c^2}} = \frac{a^2c^3}{b}$$

and since P' lies on ω we have $\overline{p'} = \frac{1}{p'}$ so

$$\frac{p' - a^2}{\overline{p'} - \overline{a}^2} = \frac{p' - a^2}{\frac{1}{p'} - \frac{1}{a^2}} = -a^2p'$$

so

$$p' = -\frac{c^3}{b}.$$

Therefore

$$p = a^2 + c^2 - a^2c^2\overline{p'} = a^2 + c^2 + \frac{a^2b}{c}.$$

Now the circumcenter of triangle APC is clearly the reflection of the circumcenter of triangle ABC over line AC - hence, it has complex coordinate a^2+c^2. Thus we have

$$y + 2(a^2 + c^2) = a^2 + c^2 + p \Longrightarrow y = \frac{a^2b}{c}.$$

Now there are two ways to proceed - it is possible to guess the coordinate of x by using the fact that our goal is to prove $|x - y| = 1$. However, we provide an explicit calculation. Let $A_1B_1P_1$ be the translation of triangle ABP by $-a^2$ in the complex plane so that

$$\begin{aligned} a_1 &= 0 \\ b_1 &= b^2 - a^2 \\ p_1 &= \frac{c^3 + a^2b}{c} \end{aligned}$$

Then the circumcircle of triangle $A_1B_1P_1$ has complex coordinate

$$\begin{aligned} \frac{b_1p_1(\overline{b_1} - \overline{p_1})}{\overline{b_1}p_1 - b_1\overline{p_1}} &= \frac{\left(\frac{c^3+a^2b}{c}\right)(b^2 - a^2)\left(\frac{c^3+a^2b}{a^2bc^2} - \frac{a^2-b^2}{a^2b^2}\right)}{\left(\frac{c^3+a^2b}{a^2bc^2}\right)(b^2 - a^2) - \left(\frac{c^3+a^2b}{c}\right)\left(\frac{a^2-b^2}{a^2b^2}\right)} \\ &= \frac{(c^3 + a^2b)(b^2 - a^2)(b + c)(a^2b + bc^2 - ca^2)}{c(c^3 + a^2b)(b^2 - a^2)(b + c)} \\ &= \frac{a^2b}{c} + bc - a^2 \end{aligned}$$

and translating back by a^2 we see that

$$x = \left(\frac{a^2b}{c} + bc - a^2\right) + a^2 = \frac{a^2b}{c} + bc.$$

Hence

$$XY = |x - y| = |bc| = 1 = R$$

as desired. □

Delta 24.8. (EGMO 2015) Let H be the orthocenter and G be the centroid of acute-angled triangle ABC with $AB \neq AC$. The line AG intersects the circumcircle of triangle ABC again at P. Let P' be the reflection of P over the line BC. Prove that $\angle CAB = 60°$ if and only if $HG = GP'$

Proof. Without loss of generality let the circumcircle of triangle ABC be the unit circle. Then since line AG passes through the midpoint of side BC which has complex coordinate $\frac{b+c}{2}$ we have that

$$\frac{p-a}{\overline{p}-\frac{1}{a}} = \frac{a-\frac{b+c}{2}}{\frac{1}{a}-\frac{b+c}{2bc}} = \frac{abc(2a-b-c)}{2bc-ab-ac}.$$

Since P lies on the circumcircle of triangle ABC we have that $\overline{p} = \frac{1}{p}$ hence

$$\frac{p-a}{\overline{p}-\frac{1}{a}} = \frac{p-a}{\frac{1}{p}-\frac{1}{a}} = -ap.$$

Therefore

$$p = -\frac{bc(2a-b-c)}{2bc-ab-ac}$$

so we can compute

$$p' = b+c-bc\overline{p} = b+c+\frac{2bc-ab-ac}{2a-b-c} = \frac{ab+ac-b^2-c^2}{2a-b-c}.$$

Now let M be the midpoint of segment $P'H$. Since $h = a+b+c$ we have

$$m = \frac{p'+h}{2} = \frac{a^2-b^2-c^2+ab+ac-bc}{2a-b-c}$$

so since $g = \frac{a+b+c}{3}$ we obtain

$$g-m = \frac{2b^2+2c^2-a^2+bc-2ab-2ac}{3(2a-b-c)}.$$

Another simple computation yields

$$h-p' = \frac{2(a^2-bc)}{2a-b-c}.$$

We know that

$$GP' = GH \iff GM \perp P'H \iff \frac{h-p'}{g-m} = -\frac{\overline{h}-\overline{p'}}{\overline{g}-\overline{m}}.$$

But our previous calculations show that this is equivalent to

$$\frac{a^2-bc}{2b^2+2c^2-a^2+bc-2ab-2ac} = -\frac{bc(bc-a^2)}{2a^2c^2+2a^2b^2-b^2c^2+a^2bc-2abc^2-2ab^2c}.$$

After factoring out the a^2-bc term and cross multiplying, lots of things cancel and we are left with a

$$2(b^3c+bc^3+b^2c^2-a^2b^2-a^2c^2-a^2bc) = 2(bc-a^2)(b^2+bc+c^2)$$

term. Hence

$$GP' = GH \iff (a^2 - bc)^2(b^2 + bc + c^2) = 0.$$

But

$$bc - a^2 = \frac{abc\left(\frac{(a-c)^2}{ac} - \frac{c(a-b)^2}{ab}\right)}{c-b} = \frac{abc}{c-b} \cdot (AC - AB) \neq 0$$

and

$$b^2 + bc + c^2 = 0 \iff \angle BAC = 60^\circ$$

so we are done. Also, note that division by $2a - b - c$ earlier in the proof was acceptable because if $2a - b - c = 0$ then the midpoint of side BC would be the circumcenter of triangle ABC and hence triangle ABC would not be acute. This completes the proof. □

The next problem was actually G9 on the 2006 IMO Shortlist. However, complex numbers make the problem trivial!

Delta 24.9. (IMO Shortlist 2006) Points A_1, B_1, C_1 are chosen on the sides BC, CA, AB of a triangle ABC respectively. The circumcircles of triangles AB_1C_1, BC_1A_1, CA_1B_1 intersect the circumcircle of triangle ABC again at points A_2, B_2, C_2 respectively. Points A_3, B_3, C_3 are symmetric to A_1, B_1, C_1 with respect to the midpoints of the sides BC, CA, AB respectively. Prove that the triangles $A_2B_2C_2$ and $A_3B_3C_3$ are similar.

Proof. A_2 is the center of the spiral similarity taking segment BC to segment C_1B_1 so

$$a_2 = \frac{bb_1 - cc_1}{b + b_1 - c - c_1}$$

and similarly

$$b_2 = \frac{cc_1 - aa_1}{c + c_1 - a - a_1}$$

$$c_2 = \frac{aa_1 - bb_1}{a + a_1 - b - b_1}$$

Also it's clear that $b_1 + b_3 = c + a$ and $c_1 + c_3 = a + b$ so

$$b_3 - c_3 = (c + a - b_1) - (a + b - c_1) = c + c_1 - b - b_1$$

and similarly

$$c_3 - a_3 = a + a_1 - c - c_1$$

$$a_3 - b_3 = b + b_1 - a - a_1$$

so

$$a_2(b_3-c_3)+b_2(c_3-a_3)+c_2(a_3-b_3)=(cc_1-bb_1)+(aa_1-cc_1)+(bb_1-aa_1)=0$$

hence triangles $A_2B_2C_2$ and $A_3B_3C_3$ are similar as desired. □

Delta 24.10. Let $ABCD$ be a convex quadrilateral with $AB = AC = BD$ and let $P = AC \cap BD$. If O and X are the circumcenter and incenter of triangle APB respectively, show that $XO \perp CD$.

Proof. Assume without loss of generality that the circumcircle of triangle ABP is the unit circle and let A, B, P have complex coordinates a^2, b^2, p^2 respectively for some complex numbers a, b, p. Since $AC = AB$ we have that $|c - a^2| = |b^2 - a^2|$ so

$$c - a^2 = e^{i\angle PAB}(b^2 - a^2).$$

But we also have

$$\frac{x-a^2}{\overline{x}-\overline{a}^2} = e^{2i\angle XAB}\frac{b^2-a^2}{\overline{b}^2-\overline{a}^2} = -a^2b^2e^{i\angle PAB}.$$

Since $x = -bp - pa - ab$ we have

$$\frac{x-a^2}{\overline{x}-\overline{a}^2} = \frac{-(a+b)(a+p)}{-\frac{(a+b)(a+p)}{a^2bp}} = a^2bp$$

so

$$e^{i\angle PAB} = -\frac{p}{b} \Longrightarrow c = \frac{a^2b+a^2p-b^2p}{b}$$

and analogously

$$d = \frac{b^2a+b^2p-a^2p}{a}.$$

Therefore

$$c-d = \frac{a(a^2b+a^2p-b^2p)-b(b^2a+b^2p-a^2p)}{ab} = \frac{(a^2-b^2)(bp+pa+ab)}{ab}$$

so

$$\frac{c-d}{\overline{c}-\overline{d}} = \frac{\frac{(a^2-b^2)(bp+pa+ab)}{ab}}{\frac{(b^2-a^2)(a+b+p)}{a^2b^2p}} = -\frac{abp(bp+pa+ab)}{a+b+p}.$$

But

$$\frac{x-o}{\overline{x}-\overline{o}} = \frac{-bp-pa-ab}{-\frac{a+b+p}{abp}} = \frac{abp(bp+pa+ab)}{a+b+p}$$

.

so

$$\frac{c-d}{\overline{c}-\overline{d}} = -\frac{x-o}{\overline{x}-\overline{o}}$$

which implies the desired perpendicularity. □

Delta 24.11. (China 1996) Let H be the orthocenter of triangle ABC. Let ω be the circle with diameter BC and let the tangents from A to ω intersect ω at P and Q. Show that points P, Q, H are collinear.

Proof. Let O be the center of ω. Assume without loss of generality that ω is the unit circle and let $b = -1$ and $c = 1$. Then since P lies on ω and $AP \perp OP$ we have

$$\frac{a-p}{\overline{a}-\overline{p}} = -\frac{p-o}{\overline{p}-\overline{o}} = -p^2$$

so upon expanding we find

$$\overline{a}p^2 - 2p + a = 0.$$

Treating this as a quadratic in p it's easy to see that its roots are precisely p and q - hence, by Vieta's formulas we have

$$p+q = \frac{2}{\overline{a}}$$
$$pq = \frac{a}{\overline{a}}$$

Now let H' be the intersection of the A-altitude of triangle ABC with line PQ. Since H' lies on PQ we have

$$h' = p + q - pq\overline{h'} = \frac{2 - a\overline{h'}}{\overline{a}}$$

and since $AH' \perp BC$ we have

$$\frac{h'-a}{\overline{h'}-\overline{a}} = -\frac{b-c}{\overline{b}-\overline{c}} = -1 \Longrightarrow h' = a + \overline{a} - \overline{h'}.$$

Therefore

$$\frac{2-a\overline{h'}}{\overline{a}} = a + \overline{a} - \overline{h'} \Longrightarrow \overline{h'} = \frac{a\overline{a} + \overline{a}^2 - 2}{\overline{a} - a}$$

so

$$h' = \frac{a\overline{a} + a^2 - 2}{a - \overline{a}}.$$

We want to show that $H' = H$ so it suffices to show that $CH' \perp AB$. But we have

$$h' - c = h' - 1 = \frac{a\overline{a} + a^2 - 2 - a + \overline{a}}{a - \overline{a}} = \frac{(a+1)(a + \overline{a} - 2)}{a - \overline{a}}$$

so

$$\frac{h' - c}{\overline{h'} - \overline{c}} = \frac{\frac{(a+1)(a+\overline{a}-2)}{a-\overline{a}}}{\frac{(\overline{a}+1)(a+\overline{a}-2)}{\overline{a}-a}} = -\frac{a+1}{\overline{a}+1} = -\frac{a-b}{\overline{a}-\overline{b}}$$

which implies the desired perpendicularity. This completes the proof. $\square$

Delta 24.12. (USA TST 2014) Let $ABCD$ be a cyclic quadrilateral, and let E, F, G, and H be the midpoints of AB, BC, CD, and DA respectively. Let W, X, Y and Z be the orthocenters of triangles AHE, BEF, CFG and DGH, respectively. Prove that the quadrilaterals $ABCD$ and $WXYZ$ have the same area.

Proof. Let O be the origin and assume without loss of generality that the circumcircle of $ABCD$ is the unit circle. Then we have

$$e = \frac{a+b}{2}$$
$$h = \frac{d+a}{2}$$

and now it's easy to see that the circumcenter of triangle AHE has complex coordinate $\frac{a}{2}$. Hence

$$w + 2 \cdot \frac{a}{2} = a + h + e = \frac{4a + b + d}{2} \implies w = \frac{2a + b + d}{2}.$$

Similarly

$$x = \frac{2b + c + a}{2}$$
$$y = \frac{2c + d + b}{2}$$
$$z = \frac{2d + a + c}{2}$$

Now, translate quadrilateral $WXYZ$ by $-\frac{a+b+c+d}{2}$ in the complex plane to quadrilateral $W'X'Y'Z'$. It's clear that $[WXYZ] = [W'X'Y'Z']$. We have

$$\begin{aligned} w' &= \frac{a-c}{2} \\ x' &= \frac{b-d}{2} \\ y' &= \frac{c-a}{2} \\ z' &= \frac{d-b}{2} \end{aligned}$$

Since $w' + y' = x' + z' = 0$ we have that $W'X'Y'Z'$ is a parallelogram with center O so

$$\begin{aligned} [W'X'Y'Z'] = 4[W'X'O] &= \frac{i}{4}\begin{vmatrix} a-c & \overline{a}-\overline{c} & 1 \\ b-d & \overline{b}-\overline{d} & 1 \\ 0 & 0 & 1 \end{vmatrix} \\ &= \frac{i}{4}(a\overline{b} + b\overline{c} + c\overline{d} + d\overline{a} - \overline{a}b - \overline{b}c - \overline{c}d - \overline{d}a) \end{aligned}$$

Moreover, we have

$$\begin{aligned} [ABCD] = [ABC] + [ADC] &= \frac{i}{4}\begin{vmatrix} a & \overline{a} & 1 \\ b & \overline{b} & 1 \\ c & \overline{c} & 1 \end{vmatrix} + \frac{i}{4}\begin{vmatrix} a & \overline{a} & 1 \\ d & \overline{d} & 1 \\ c & \overline{c} & 1 \end{vmatrix} \\ &= \frac{i}{4}(a\overline{b} + b\overline{c} + c\overline{d} + d\overline{a} - \overline{a}b - \overline{b}c - \overline{c}d - \overline{d}a) \end{aligned}$$

so $[ABCD] = [W'X'Y'Z'] = [WXYZ]$ as desired. □

Delta 24.13. (APMO 2010) Let ABC be an acute angled triangle satisfying the conditions $AB > BC$ and $AC > BC$. Denote by O and H the circumcenter and orthocenter, respectively, of triangle ABC. Suppose that the circumcircle of triangle AHC intersects the line AB again at M, and the circumcircle of triangle AHB intersects the line AC again at N. Prove that the circumcenter of triangle MNH lies on the line OH.

Proof. Assume without loss of generality that the circumcircle of triangle ABC is the unit circle. Let D be the foot of the C-altitude of triangle ABC. Then we have that

$$\angle CMB = 180^\circ - \angle AMC = 180^\circ - \angle AHC = \angle CBM$$

so triangle BCM is isosceles and hence M is the reflection of B over D. Since

$$d = \frac{1}{2}\left(a + b + c - \frac{ab}{c}\right)$$

we have that

$$m = 2d - b = a + c - \frac{ab}{c}$$

and similarly

$$n = a + b - \frac{ac}{b}.$$

Now translate triangle MNH by $-a - b - c$ in the complex plane to triangle $M'N'O$ so that

$$m' = -\frac{b(a+c)}{c}$$
$$n' = -\frac{c(a+b)}{b}$$

Since $h = a + b + c$ this translation sends H to O and O to the reflection of H over O so it preserves the line OH. Thus, it suffices to show that the circumcenter P of triangle $M'N'O$ lies on line OH. But we have

$$p = \frac{\left(\frac{b(a+c)}{c}\right)\left(\frac{c(a+b)}{b}\right)\left(\frac{a+b}{ac} - \frac{a+c}{ab}\right)}{\left(\frac{a+c}{ab}\right)\left(\frac{c(a+b)}{b}\right) - \left(\frac{a+b}{ac}\right)\left(\frac{b(a+c)}{c}\right)} = -\frac{bc(a+b+c)}{b^2+bc+c^2}.$$

Conjugation then yields that

$$\frac{p-o}{\overline{p}-\overline{o}} = \frac{-\frac{bc(a+b+c)}{b^2+bc+c^2}}{-\frac{bc+ca+ab}{a(b^2+bc+c^2)}} = \frac{abc(a+b+c)}{bc+ca+ab} = \frac{h-o}{\overline{h}-\overline{o}}$$

which implies the desired collinearity. □

We end the section with an incredibly difficult IMO problem 6 which you already saw in **Section 14**.

Delta 24.14. (IMO 2011) Let ABC be an acute triangle with circumcircle Γ. Let ℓ be a tangent line to Γ, and let ℓ_a, ℓ_b and ℓ_c be the lines obtained by reflecting ℓ over the lines BC, CA and AB, respectively. Show that the circumcircle of the triangle determined by the lines ℓ_a, ℓ_b and ℓ_c is tangent to Γ.

Proof. Let ℓ be tangent to Γ at P and assume without loss of generality that P lies on minor arc BC of Γ. Let $A_1 = \ell_b \cap \ell_c$ and define B_1 and C_1 similarly. We want to prove two circles are tangent, and one way of doing so is to find two homothetic triangles with each of these circles as circumcircles whose center of homothety lies on one of the circles.

Motivated by this, we start by finding a chord of Γ parallel to ℓ_a. The angle ℓ makes with line BC is clearly $\frac{|\widehat{PB}-\widehat{PC}|}{2}$. Hence if we let B_2 be the point on minor arc AC such that $\widehat{B_2C} = \widehat{PC}$ and C_2 be the point on minor arc AB such that $\widehat{C_2B} = \widehat{PB}$ then the angle line B_2C_2 makes with line BC is also $\frac{|\widehat{C_2B}-\widehat{B_2C}|}{2} = \frac{|\widehat{PB}-\widehat{PC}|}{2}$. Therefore $B_2C_2 \parallel \ell_a$ and defining A_2 analogously we have that triangles $A_1B_1C_1$ and $A_2B_2C_2$ are homothetic. Hence, it suffices to show that the center of the homothety S taking triangle $A_2B_2C_2$ to triangle $A_1B_1C_1$ (and thus also taking Γ to the circumcircle of triangle $A_1B_1C_1$) lies on Γ (for then S will be the desired tangency point). We use complex numbers!

Let O be the center of Γ. Assume without loss of generality that Γ is the unit circle and that $p = 1$. Since A is the midpoint of arc PA_2 we have that

$$a_2p = a^2 \Longrightarrow a_2 = a^2$$

and similarly

$$b_2 = b^2$$
$$c_2 = c^2$$

Now the equation of line ℓ is

$$z = 2 - \overline{z}$$

so since the reflection of A_1 over chord AB of Γ lies on ℓ we have

$$a + b - ab\overline{a_1} = 2 - \frac{1}{a} - \frac{1}{b} + \frac{a_1}{ab}$$

and upon multiplying both sides by ab and rearranging we find

$$a_1 = a^2b^2\overline{a_1} + 2ab - a^2b - ab^2 - a - b.$$

Similarly

$$a_1 = a^2c^2\overline{a_1} + 2ac - a^2c - ac^2 - a - c$$

so

$$a^2(b-c)(b+c)\overline{a_1} = (b-c)(-2a + a^2 + ab + ac + 1)$$

which means that

$$\overline{a_1} = \frac{1}{a} + \frac{(a-1)^2}{a^2(b+c)} \Longrightarrow a_1 = a + \frac{bc(a-1)^2}{b+c}.$$

Now, let line A_1A_2 intersect Γ again at T. We have that

$$\frac{t - a_2}{\overline{t} - \overline{a_2}} = \frac{a_1 - a_2}{\overline{a_1} - \overline{a_2}}$$

but

$$a_1 - a_2 = a(1-a) + \frac{bc(a-1)^2}{b+c} = \frac{(1-a)(bc+ca+ab-abc)}{b+c}$$

so

$$\frac{a_1 - a_2}{\overline{a_1} - \overline{a_2}} = \frac{\frac{(1-a)(bc+ca+ab-abc)}{b+c}}{\frac{(a-1)(a+b+c-1)}{a^2(b+c)}} = -\frac{a^2(bc+ca+ab-abc)}{a+b+c-1}.$$

Moreover

$$\frac{t-a_2}{\overline{t}-\overline{a_2}} = \frac{t-a_2}{\frac{1}{t}-\frac{1}{a_2}} = -a_2 t = -a^2 t$$

so

$$t = \frac{bc+ca+ab-abc}{a+b+c-1}.$$

Since the coordinate of t is symmetric in a, b, c we have that T lies on lines B_1B_2 and C_1C_2 as well and therefore T is the center of the homothety that takes triangle $A_2B_2C_2$ to triangle $A_1B_1C_1$. Since by definition T lies on Γ, we are done. □

Assigned Problems

Epsilon 24.1. (USA TST 2014) Let ABC be an acute triangle, and let X be a variable interior point on the minor arc BC of its circumcircle. Let P and Q be the feet of the perpendiculars from X to lines CA and CB, respectively. Let R be the intersection of line PQ and the perpendicular from B to AC. Let ℓ be the line through P parallel to XR. Prove that as X varies along minor arc BC, the line ℓ always passes through a fixed point.

Epsilon 24.2. (China TST 2011) Let AA', BB', CC' be three diameters of the circumcircle of an acute triangle ABC. Let P be an arbitrary point in the plane of triangle ABC, and let D, E, F be the orthogonal projections of P on sides BC, CA, AB respectively. Let X be the point such that D is the midpoint of $A'X$, let Y be the point such that E is the midpoint of $B'Y$, and similarly let Z be the point such that F is the midpoint of $C'Z$. Prove that triangle XYZ is similar to triangle ABC.

Epsilon 24.3. Prove Brokard's Theorem (**Theorem 12.2**) with complex numbers.

Epsilon 24.4. Let $ABCD$ be a convex quadrilateral and let $P = AC \cap BD$. If G_1, G_2 are the centroids of triangles DPA and BPC respectively and H_1, H_2 are the orthocenters of triangles APB and CPD respectively, show that $G_1G_2 \perp H_1H_2$.

Epsilon 24.5. Show that the area of a triangle whose vertices are the feet of the projections from an arbitrary vertex of a cyclic pentagon to its three opposing sides does not depend on the choice of the vertex.

Epsilon 24.6. (BMO 2003) Let ω be the circumcircle of triangle ABC and let the line tangent to ω at A intersect line BC at D. Let the perpendicular bisector of segment AB intersect the line through B perpendicular to BC at E and let the perpendicular bisector of segment AC intersect the line through C perpendicular to BC at F. Show that points D, E, F are collinear.

Epsilon 24.7. (ELMO Shortlist 2013) Let ABC be a triangle inscribed in a circle ω, and let the medians from B and C intersect ω at D and E respectively. Let O_1 be the center of the circle through D tangent to AC at C, and let O_2 be the center of the circle through E tangent to AB at B. Prove that O_1, O_2, and the nine-point center of triangle ABC are collinear.

Epsilon 24.8. (IMO Shortlist 2005) Let triangle ABC be an acute-angled triangle with $AB \neq AC$. Let H be the orthocenter of triangle ABC, and let

M be the midpoint of side BC. Let D be a point on side AB and E a point on side AC such that $AE = AD$ and the points D, H, E are on the same line. Prove that the line HM is perpendicular to the common chord of the circumcircles of triangles ABC and ADE.

Chapter 25

3D Geometry

Because they are not seen very often, geometry problems involving three-dimensional objects often take competitors by surprise. In many cases, 3D geometry problems are merely analogues to standard two-dimensional configurations. Sometimes, however, they require an altogether different approach.

We'll start with some novel applications of 3D geometry. Recall that we gave a slick proof of Monge's Theorem using three dimensions - we do the same with Menelaus' Theorem and Desargues' Theorem.

Delta 25.1. (Menelaus' Theorem) Let ABC be a triangle and let D, E, F be points on sides BC, CA, AB of triangle ABC respectively. Then points D, E, F are collinear if and only if

$$\frac{BD}{CD} \cdot \frac{CE}{AE} \cdot \frac{AF}{BF} = -1$$

where we use directed lengths.

Proof. We begin with the direct implication. Assume points D, E, F are collinear. Let A' be a point in space such that line AA' is orthogonal to the plane of triangle ABC. Let B' and C' be points on the plane determined by A', D, E, F such that lines BB' and CC' are orthogonal to the plane of triangle ABC. Now, note that triangles $BB'D$ and $CC'D$ are similar, hence $\frac{BD}{CD} = \frac{BB'}{CC'}$ (these lengths are undirected). Similarly we find that $\frac{CE}{AE} = \frac{CC'}{AA'}$ and $\frac{AF}{BF} = \frac{AA'}{BB'}$. Upon multiplying we obtain

$$\frac{BD}{CD} \cdot \frac{CE}{AE} \cdot \frac{AF}{BF} = -\frac{BB'}{CC'} \cdot \frac{CC'}{AA'} \cdot \frac{AA'}{BB'} = -1$$

as desired.

The converse is then easily showed by using a ghost point and applying the direct implication. □

Delta 25.2. (Desargues' Theorem) Let ABC and DEF be triangles and let $X = BC \cap EF$ and $Y = CA \cap FD$ and $Z = AB \cap DE$. Then points X, Y, Z are collinear if and only if lines AD, BE, CF intersect.

Proof. We begin with the direct implication. Assume that lines AD, BE, CF concur at a point P. Assume that triangles ABC and DEF are not in the same plane. Let line ℓ be the intersection of the planes determined by triangles ABC and DEF. Because lines BE and CF intersect at P, they lie in the same plane and hence lines BC and EF intersect on ℓ. Similarly lines CA and FD intersect on ℓ and lines AB and DE intersect on ℓ so we are done.

Now consider the case where triangles ABC and DEF lie in the same plane. Let G be a point not on this plane, and let G' be a point on line GA. Then line DG meets line PG' at a point A'. Then we use the case where the triangles are in different planes on triangles $G'BC$ and $A'EF$, and project from E to obtain the desired result. Note that if points B, C, E, F are collinear then triangles $G'BC$ and $A'EF$ are still coplanar, but this is easily remedied by repeating the proof using B instead of A. We leave the converse as an exercise to the reader. □

Now that we're warmed up, let's do some Olympiad problems!

Delta 25.3. (USAMO 1985) Let A, B, C, D be four points in space such that at most one of the distances AB, BC, CD, DA, AC, BD is greater than 1. Determine the maximum value of the sum of the six distances.

Proof. Assume WLOG that $AB > 1$. Now consider two spheres, centered at A and B, each with radius 1. Points C and D must be inside both spheres. Standard methods then yield that $AC + BC$ is maximized when C is on the circle formed by intersecting the two spheres. Similarly D, is on this circle as well. And it is clear that to maximize CD, these points must be antipodal on the circle. Hence we have shown that to obtain the largest sum, $ACBD$ is a rhombus with side length 1. Now, note that $AB^2 + CD^2 = 4$ and $CD \leq 1$. Your inequality of choice then yields that $AB+CD$ is maximized when $CD = 1$ and $AB = \sqrt{3}$, and so the maximum value is $5 + \sqrt{3}$ obtained when $ACBD$ is a rhombus made of two equilateral triangles. □

While it turned out that that the solution to the last problem actually came from a two-dimensional configuration, considering the problem in 3D was crucial.

Delta 25.4. (Bulgarian NMO 2014) A real number $f(X) \neq 0$ is assigned to each point X in the space. It is known that for any tetrahedron $ABCD$ with O the center of its inscribed sphere, we have:

$$f(O) = f(A)f(B)f(C)f(D).$$

Prove that $f(X) = 1$ for all points X.

Proof. Let P be an arbitrary point in space, and let $ABCD$ be a regular tetrahedron with center P. Let A', B', C', D' be the centers of the inscribed spheres of tetrahedrons $BCDP$, $CDAP$, $DABP$, and $ABCP$ respectively. Note that regular tetrahedron $A'B'C'D'$ has center P. Using the formula we get

$$f(P) = f(A)f(B)f(C)f(D)$$
$$f(A') = f(P)f(B)f(C)f(D)$$

Therefore $f(A)f(A') = f(P)^2$ and similarly we obtain

$$f(A)f(A') = f(B)f(B') = f(C)f(C') = f(D)f(D') = f(P)^2.$$

Now, noting that

$$f(P) = f(A')f(B')f(C')f(D')$$

and multiplying the four expressions for $f(P)^2$ together, we obtain

$$f(P)^2 = f(P)^8 \implies f(P) = \pm 1$$

Now, assume for the sake of contradiction $f(P) = -1$. Since $|f(A)| = |f(B)| = |f(C)| = |f(D)| = 1$ we can assume WLOG that $f(A) = -1$ and $f(B) = f(C) = f(D) = 1$. Let A_1, B_1, C_1, D_1 be the reflections of A, B, C, D over the planes determined by triangles BCD, CDA, DAB, ABC. Let A_2, B_2, C_2, D_2 be the centers of regular tetrahedrons A_1BCD, AB_1CD, ABC_1D, $ABCD_1$. Note that the P is the center of regular tetrahedrons $A_1B_1C_1D_1$ and $A_2B_2C_2D_2$. Multiple applications of the formula then yield

$$f(A_2) = f(A_1)$$

$$f(B_2) = -f(B_1)$$
$$f(C_2) = -f(C_1)$$

$$f(D_2) = -f(D_1)$$

and multiplying yields

$$f(P) = f(A_2)f(B_2)f(C_2)f(D_2) = -f(A_1)f(B_1)f(C_1)f(D_1) = -f(P),$$

contradiction! Hence $f(P) = 1$ and we are done. $\square$

Delta 25.5. (USAMO 1978) Prove that if the six dihedral (i.e. angles between pairs of faces) of a given tetrahedron are congruent, then the tetrahedron is regular. Is a tetrahedron necessarily regular if five dihedral angles are congruent?

Proof. Let the inscribed sphere of the tetrahedron have center O and touch faces BCD, CDA, DAB, ABC at points W, X, Y, Z respectively. Now, note that the angles between segments OW, OX, OY, OZ are supplementary to the dihedral angles of tetrahedron $ABCD$ and hence are all equal. Also, since by definition $OW = OX = OY = OZ$, by the Law of Sines we have that

$$\frac{\sin\frac{\angle WOX}{2}}{WX} = \frac{\sin\frac{\angle WOY}{2}}{WY} = \frac{\sin\frac{\angle WOX}{2}}{WZ} = \frac{\sin\frac{\angle XOY}{2}}{XY} = \frac{\sin\frac{\angle ZOX}{2}}{ZX} = \frac{\sin\frac{\angle YOZ}{2}}{YZ}$$

and so $WXYZ$ is a regular tetrahedron. Then, by symmetry, tetrahedron $ABCD$ is regular as well.

For the second part of the problem, the answer is surprisingly negative. Carrying over the notation from the first part of the problem, if $WX = WY = XY = ZX = YZ \neq WZ$ (convince yourself that this is possible by imagining two non-coplanar equilateral triangles joined at an edge) then by the relationship found in the proof of the first part, every dihedral angle is equal except for the one determined by planes BCD and ABC. $\square$

Delta 25.6. (MOP 2014) The insphere and one of the exspheres of tetrahedron $ABCD$ touch face BCD at points X and Y respectively. Prove that triangle AXY is obtuse.

Proof. Let X' be the point on the inspehere of tetrahedron $ABCD$ antipodal to X. Then by the natural 3D extension of **Theorem 14.2** we have that A, X', Y are collinear (the proof of this in 2D only relies upon a homothety, which can be utilized in the exact same way in 3D). Hence we have that

$$\angle AXY = \angle AXX' + \angle YXX' = \angle AXX' + 90^\circ > 90^\circ$$

and hence triangle AXY is obtuse as desired. $\square$

Now we consider some non-standard problems, all of which involve 3D geometry.

Delta 25.7. Let a "strip of width w" be an infinitely long rectangle with width w. Prove that if a disk can be covered with n strips of width $w_1, w_2, \ldots, w_n$ then it can be covered by a strip of width $\sum_{i=1}^n w_i$.

Proof. Consider a sphere such that the disk divides it into two hemispheres. Associate to each strip the spherical band formed by orthogonally projecting the strip onto the two hemispheres. If the strips cover the disk, then the bands cover the sphere. We claim that the surface area of a spherical band is proportional to its width. To show this, consider the unit sphere centered about the origin. The surface area of the band formed by rotating the function $f(x) = \sqrt{1-x^2}$ between $x = a$ and b over the x-axis is given by

$$\begin{aligned} 2\pi \int_a^b f(x)\sqrt{1+f'(x)^2}\,dx &= 2\pi \int_a^b \sqrt{1-x^2}\sqrt{1+\frac{x^2}{1-x^2}}\,dx \\ &= 2\pi(b-a), \end{aligned}$$

so the surface area is based solely on $b-a$ as desired.

Therefore, if s_i denotes the surface area of the band associated to the strip with width w_i, s denotes the surface area of the sphere, and d denotes its diameter, then

$$\sum_{i=1}^n \frac{w_i}{d} = \sum_{i=1}^n \frac{s_i}{s} \geq 1.$$

Consequently, $\sum_{i=1}^n w_i \geq d$, and so a strip of width $\sum_{i=1}^n w_i$ covers the disk. $\square$

Next, we'll look at a difficult 3D problem involving something you don't see too often - a square pyramid.

Delta 25.8. A circumscribed pyramid $ABCDS$ is given. Let $P = AB \cap CD$ and $Q = AD \cap BC$. The inscribed sphere of the pyramid touches faces ABS and BCS at points K and L respectively. Prove that if PQ and KL are coplanar, then the tangency point between the inscribed sphere and base $ABCD$ lies on line BD.

Proof. Let $\mathcal{S}$ denote the insphere of pyramid $ABCDS$ and let it touch planes SCD, SDA, $ABCD$ at M, N, R respectively. By the equal tangents lemma, we have that $SK = SL = SM = SN$ and hence points K, L, M, N lie on circle ω, the intersection of $\mathcal{S}$ and the sphere centered at S with radius SK. Let plane $MNKL$ intersect lines SA, SB, SC, SD at A', B', C', D' respectively. Then quadrilateral $A'B'C'D'$ has incircle ω tangent to $B'C'$, $C'D'$, $D'A'$, $A'B'$ at L, M, N, K respectively, and thus by Newton's Theorem lines $A'C', B'D', NL, MK$ concur at a point T.

Now, we extend the concept of poles and polars to 3D. Here, poles and polars are taken with respect to spheres, and the polar of a point is a plane. We also introduce the concept of a *conjugate line.* Two lines are conjugate with respect to a sphere if for any point on one line, its polar contains the other line. Take some time to figure out how the identities we proved in **Chapter 12** carry over to three dimensions. All poles and polars and conjugate lines in the rest of the proof will be with respect to $\mathcal{S}$.

It's clear that planes RNL and RMK are the polars of P and Q respectively. Since these two planes intersect at line RT, we have that PQ and RT are conjugate lines. If PQ and KL are coplanar then their conjugate lines RT and SB are coplanar as well. Therefore projecting from S to the base $ABCD$ sends $T = A'C' \cap B'D'$ to $E = AC \cap BD$ and so R lies on line BE, which is the same as line BD, as desired. □

The next exercise is an unbelievable problem from a recent MOP.

Delta 25.9. (MOP 2014) Let a *smushed box* be a convex polyhedra in space with 6 faces and 8 vertices. Show that if 7 of the vertices of a smushed box lie on a sphere, then the eighth vertex lies on the sphere as well.

Proof. Denote the eight vertex by Q. Denote by P the vertex opposite to Q, and denote by A_1, A_2, A_3 the vertices connected to P by an edge. Denote by B_1 the remaining vertex on the same plane as points P, A_2, A_3, and denote the remaining vertices by B_2 and B_3 in a similar fashion. Consider the inversion about a circle centered at P - we'll denote the inverses of points by adding an apostrophe to their letters. It suffices to show that points A_1', A_2', A_3', B_1', B_2', B_3', Q' are coplanar. Because quadrilateral $PA_2B_1A_3$ is cyclic, we have that points A_2', A_3', B_1' are collinear. Similarly, points A_3', A_1', B_2' are collinear and points A_1', A_2', B_3' are collinear. Now, Q is the intersection of the spheres circumscribing the tetrahedrons $PA_1'B_2'B_3'$, $PA_2'B_3'B_1'$, $PA_3'B_1'B_2'$. But by Miquel's Theorem on triangle $A_1'A_2'A_3'$ with points B_1', B_2', B_3' on its sides, we have that the circumcircles of triangles $A'_1B'_2B'_3$, $A'_2B'_3B'_1$, and $A'_3B'_1B'_2$ concur at a point X. Thus, X lies on the three spheres discussed earlier, and so $Q' = X$. Therefore points $A'_1, A'_2, A'_3, B'_1, B'_2, B'_3, Q'$ are coplanar, which implies the desired result. □

Last but not least we end the chapter (and the book) with two terrific problems that ask for similar things: to prove that three lengths represent the sidelengths of a triangle.

Delta 25.10. (Lev Emelyanov, Tuymaada Yacut Olympiad 2005) In a triangle ABC, let A_1, B_1, C_1 be the points where the excircles touch the sides BC, CA and AB respectively. Prove that AA_1, BB_1 and CC_1 are the side lengths of a triangle.

Proof. The straightforward solution is of course computational. One can confidently compute the precise lengths of segments AA_1, BB_1, CC_1 in terms of the sides of triangle ABC, using complex numbers or Stewart's theorem, after which the problem just turns into algebraic manipulations.

However, we present an argument "from the book" that involves 3D geometry! Consider the line through A parallel to BC, the line through B parallel to CA and the line through C parallel to AB. Let $A'B'C'$ be the triangle thus obtained - this is usually called the *anticomplementary triangle* of ABC (because ABC is the medial triangle of $A'B'C'$). Furthermore, let D, E, F be the tangency points of the incircle of ABC with BC, CA, AB respectively. Since the tangency points of the excircles with the sides are just the reflections of D, E, F with respect to the midpoints of BC, CA, AB, respectively, we notice that $A'D = AA_1$, $B'E = BB_1$ and $C'F = CC_1$. Now, fold the triangles $A'BC$, $B'CA$, $C'AB$, which are congruent to ABC, to form a tetrahedron whose base is ABC in such a way that A', B' and C' become the same point, say P. The previous equalities become $PD = AA_1$, $PE = BB_1$, and $PF = CC_1$, respectively. By the triangle inequality, $EF + PE > PF$; on the other hand, however, $PD > EF$, since EF is a chord of the incircle, while PD is at least the length of an altitude of ABC. Hence,

$$PD + PE > EF + PE > PF.$$

Analogously, we get that $PE + PF > PD$ and $PF + PD > PE$. This proves that the lengths $PD = AA_1$, $PE = BB_1$, $PF = CC_1$ can be the side lengths of a triangle, as claimed.

□

The above solution is taken from our previous work, 110 Geometry Problems for the International Mathematical Olympiad; we urge the reader to go pick up that book if you have been enjoying this one so far, since we are almost done. We have one more problem to show you.

Delta 25.11. In the regular tetrahedron $ABCD$, let M be a point on the face ABC, and let N be a point on the face ACD. Prove that BN, DM, and MN represent the side lengths of a triangle.

Proof. The idea is to consider a point $E \in \mathbb{R}^4$, the four dimensional Euclidean space, such that the simplex $EABCD$ is regular. In other words, choose a point $E \in \mathbb{R}^4$ such that the segments EA, EB, EC, ED, AB (and also AC, AD, BC, BD, CD) are all equal in length. We leave as an exercise to the careful reader the question of figuring out why such a point E must exist (hint: think about the distance formula in $\mathbb{R}^4$ and rewrite the claim in terms of a system of equations). Then, tetrahedra $EABC$ and $DABC$ are similar, so $EM = DM$; likewise tetrahedra $EACD$ and $BACD$ are similar, so $EN = BN$. Consequently, EMN is a triangle with side lengths BN, DM, and MN, as desired. This completes the proof.

□

A bit surprising, right? Here's a baby version of the above problem, which in some sense should provide some intuition for the above solution.

Baby. Let ABC be an equilateral triangle with M on the side AB, and N on the side AC. Is it true that BN, CM, and MN represent the side lengths of a triangle?

Here take of course a point E in $\mathbb{R}^3$ such that $EABC$ is a regular tetrahedron (why does this exist again? Is this point unique? Does the point $E \in \mathbb{R}^4$ in the above proof have to be unique?); likewise, triangles ABC and ABE are similar, and so are triangles ACB and ACE. Consequently, $BN = EN$ and $CM = EM$, so a triangle with side lengths BN, CM, and MN is in fact triangle EMN.

Bibliography

[1] A. Bogomolny, Fagnano's Problem: What is it?
http://www.cut-the-knot.org/Curriculum/Geometry/Fagnano.shtml.

[2] A. Bogomolny, Fagnano's Problem: Morley's Miracle
http://www.cut-the-knot.org/triangle/Morley/index.shtml.

[3] A. Bogomolny, An Old Japanese Theorem:
http://www.cut-the-knot.org/proofs/jap.shtml.

[4] N. A. Court, *College Geometry*, Dover reprint, 2007.

[5] H. S. M. Coxeter, *Introduction to Geometry*, Wiley, 1969.

[6] R. Courant and H. Robbins, *What is Mathematics?: An Elementary approach to Ideas and Methods*, Oxford Univesity Press, 1941.

[7] H. Dorrie, *Mathematische Miniaturen*, Wiesbaden, 1969.

[8] E. Daneels, N. Dergiades, A theorem on orthology centers, *Forum Geom.*, **4** (2004), 135-141.

[9] J.-P. Ehrmann, Hyacinthos message 95, January 8, 2000.

[10] R. Gologan, M. Andronache, M. Balună, C. Popescu, D. Schwarz, D. Şerbanescu, Problem 1, the 5th Romanian IMO Team Selection Test, *Romanian Mathematical Competitions 2007*, Romanian Mathematical Society, 88-89.

[11] D. Grinberg, New Proof of the Symmedian Point to be the centroid of its pedal triangle, and the Converse.

[12] D. Grinberg, The Lamoen Circle.

[13] N. M. Ha, Another proof of van Lamoen's Theorem and its converse, *Forum Geom.*, **5** (2005), 127-132.

[14] M. Hajja, A short trigonometric proof of the Steiner-Lehmus theorem, *Forum Geom.*, **8** (2008), 39-42.

[15] A. Henderson, A classic problem in Euclidean geometry, *J. Elisha Mitchell Soc.*, (1937), 246-281.

[16] A. Henderson, The Lehmus-Steiner-Terquem problem in global survey, *Scripta Mathematica*, **21** (1955), 309-312.

[17] F. Holland, Another verification of Fagnano's theorem, *Forum Geom.*, **7** (2007), 207-210.

[18] R. Honsberger, *Episodes of 19th and 20th Century Euclidean Geometry*, Math. Assoc. America, 1995.

[19] R. A. Johnson, *Advanced Euclidean Geometry*, Dover reprint, 2007.

[20] N. D. Kazarinoff, *Geometric Inequalities*, Random House, New York, 1961.

[21] D. C. Kay, Nearly the last comment on the Steiner-Lehmus theorem, *Crux Math.*, **3** (1977), 148-149.

[22] C. Kimberling, Hyacinthos message 1, December 22, 1999.

[23] C. Kimberling, Triangle centers and central triangles, *Congressus Numeratium*, **129** (1998), 1-285.

[24] A. Letac, Solution to Problem 490, *Sphinx*, **9** (1939), 46.

[25] J. S. Mackay, History of a theorem in elementary geometry, *Edinb. Math. Soc. Proc.*, **20** (1902), 18-22.

[26] A. Myakishev and Peter Y. Woo, On the Circumcenters of Cevasix Configurations, *Forum Geom.*, **3** (2003) 57-63.

[27] M. H. Nguyen, Another proof of Fagnano's inequality, *Forum Geom.*, **4** (2004) 199-201.

[28] V. Nicula, C. Pohoaţă, On the Steiner-Lehmus theorem, *Journal for Geometry and Graphics*, 2008.

[29] C. Pohoaţă, A short proof of Lemoine's theorem, *Forum Geom.*, **8** (2008), 97-98.

[30] L. Panaitopol, M. E. Panaitopol, *Probleme de Geometrie Plană* (in Romanian), GIL reprint, 2007.

[31] H. Rademacher and O. Toeplitz, *The Enjoyment of Mathematics*, Princeton University Press, 1957.

[32] K. R. S. Sastry, A Gergonne analogue of the Steiner-Lehmus theorem, *Forum Geom.*, **5** (2005), 191-195.

[33] L. Sauve, The Steiner-Lehmus theorem, *Crux Math.*, **2** (1976), 19-24.

[34] D. O. Shklyarsky, N. N. Chentsov, Y.M.Yaglom, *Selected Problems and Theorems of Elementary Mathematics*, vol. 2, Moscow, 1952.

[35] C. W. Trigg, A bibliography of the Steiner-Lehmus theorem, *Crux Math.*, **2** (1976), 191-193.

[36] B. Work, Hyacinthos message 19, December 27, 1999.

[37] P. Yiu, *Introduction to the Geometry of the Triangle*, Florida Atlantic University Lecture Notes, 2001.

Printed by “Combinatul Poligrafic”
Com. nr. 60586